D1198066

CHANNELIZED RIVERS

Frontispiece. Construction of a comprehensive flood alleviation scheme on the River Usk at Brecon, Wales (UK), completed in 1983. Widening of the bed resulted in reduced flow velocities. Between 5–8000 tonnes of gravel has to be removed at least once per year from a total channel length of about 500 metres.

CHANNELIZED RIVERS

PERSPECTIVES FOR ENVIRONMENTAL MANAGEMENT

ANDREW BROOKES

Environmental Consultant, Reading, UK

A Wiley–Interscience Publication

JOHN WILEY & SONS
Chichester · New York · Brisbane · Toronto · Singapore

Copyright © 1988 by John Wiley & Sons Ltd.

Reprinted June 1990
Reprinted June 1992

All rights reserved.

No part of this book may be reproduced by any means, or transmitted, or translated into a machine language without the written permission of the publisher.

Library of Congress Cataloging in Publication Data

Brookes, Andrew.
Channelized rivers : perspectives for environmental management / Andrew Brookes.
p. cm.
'A Wiley–Interscience publication.'
Bibliography: p.
Includes indexes.
ISBN 0 471 91979 9
1. Stream channelization. 2. Stream channelization—Environmental aspects. I. Title.
TC529.B76 1988 88-5647
627'.12—dc19 CIP

British Library Cataloguing in Publication Data available

Printed and bound in Great Britain by
Biddles Ltd, Guildford and King's Lynn

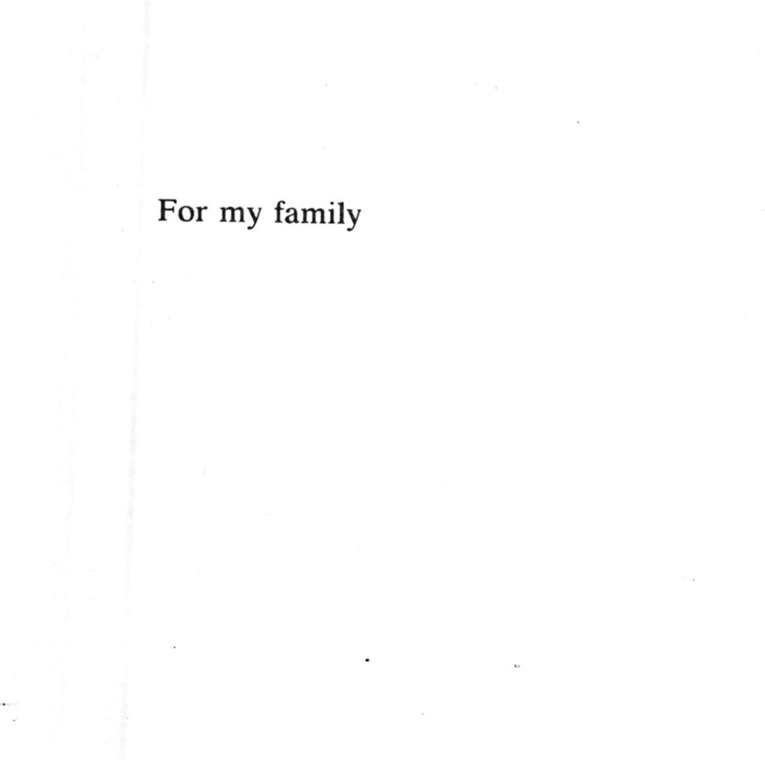

For my family

Contents

POSTSCRIPT AND PROSPECTS

Preface

River channelization encompasses those methods of engineering which modify river channels for the purposes of flood control, land drainage, navigation or the prevention of erosion, and has been one of the most dramatic and widespread forms of human impact. In recent years, many specialists in such disciplines as engineering, geomorphology, hydrology, biology and conservation have taken up the challenge of applying their knowledge to the avoidance or solution of environmental problems connected with river channelization.

This book was originally conceived whilst working as a University of Wales Research Fellow on the geomorphological effects of restraining gravel-bed rivers, but the need for a summary of the environmental implications of river channelization became even more apparent whilst appraising river works as part of an interdisciplinary team within Thames Water. The majority of the book was written during the 'European Year of the Environment' (1987–1988). During the past two decades many river managers and conservation bodies throughout the world have expressed an interest in the effects of river channelization, but progress has been hindered by the lack of a summary of published scientific research. This research appears in a wide variety of publications, ranging from international periodicals to the more obscure journals and reports of local wildlife organizations, and a large number of these articles are listed in the bibliography.

Although this book summarizes data from a number of disciplines, it is demonstrated throughout that fluvial geomorphology has a fundamental role to play in the maintenance or reclamation of river quality. The value of geomorphology is in understanding the factors contributing to the stability of natural river channels and in anticipating the consequences of a particular action. Construction can have repercussions in contiguous areas, and often there are longer term implications. Recovery of biological populations in modified channels is often dependent upon adjustments or alterations of the channel morphology. Geomorphology has also proved successful in developing alternative designs and strategies which work with nature rather than against it. Thus it is imperative that geomorphologists continue to be involved in the decision-making processes that plan and manage the river environment.

The text is organized into five parts. The first, 'Introduction and Problem', explains the need for river channelization and the historical background (Chapter I) before reviewing the methods and limitations of conventional engineering design (Chapter II). Many problems require specialized engineering design not covered by this book, and the objective is not to replace texts dealing with these topics, although material is included which is usually found in books on engineering.

Following the realization that river channels have been extensively modified by such engineering designs, new or revised legislation has recently been implemented with the intention of minimizing the environmental impacts. Several new management techniques have evolved to meet the requirements of this legislation (Chapter III; 'Environmental Legislation'). Undesirable physical and biological consequences of river channelization are detailed in the third part (Chapters IV, V and VI). Much of the published literature emphasizes the adverse effects, although it should be appreciated that there are also situations where river channelization is not so detrimental.

The fourth part, 'Recommendations', covers alternative procedures, techniques and designs which are intended to mitigate the adverse effects or to restore natural characteristics to engineered channels (Chapters VII and VIII). Examples of solutions taken from various parts of the World demonstrate how a balance can be achieved between hydraulic efficiency and environmental acceptability. The need for further research into the environmental repercussions of river channelization and the formulation of alternative strategies is emphasized in the final part (Chapter IX, 'Postscript and Prospects').

This book is therefore intended for a wide readership, including civil engineering, geomorphology, hydrology, biology and conservation. Although not a design handbook, it is written for both the researcher and the practical river manager. The S.I. system of metric units is used throughout.

Reading
February 1988

Andrew Brookes

Acknowledgements

This book is the result of eight years involvement in research and practical management of rivers. Many individuals throughout the world have helped either directly or indirectly. Particular thanks must go to Professor K.J. Gregory of the University of Southampton who trained me as a fluvial geomorphologist and who has remained a constant source of encouragement. Research under a Natural Environment Research Council (CASE) studentship, held jointly at the Freshwater Biological Association River Laboratory in Dorset, provided a valuable insight into the ecology of watercourses under the guidance of Dr F.H. Dawson. The Royal Society in London awarded a European Science Exchange Programme Fellowship at the National Agency of Environmental Protection in Denmark (1983/4) and I am extremely grateful for the experience obtained in restoring rivers. My colleagues in Denmark included Ulrich Kern-Hansen, Mogens Bjørn Nielsen and Erik Nielsen. Thanks must also be extended to Professor Ed Keller of the University of California, Santa Barbara, for his continued discussions throughout my research on channelization and to Professor John Lewin of the University College of Wales, Aberystwyth, who provided guidance on my work on Welsh gravel-bed rivers. The author is currently working in the Project Development and Appraisal Section of the Rivers Division of Thames Water at Reading. I am also grateful to my family who helped in many ways with this book.

For permission to use and reproduce material I express gratitude to the following: The American Society of Civil Engineers; The US Army Corps of Engineers; The US Geological Survey; The Geological Society of America; Blackwell Scientific Publications Ltd; Elsevier Applied Science Publishers Ltd; Elsevier Science Publishers B.V.; Thomas J. Hassler; North Carolina Wildlife Resources Commission; Springer-Verlag, New York Inc.; John Wiley & Sons Ltd; *Science*; US Fish and Wildlife Service; US Soil Conservation Service; Virginia Department of Game and Inland Fisheries; American Fisheries Society; Paul Parey Publishers; Ulrich Kern-Hansen; Pitman Publishing; Professor M. de Vries; Akademie-Verlag; *American Journal of Science*; American Society of Agricultural Engineers; American Water Resources Association; Soil and Water Conservation Society; Professor E.A.

Keller; Academic Press Inc. (London) Ltd; Purdue University Water Resources Research Center; Thames Water Authority; Martinus Nijhoff/Dr W. Junk Publishers; George Allen and Unwin; The Wildlife Society; Professor D.R. Coates; Water Resources Publications (Colorado).

Reading Andrew Brookes

Introduction and Problem

CHAPTER I

Introduction

> In the course of my field-work in the rural districts I am constantly struck with the effect of human culture upon the streams. Hardly in any particular has Man in a settled country set his mark more conspicuously on the physical features of the land.
>
> (G.W. Lamplugh, 1914, p.651.)

G.W. Lamplugh is referring to the way in which river channels become modified or restrained (channelized) following settlement by man and this is the focus of the current book. The purpose of this introductory chapter is to precisely define what is encompassed by the term river channelization, to describe the historical background and spatial extent of channelization, and then to indicate why the subject has recently become such a controversial issue and the environmental ramifications a topic of scientific research. However it is also useful to place river channelization in the context of other human impact studies and these are reviewed first.

Human Impact

The physical landscape has been influenced by gradual economic and social change over a period of more than 6000 years but only recently has there been an awareness of the need for an understanding of the impact of man. It is now apparent that the effects of human activities are significant at the global scale, whereas few areas of the environment escape at least some form of impact. One of the earliest reviews, *Man and Nature*, was written by George Perkins Marsh in 1864 with the objective of revealing the character and extent of the changes produced by human action. Subsequently, this book became a fundamental basis for the conservation movement (Mumford, 1931). In the ensuing century few books were written on the subject, although the theme of *Man as a Geological Agent* was taken up by Sherlock in 1922 and this work concentrated on aspects such as denudation, subsidence, accumulation, coastal and hydrological changes, and man's impact on climate and scenery. In a later article these topics were extended to include the effects of human activity on plant and animal species (Sherlock, 1923).

An important landmark was the international and interdisciplinary symposium held at Princetown in New Jersey in 1955, entitled 'Man's Role in Changing the Face of the Earth', which was subsequently published

(Thomas, 1956). Recent concern for a better understanding of human impacts on all facets of the landscape has arisen for several reasons, including the increased need for the planning and exploitation of the earth's resources, fears over long-term and large-scale effects of environmental pollution and because of increased sensitivity to the issues of environmental quality and landscape aesthetics (Park, 1981). Recent research concerned with the effects of man on environmental processes has been collated by a number of authors (Coates, 1972; Detwyler, 1971; Manners and Mikesell, 1974; Hails, 1977; Goudie, 1981; Gregory and Walling, 1986). The effects of man now appear to be far more complex than previously thought, and an impact in one part of the environment can trigger serious and in certain instances irreversible changes in another.

The impact of man on rivers and river channels has been widespread throughout the period of habitation of the planet. Water supply and land drainage schemes were implemented as early as 3200 BC (Drower, 1954) and the Hwang He (Yellow River) in China has been regulated for at least 4000 years. Prior to the 11th century in Europe, embankments were systematically built for flood control and land reclamation, and primitive weirs were constructed to regulate river flows for water power, whilst in Britain river regulation was widely used by Domesday times (Cole, 1976). Concern for the morphological, hydrological and biological impacts of man on rivers has been reported in an abundance of research papers, reviewed or edited by Hynes (1960, 1970); Moore and Morgan (1969); Whitton (1975); Gregory (1977a) and Hollis (1979). Mrowka (1974) identified four groups of effect on watercourses, namely direct channel manipulation, rural watershed alterations, the effects of urbanization and water pollution activities. Connectivity in the fluvial system means that repercussions of any man-induced change at any given location can be transmitted over a wide area, especially in the downstream direction. This interest has arisen because of the extensive distribution and in many cases the intensive nature of environmental problems. For example, Beaumont (1978) reported that 20% of the total runoff in North America and Africa is now regulated by reservoirs, 15% in Europe and 14% in Asia. Although the first dam was probably constructed

in Egypt some 5000 years ago, there was a marked increase in construction between 1945 and 1971. These have been built either as sources of water for irrigation or domestic supply, as a method of flood control, or to generate power (Petts, 1984). There has also been a realization that environmental scientists can extend the understanding of the complexities of the fluvial system and recommend alternative strategies for management which work with Nature rather that against it.

A major component of the research effort during the past two decades has concentrated upon establishing the ways in which river channel stability has been affected by a range of impacts, including land use changes such as deforestation, overgrazing, cultivation and urbanization, and the construction of dams, bridges, channelization works and irrigation diversions. Although the direct effects of engineering structures have been acknowledged by design engineers for many years, the more widespread effects of these structures were less well appreciated. For example, the potential occurrence of scour immediately below dams had to be allowed for at the design stage and equations were developed to predict scour in relation to local environmental and hydraulic conditions (e.g. Komura and Simons, 1967). However the downstream effects of dams were less immediately obvious but research by environmental scientists during the past decade has revealed a series of effects upon the morphology and biology of river channels, which can persist for very considerable distances downstream. Gregory and Brookes (1983) identified a range of morphological effects occurring downstream from road bridges in south-central England.

A second component of research has investigated the magnitude and spatial extent of the effects which can be identified along individual river channels and this has produced valuable results. For example, below reservoirs in Britain, channel capacities may be reduced to as little at 30% of those expected (Petts, 1980), whilst the effect of the Aswan High Dam on the channel morphology persists downstream for nearly 1000 km (Kashef, 1981). Effects are difficult to predict because a particular situation may be affected by a complex of variables (Schumm, 1977). For example, downstream of urban areas an increased peak discharge may not necessarily lead to channels which are greatly enlarged in size if the calibre of the sediment released from new sediment sources is at variance with the sediment already exposed in the channel bed and banks (Gregory, 1977b).

A third component of research has focused on the realization that there are a range of strategies available for the management of changes (e.g. Whipple and Dilouie, 1981). It has been argued that the work of the geomorphologist is complementary to that of the engineer, the latter concerned with site applications and implications, whereas the geomorphologist is concerned with the more widely distributed spatial impacts (Gregory, 1979). Research has also shown how adjustments of the channel morphology relate to biological changes (e.g. Brooker, 1981; Petts and Greenwood, 1981).

River Channelization

This book is concerned with one of the most dramatic aspects of man's impact on the fluvial system, namely river channelization, which involves the direct modification of river channels. Channelization is the term used to embrace all processes of river channel engineering for the purposes of flood control, drainage improvement, maintenance of navigation, reduction of bank erosion or relocation for highway construction (Herr, 1973; Funk and Ruhr, 1971; Keller, 1976; Brookes, 1985a). River channelization may also be associated with programmes of forest or field drainage (Marshall *et al.*, 1978; Stewart and Lance, 1983). In Scandinavia watercourses have been cleared and straightened in order to float logs out from forests (Hynes, 1960).

Although the term channelization has been used extensively to denote channel works in many other countries, equivalent terms for the same group of engineering methods are 'kanalisation' employed in Germany (Statzner and Stechman, 1977), 'chenalisation' in France (Cuinat, 1983) and 'canalization' in the U.K.

Table I lists some of the principal methods of channelizing rivers, although the use or connotation of the term for a particular method may differ between countries. Channelization may be undertaken by engineering procedures which either enlarge, straighten, embank, or protect an existing channel, or which involve the creation of new channels. Other channelization procedures may be classified as river channel maintenance, including dredging, clearing and snagging, or the removal of obstructions from urban channels. The technical details of how these procedures are used to modify river channels are covered in Chapter II and it is the intention of the following discussion to describe the historical context.

Table I. Terminologies for the methods of channelization (based on terms given in Nunnally and Keller, 1979* and Thorn, 1966†)

American term*	British equivalent†	Procedure
Widening Deepening	Resectioning Resectioning	Manipulating width and/or depth variable to increase the channel capacity
Straightening	Realigning	Steepening the gradient to increase the flow velocity
Levee construction	Embanking	Confining floodwaters by raising the height of the channel banks
Bank stabilization	Bank protection	Use of structures such as gabions and steel piles to control bank erosion
Clearing and snagging	Pioneer tree clearance Weed control Dredging of silt Clearing trash from urban areas	Decreasing the hydraulic resistance and increasing the flow velocity by removing obstructions

The principal purpose of much of the earliest channelization was to increase arable land and this procedure has been applied extensively. The more easily accomplished and economically feasible channelization for this purpose has already been carried out in the majority of intensively cultivated areas. Flood control is essential to protect buildings from damage in both urban and rural areas. Rivers of all sizes are involved, but generally the larger the watercourse the greater the danger of flooding and the more extensive the work involved. Navigation is limited to larger watercourses, but again it has often been necessary to modify these channels to provide sufficient depth of water to enable movement of boats.

Clearly channelization may be desired to restore channels to their 'former state', involving the removal of sediment and debris which may have accumulated following land use changes in the basin upstream, such as the agricultural changes which have occurred in North Carolina (Canterbury, 1972).

Emergency Channel Works

Immediately following a major flood, works may be undertaken on a 'reactive' basis and these are essentially unplanned. In the United States, in particular, emergency works have often been implemented following catastrophic storms without engineering design and environmental impact assessment. For example, following severe floods in Virginia after Hurricane Camille in 1969, emergency federal money was used to straighten and enlarge channels (Keller, 1976). Flood damage caused by Hurricanes Camille (1969) and Agnes (1972) prompted the channelization of more than 1100 km of Virginia streams alone (Corning, 1975). The works were often done by local contractors with little or no knowledge of streams. Keller (1976) suggested that emergency funds should not be considered as a licence for the wholesale modification of any stream at the request of property owners.

Emergency channel work is also required in the USA where streams have become choked with sediment and debris. North Carolina has experienced a proliferation of state-based stream alteration operations during the past 25 years for reasons of public health, undertaken by the North Carolina Department of Human Resources, Division of Health Services. Unlike Soil Conservation Service Projects, stream excavation for public health reasons can be initiated immediately without the long delays and environmental impact statements required by the Watershed Protection and Flood Prevention Act of 1954 (Public Law 566). Hazards include high mosquito populations as well as contamination of wells and malfunctions of septic tanks. However in North Carolina, at least, projects are undertaken in conjunction with the North Carolina Wildlife Resources Commission and involve minimal impact by removing only the accumulated debris and sediment from the channel. The original course and dimensions are therefore retained.

The Proliferation of Channelization

There is a long history of channelization throughout the world. Examples of the earliest forms of channelization come from the Old World, where canals and ditches were built to carry water into and sewage out of ancient cities, and the diversion of streams for irrigation. The construction of embankments is recorded in the histories of most early civilizations. Flood banks were constructed on the Yellow River in China as early as 600 BC, and their construction was brought under unified control by the Han Dynasty in 69 BC. In Britain the Romans first built embankments to control flooding and subsequently many kilometres of banks were built to protect low-lying marsh areas in the Fens (Darby, 1956) and the Somerset Levels (Williams, 1970). Channelization has been most prolific in developed countries in both urban and rural areas (e.g. Neller and Broughton, 1981).

> In uninhabited regions the rivers are wayward and restless, ever shifting from place to place within the bounds of their valleys, that are theirs to sprawl across at will. If a flood should heap up a bar in the channel; or fallen timber gather into a dam; or swamp-vegetation block the fairway in a sluggish reach; the stream swings easily aside into a fresh course...
>
> But as soon as a country acquires a settled population this unstable habit of running water is corrected. For many reasons, human interests demand that a stream shall have a fixed course. When tribal or individual ownership of land was established, the rivers and streams often afforded the best natural boundaries. The convenience of sites chosen for dwellings depended upon the constancy of the waters; and every cattle-enclosure required a permanent drinking-place....
>
> The control was in most cases first established on the brooks, and was extended gradually from them to the larger streams and rivers."
>
> (from G.W. Lamplugh (1914) 'Taming of streams', p.651.)

Man initially lived along watercourses for various reasons: as a supply of surface and ground water, because sand and gravel in floodplain deposits provided a readily available supply of building materials, for disposal of waste and for the irrigation of fields. Floodplains are also the most fertile areas and are flat and easy to develop. A floodplain is also the preferred location for land transportation routes because of the gentle gradients.

The environmental impact of the earliest work was minimal due to the limitations of hand and animal labour. The revolution in the practice of channelization came as a result of several factors. These were the availability of heavy equipment such as bulldozers and draglines, the involvement of government agencies and the increased demand placed on floodplain lands either for increased agricultural productivity or for urban development.

Although river channelization is extensive in many countries throughout the world, the historical background and geographical distributions are well documented for only a relatively few countries. The three contrasting examples described here are the United States of America, which has undergone an intense period of channelization during the past 150 years; Britain, which has had a much longer history of channelization extending back for at least 2000 years; and Denmark, where almost 100% of the drainage network has been modified. In each of these countries channelization has only recently become a controversial issue.

United States of America

It was only in the 19th century that channel improvements became widespread in the United States following early settlement. In a 150 year period at least 320,000 km of rivers have been modified (Little, 1973). The primary purposes of this effort were to drain land for agriculture, to control floods and to provide for the waterborne transportation of goods.

In the early 19th century in the midwest, particularly in Iowa and east-central Illinois, much of the land suffered from poor drainage, causing swampy conditions, crop failures and encouraging mosquitoes (Ames, 1970). The land was considered worthless and an early resident refused to trade his saddle and horse for 260 hectares which was valued in 1974 at about $1 million (Hay and Stall, 1974). As a result of 150 years of channel improvements and the construction of drainage ditches, swamp was converted into rich cropland. In the Vermilion River watershed (3240 sq kilometres) in east-central Illinois 105 legal drainage districts were formed between 1880 and 1974 to provide for surface drainage. Early drainage was by horse-drawn slip scrapers, gangs of men with shovels, or by enormous ditching ploughs which required 68 oxen driven by 8 men. Later work was undertaken by floating dredgers, which were assembled on site. Many of the 'man-made' channels require regular maintenance.

In California farmers and ranchers in the Coastal Ranges straightened and enlarged streams for flood prevention by means of horse-drawn slip scrapers and Chinese labourers as early as 1871 (Keller, 1976).

Most of the early channel modification was highly fragmented, being carried out by a variety of bodies including States, counties, towns, local improvement districts, individuals and private companies (Little, 1973). Whilst this enormous effort achieved the drainage of about 53 million hectares of wetlands, most of this effort was not properly planned, engineered or financed. The doctrine of comprehensive, coordinated, integrated and unified river basin and watershed planning and development was not advanced until the turn of the century. Early piecemeal channelization by farmers and ranchers has been replaced by better planning and engineered projects, sometimes involving the entire basin (Keller, 1976).

This doctrine has been declared in nearly all national Acts dealing with river management this century. The principal agencies now engaged in sponsoring channelization in the United States are the Corps of Engineers, empowered under the Flood Control Acts of 1936 and 1944, and the Soil Conservation Service, carrying out small watershed projects under the Watershed Protection and Flood Prevention Act of 1954 (Ash, 1973; Buie, 1973). As an example of this legislation, the Watershed Protection and Flood Prevention Act (Public Law 566), as subsequently amended on September 27, 1962, stated:

> erosion, floodwater, and sediment damages in the watersheds of rivers and streams of the United States... constitute a menace to the national welfare... that the Federal Government should cooperate with States

> and their political subdivision, soil or water conservation districts, flood prevention or control districts, and other local public agencies for the purpose of preventing such damages and of furthering the conservation, development, utilization, and disposal of water and thereby of preserving and protecting the Nation's land and water resources.

Other projects on a smaller scale include the flood control works of the Tennessee Valley Authority and irrigation and drainage canals (US Bureau of Reclamation, 1952). Federal and non-federal public agency bodies, together with private individuals, still undertake channel alteration activities.

It has been estimated that 26,550 km of river had been modified in the USA since the Flood Control Act of 1936 (Leopold, 1977) with a further 16,090 km predicted. The distribution of major flood protection and levee works is depicted in a map by the US Geological Survey (1969). About 50% of Corps and SCS works already undertaken or predicted on 1630 projects is in five states; 65% of channel alteration works is most heavily concentrated in eight southern States, particularly Illinois, Indiana, North Dakota, Ohio and Kansas and 51% of levee work in California, Illinois and Florida (Little, 1973). Over 3200 km of river levees have been built along a 1520 km length of the Lower Mississippi. Table II lists some of the estimated lengths of works undertaken in various regions and States by Government and other bodies.

Table II. Examples of the extent of channel works in the United States

Location	Extent	Source
Missouri, north of the Missouri River	2947 km of major streams; 1600 km already channelized	Funk (1968)
Pennsylvannia State	480 km channelized as emergency works following Hurricane Agnes in 1972	Duvel *et al.* (1976)
Seven mid-western States	46,530 km already undertaken	Thrienen (1971)
Soil Conservation Service projects in US	12,366 km completed and 21,403 km planned	Committee on Government Operations (1973b)
Twelve south-eastern States	40,000 km planned for channelization	Martin (1969)
Mississippi River (SCS and US Army projects)	1750 km completed (142 km in Tombigbee basin alone)	Freeman (1972)
Minnesota	34,720 km in total	Funk and Ruhr (1971)

About 7% (541,310 sq kilometres) of the USA, excluding Alaska, is subject to inundation by the 100 year flood (Goddard, 1976). Using the 1970 census and land data, Goddard (1976) estimated there to be about 22,000 flood-prone communities and 42,730 sq kilometres of urban floodplain in the USA. This covers a total of 6.4 million single family homes. Since 1953 there have been about 375 national disasters caused by flooding and an average of 100 people die each year from flooding (Costa, 1978). In spite of increasing annual expenditures for flood control, losses from flooding continue to rise in the United States. This results from a false sense of security arising from Federally supported structural solutions to flood problems, which encourage further development. It is inevitable that unless development is more strictly controlled the only alternative may be to modify every stream in the country (Costa, 1978).

Britain

In the UK there is a complex history of channelization, either for the purpose of making rivers navigable or to achieve the engineering objectives of flood alleviation or agricultural drainage (e.g. Vaughan, 1610; Taylor, 1864; Grantham, 1859; Smith, 1910). Lamplugh, in 1914, identified changes to the natural drainage network from maps of a scale 6 inches to a mile, which extend from the earliest cultivation to the late 19th century. This sequence of events may be typical of the majority of developed countries. Small streams had been aligned to run parallel with straight fencelines along old field systems, either for convenience of cultivation, or as boundaries between individual farms. Occasionally original bends had been retained on each side of the fence to allow access by cattle to the watercourse. Where mechanization had caused old field systems to be replaced by new, larger fields, then brooks were found to have been completely diverted away from their original course to follow a fenceline. The only trace of an old course was often a partly obliterated depression within a cultivated field.

Alluvial streams of a larger size were originally bordered by swamp and tangled thicket, but with gradual clearing they were converted to rough pasture or meadow. Typically the sinuous course was eliminated and the channel fixed artificially. Where the valley flats of larger watercourses were required for cultivation the rivers had been diverted or impounded over short lengths for irrigation of water meadows, to feed ornamental lakes, or most often to provide power for small water mills. Within urban cities and towns, brooks had often been confined within a conduit. It was possible to follow the course of a stream from the open country to the edge of the town, where it 'disappeared'.

Major rivers offered greater resistance to human interference because modification was both costly and difficult. However major rivers were changed in the late 19th century for agricultural or navigation purposes. These changes had occurred since the establishment of parish or county boundaries. At this time large rivers were deepened and stabilized, fords were replaced by bridges and weirs were used to check the gradient (Lamplugh, 1914). Finally, on estuaries the most impressive works were built: large embankments or walls being used to confine tidal waters and protect from flooding.

Navigable Rivers in Britain

The conversion of natural rivers into canals for the purpose of navigation is well documented (Priestly, 1831; Willan, 1936; Rolt, 1950, 1969). The raids carried out by Danish invaders of Saxon England represent some of the first historical references to extensive inland navigation, although there is less evidence to indicate that physical improvements were made to river channels at that time. Rolt (1950) suggests that much of this early navigation depended on a good tide; when boats became grounded on a shoal, temporary dams of turf and brushwood were sometimes thrown astern of the boat to raise the level of the water. During the 14th and 15th centuries extensive silting of rivers is generally thought to have occurred, at least partly as a result of changing land use, and several acts were passed to aid navigation.

During the period from 1600 to 1750 major improvements such as widening and deepening became inevitable following the substantial growth of population in England and the increased demands for essential commodities. The chief instrument of the administration of rivers before the 17th century was the Commission of Sewers, consisting of the chief landowners of a district and, although the primary function was that of drainage and flood control, support was often given to the defenders of navigation. Later, the authorization of river improvement by the Crown was through Letters Patent, which gave the undertakers the right to negotiate with the landowners concerned.

Examples of the earlier methods of improving channels are listed in Table III. An elementary technique used on the upper Thames to improve the channel was referred to as 'ballasting' (Robertson, 1875). A long-handled scoop was outrigged from a simple wooden punt, typically with a two-man crew. One man would hold down the scoop with a vertical pole, whilst the other would wind it in by means of a winch, and the spoil thus obtained from the channel bed would be tipped into the boat. More mechanized methods were adopted in later years, and in the 17th century many patents were taken out for dredgers. On July 16, 1618, John Gilbert took out a patent for a water plough for taking up sand and gravel and there is evidence that it made great holes in the bed of the River Thames.

Cutoffs were long recognized as a means of improving navigation by shortening the length of a channel, although this method was not widely used until the 17th century. The River Wey navigation, built in the 1600s, involved extensive cuts and of the total length of 24 km, 11 km consisted of

Table III. Selected examples of improvement for navigation (examples taken from Priestly, 1831)

River	Method	Date
Bristol Avon	removing obstacles	1745
Warwickshire Avon	dredging of shoals	1636
Beverley Brook	bank piling	C.1726
Kennet	artificial cuts	1715–32
Lee	scouring	C.1425
Medway	enlarging and widening	C.1792
Nene	embanking and straightening	C.1650
Great Ouse	bank protection with wood piles	1749
Yorkshire Ouse	contraction and scour	1726–32
Trent	dredging of shoals	1777
Wey	artificial cuts	1651–53
Witham	straightening	C.1600s

artificial cuts. Elementary bank protection involved the driving of alder piles against the bank with mallets, and typically a horizontal walling of poles would be attached to these piles, and the areas between the banks and the wall infilled with marl.

Many of the rivers made navigable needed constant maintenance, probably because disrupted river systems attempted to regain equilibrium. The works carried out to improve the Hampshire Avon between Christchurch and Salisbury in the 1660s were destroyed by a flood soon after completion (Priestly, 1831).

Later works for navigation were more substantial following improvements in dredging technology. Improvements on the River Severn were commenced in 1856 (Williams, 1860). From Stourport to Tewkesbury beds of marl and compacted gravel intersected the channel at various points, causing shallow rapids at low water and these were removed by dredging. Between Tewkesbury and Gloucester shoals of sand were deposited by the winter floods and spring tides due to decreased velocities in wide sections of channel. This problem was overcome by the construction of fascine embankments, which provided a uniform cross-sectional area for the channel, and the accumulations of sand and gravel removed by steam dredgers were deposited behind the fascines.

After 1760 the history of river transport in England and Wales became bound with that of artificial canals rather than the improvement of rivers. In 1942 the waterways were brought under unified government control, being passed to the British Transport Commission in 1948. The Commission was superseded by the British Waterways Board under the Transport Act of 1962 (British Waterways Board, 1964, 1973). Several rivers have undergone improvement and maintenance schemes and these organizations have been responsible for protecting many kilometres of banks of navigable rivers with steel and concrete piling.

Flood Alleviation and Agricultural Drainage in Britain

Flood and agricultural drainage schemes have been carried out by a number of organizations in Britain during the past 550 years. The need for flood control and land drainage, arising primarily from the growth of population, was recognized by various piecemeal legislations, commencing with the Statute of Sewers in 1427, and subsequently the Court of Sewers Act of 1531, which could compel landowners to maintain a river adjacent to their land, or levy a rate for such maintenance. However Commissioners could not 'make a new river, or try Inventions at the charge of the Country', although they could make new cuts for drainage purposes.

The Land Drainage Act of 1861 was perhaps the most notable landmark in the history of land drainage, permitting the maintenance and improvement of existing works, and the construction of new works. Elective Drainage Boards, introduced by the Act, covered areas usually smaller than the catchment area of a river and in many cases the Boards were inefficient.

A Royal Commission on Land Drainage in England and Wales reported in 1927 that rivers under 'modern' conditions of roofing, paving, roadmaking, sanitation and agricultural underdrainage were called upon to 'discharge functions for which they were not designed by Nature'. Subsequently the Land Drainage Act of 1930 put land drainage and flood control works under the integrated control of 46 Catchment Boards. Finance came principally from the County and Borough Councils within the catchments, whilst central government grants were made available to meet the costs of new drainage works or the improvement of existing works.

Under the River Boards Act of 1948, Catchment Boards and Catchment Areas were superseded by 32 River Boards and River Board Areas, except for the Thames and Lee catchments. Their functions were taken over in 1965 when the Water Resources Act of 1963 established 27 River Authorities. Finally in 1974 ten Regional Water Authorities took over the functions of the river authorities under the Water Act of 1973.

Under the Land Drainage Act of 1976 the main rivers in England and Wales are managed solely by the Water Authorities and Internal Drainage Boards (Wisdom, 1979). Drainage authorities acting within their areas can maintain or improve existing works or construct new works. Schemes can be grant aided by the Ministry of Agriculture, Fisheries and Food, whilst other works may be carried out as part of an authority's normal maintenance programme. On smaller non-main watercourses the councils are empowered to carry out drainage works for the purposes of preventing flooding or mitigating the damage caused by flooding, but the consent of water authorities is required for major works. Local authorities may obtain government grants for works. The Forestry Commission also carry out works on smaller watercourses. Under the 1976 Act Internal Drainage Districts continue for the purpose of draining low-lying areas adjacent to larger rivers and fen districts. On the smallest watercourses, field works are undertaken by farmers for the purpose of improving agricultural efficiency and a Farm Capital Grant Scheme exists to provide grants.

The extent of channel works in England and Wales has been established and mapped for the period 1930 to 1980 (Brookes, 1981; Brookes *et al.*, 1983). Figure 1 depicts a two-fold classification using data collected from individual water authority engineers and from the annual reports and accounts of organizations which document the works. The first category includes those types of channelization which may be regarded as having a

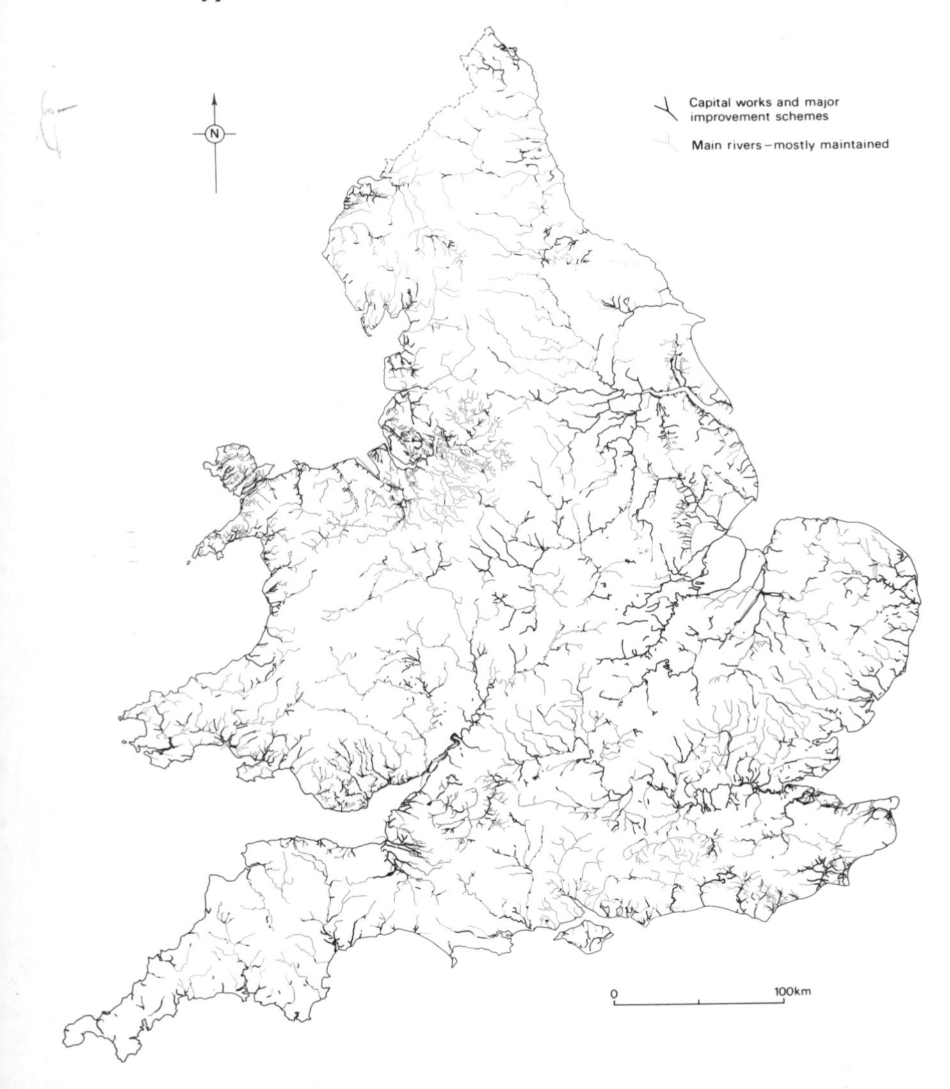

Figure 1. Rivers channelized in England and Wales 1930–1980. (From Brookes, A. *et al.*, 1983. Reproduced by permission of Elsevier Science Publishers.)

major and lasting impact on channel morphology such as embanking and embankment improvement, channel enlargement by widening or deepening, the realignment of channels, bank protection and the lining of channels with concrete. No distinction is made between channelization of one or both banks. This category includes those which are grant aided by central government and also those major works funded from a different source. Works have been undertaken on a total of 8500 km of main river.

A second category (Figure 1) includes the lengths of river in each district for which individual organizations are responsible for maintenance. Such maintenance has frequently been undertaken on a routine basis over the period, has often involved considerable lengths of channel, and has sometimes occurred two or three times per year. The maintenance activities include weed control, plant removal including weed cutting, the removal of accumulated shoals, and the removal of rubbish from urban channels. It is not possible to show the precise length of river maintained because the works are often unplanned and frequently undocumented, and therefore the main river length has been adopted as a compromise. This involves a length of 35,500 kilometres of river channel.

The lengths of capital works and major improvement schemes have been determined and presented for individual Water Authority areas (Figure 2).

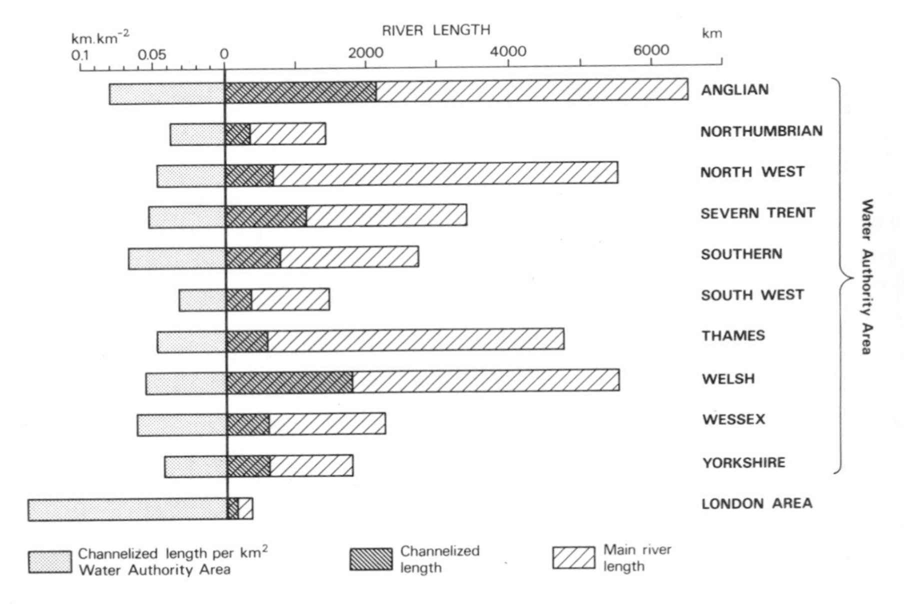

Figure 2. Lengths of channelized river in Water Authority areas in England and Wales. The works included those undertaken by Catchment Boards, River Boards, River Authorities and Regional Water Authorities 1930–1980. (From Brookes, A. *et al*., 1983. Reproduced by permission of Elsevier Science Publishers.)

The percentage of main river which is channelized varies from 41% in the London area, and 33.7% in the Severn Trent and Yorkshire Water Authority areas, to 12.3% in the North West and Thames Water Authority areas. To facilitate further comparison between the Water Authority areas the lengths of channelized river have been standardized per unit area. Clearly, those areas with a large residential or industrial area in close proximity to a major river or in lowland agricultural areas rank highest.

Denmark

In many West European lowland countries which are intensively cultivated, such as Denmark, the Netherlands and northern Belgium, channelization has been extensive. The notion of modifying channels for the purpose of agricultural drainage in Denmark was implicit in early Parliamentary legislation, including the Watercourse Law of 29 July 1846 (Andersen, 1977). The Watercourse Act of 11 April 1949, with amendments in 1963, 1965, 1969 and 1973, stated that 'drainage shall have priority over any other use'. Uses in conflict with this rule required the special permission of a Watercourse Tribunal. The natural bedforms, accumulated sediments and aquatic vegetation on the bed and banks of a river channel were regarded as obstructions which impeded drainage. Under the Act a section of watercourse was designated as either public or private, the former being managed by municipal and county authorities, the latter by riparian landowners. Rules relating to the maintenance of public rivers were laid down in a series of 'Regulatives' which defined the course, its dimensions (width and depth) and the elevation above sea-level. These regulations also specified the times of the year when dredging and weed cutting were to be undertaken to control the water level.

A survey of sinuous and straightened stream channels in Denmark, collated by Brookes (1987c) using maps, field surveys and engineering documents, revealed that there are only 880 km of sinuous channel. If the total network of public and private streams depicted on 1 : 25,000 maps is taken to be 40,000 km, this value represents only 2.2% of the total. The remaining 97.8% has been straightened, which is equivalent to a density of modified watercourses of 0.9 km km^{-2}. Although not strictly comparable, this compares with a density of channelized river in England and Wales of only 0.06 km km^{-2} (Brookes *et al.*, 1983) and 0.003 km km^{-2} for the United States of America (Leopold, 1977). Denmark thus has a density 15 times greater than England and Wales and 300 times greater than the USA. These differences can be attributed mainly to the intensities of land use: the majority of the surface area in Denmark is either intensively farmed or developed for residential or industrial purposes.

Forty-three per cent of the individual lengths of the remaining sinuous channels are less than 1000 metres and 73% are less than 3000 m. The sinuous channels which survive are either very large rivers which may have been totally difficult to regulate; streams in areas which have not been intensively farmed because they are used for other purposes such as military training; streams in forested areas; streams confined within steep-sided valleys unsuited to farming; or a very small minority of streams which have had sufficient energy to regain their sinuosity following straightening.

Case Study: Mississippi River

The history of channelization of the Mississippi River in the United States has been complex and various aspects have been reviewed (e.g. Ockerson, 1898; Elam, 1931; Brown, 1931; Elliot, 1932; Clemens, 1936; Tiffany, 1963; Stevens *et al*., 1975). Although the first levees were built in 1699 by individual landowners, it is only during the past 150 years that development has been extensive (Winkley and Harris, 1973; Winkley, 1976, 1982). This case study demonstrates the extent of human intervention on one of the world's largest river systems.

Prior to the 19th century the principal channel works involved the removal of snags, bars and caving banks. However, major earthquakes in 1811 and 1812 caused excessive bank caving and increased sedimentation, causing bar growth and navigation problems. Commencing in 1719, the construction of levees confined floodwaters and increased sediment movement downvalley. Steamboats directly caused the clearance of thousands of hectares of streambank vegetation in the 1800s. This, together with the clearance for agriculture, caused bank instability. Increased sediment loads and runoff from agricultural land caused more bank caving. In 1879 Congress established the Mississippi River Commission.

As a direct consequence of the bank instability, a large-scale protection programme was initiated in 1884 to restrict normal migration of the channel and prevent cutoffs. However, limited funding meant piecemeal protection to those sites with the greatest need. Dredging for navigation began in 1895, causing the release of sediment downstream. Training structures (dikes) were introduced in the late 1800s, although these were mostly temporary structures of wood until the use of limestone in the 1960s. These structures affected the morphology of the channel by causing local scour and deposition.

During the 1930s and 1940s, particularly as a result of the 1927 flood which overtopped the embankments, the river was shortened by 243 km through cutoffs and a further 86 km were planned. These cutoffs were undertaken over a 13 year period (1932–1955) with the intention of eliminating erosion and making energy available for the transport of sediment. They had perhaps the most dramatic effect on the morphology of the river (Figure 3). To maintain the alignment of the river following the cutoffs, it was essential to dredge continuously for a period of more than 10 years, especially in the straightened reaches. Over 1300 million cubic metres of sediment were removed to maintain navigation. Dikes and revetments were constructed between 1945 and 1970.

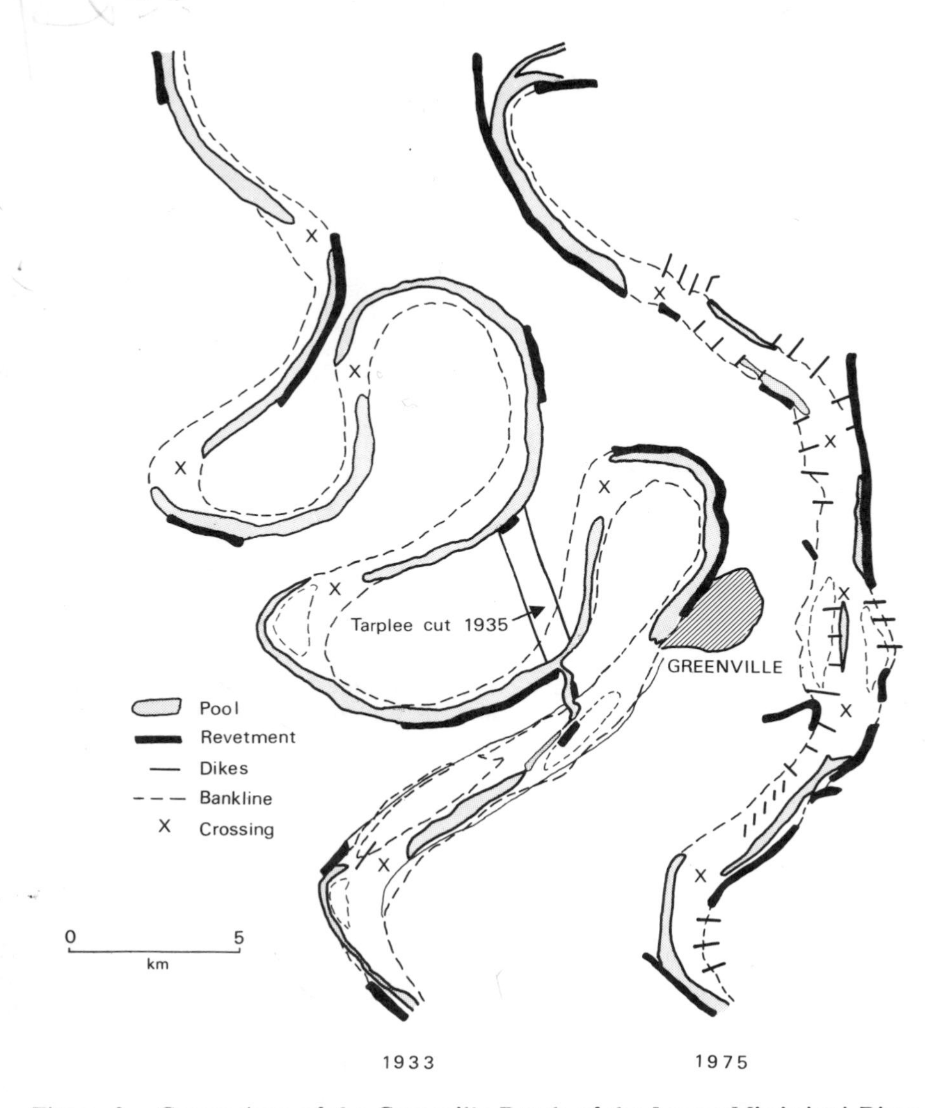

Figure 3. Comparison of the Greenville Reach of the Lower Mississippi River, 1933 and 1975. (From Winkley, B.R., 1982. Reproduced by permission of the American Society of Civil Engineers.)

Each of these types of channelization had an effect on the river and, because of the complexity over space and time, indicate that it is difficult to study cause and effect of an individual impact. Figure 3 shows the Greenville reach which was shortened from 82 to 38 km and restrained by revetments and dikes. The 1975 river had reformed pools and riffles despite the new alignment. Table IV summarizes the response of the river to the cutoff,

Table IV. Response of the Greenville Reach of the Mississippi to cutoffs

	Prior to 1933	1934–1974
Number of times dredging required to maintain navigation	0	135
Length of revetment to hold channel	23,271 m	41,776 m
Length of dikes in reach	1029 m	18,775 m
Length of river from upper end of construction to lower end	82 km	38 km

From Winkley (1976). Reproduced by permission of the American Society of Civil Engineers.

which occurred in a mobile sand-bed channel and demonstrates that a continuous and costly programme of maintenance and bank protection is required to preserve the integrity of a flood control programme.

Public Controversy

The problems surrounding channelization came to the fore in the early 1970s in the United States. It has been described as an 'insidious cancer' which contradicts many of the basic principles of water management (Bauer and East, 1970). It was also stated that 'The intent of Public Law 566 was to save and improve, but that (through channelization) it is being used to destroy' (Blackburn, 1969). Table V reflects the degree of public concern over the detrimental environmental consequences surrounding channelization.

Table V. Articles describing the controversy surrounding river channelization

Title of article	Area	Source
'Our ruined rivers'	Georgia	Bagby, 1969
'The gravediggers'	USA	Bauer and East, 1970
'Channelization: short cut to nowhere'	Virginia	Corning, 1975
'A vanishing part of Louisiana: its streams'	Louisiana	Davidson, 1969
'Crisis on our rivers'	USA	Miller and Simmons, 1970
'The vanishing stream'	USA	Gebhards, 1970
'Ravage the river'	Montana	Posewits, 1967
'Twilight for two rivers'	Montana	Robinson, 1969
'The stream that used to be'	Montana	Seaburg, 1971
'A river fights back'	Montana	Spence, 1968
'How to kill a river by "improving" it'	—	Whistleblower, 1974
'River improvement?'	Victoria (Australia)	Anon, 1977
'Wildlife down the drain'	Wales (UK)	Davies, 1982

In the United States intense opposition by conservationists to the process of channelization was witnessed between 1970 and 1975, although it is unrealistic to assume that all conventional channelization is environmentally damaging (Anon, 1972). Before the 1970s there was sparse factual data to support either side of the channelization argument; arguments being based on observations or limited studies. A Committee on Government Operations of the House of Representatives reported in 1973 that:

> A common thread running through the Subcommittee's hearings, correspondence, and subsequent studies was not that channelization, per se, was evil, but rather that inadequate consideration was being given to the adverse environmental effects of channelization. Indeed there is considerable evidence that little was known about the effects and, even more disturbing, little was done to ascertain them.

One of the outcomes of the Committee on Government Operations report was that the Fish and Wildlife Services's Office of Biological Services initiated a large-scale programme of studies throughout the USA to investigate instream alteration (Fox, 1975). The intention was to provide the Fish and Wildlife Service with a sound data base on which to make decisions. River channelization became a topic of research at various universities, including Ohio State (Edwards, 1977; Woods, 1977; Porter, 1977; Weber, 1977), Wisconsin (Vandre, 1975; Sanders, 1976) and others (Burnside, 1967; Holtz, 1969; Wyche, 1972; Blair, 1973; Perry, 1974; Ellis, 1976).

In December 1971 the US Fish and Wildlife Service issued a general policy statement which said that the Service would cooperate fully in the planning, formulation and implementation of proposals that are environmentally sound; minimize harmful effects on fish and wildlife and maximize enhancements. It was also stated that the Service would oppose any proposal for development or operation of a water project *that was not environmentally sound*; every effort would be made to avoid and prevent damage and loss to fish and wildlife. Compensation would be required for unavoidable damages.

Since the early 1970s controversy surrounding the effects of conventional engineering practices has extended to other countries, particularly in Europe, and this has been followed by efforts to elucidate the effects and implement alternative channel designs which attempt to work with Nature rather than against it.

The Impacts of Channelization

A Task Committee of the Hydraulics Division of the American Society of Civil Engineers reported in 1978 on the environmental effects of a range of hydraulic structures. The aim was to produce an overview of environmental effects for planning and design engineers. Table VI categorizes the environmental effects as physical, chemical and biological effects on water, land and air systems and lists those which are most affected by channelization (streamflow structures). The Committee also recommended that site-specific matrices should be developed for assessment of the environmental impact of hydraulic structures.

Table VI. Selected environmental effects of hydraulic structures (after Task Committee, 1978)

Water	1.	chemical (e.g. dissolved solids, dissolved oxygen, pH, organics)
	2.	biological categories (phytoplankton, zooplankton, warm and cold water fish, benthic organisms, waterfowl, aquatic plants—algae and macrophytes)
	3.	physical (turbidity, light penetration, temperature, sedimentation)
Land	1.	biological (terrestrial animals, amphibians, birds, insects, floodplain plants)
	2.	physical (sedimentation, erosion and aesthetics)
Air		e.g. odours

From Task Committee (1978). Reproduced by permission of the American Society of Civil Engineers.

The physical and ecological consequences of 42 channelization schemes in North America were assessed by the Arthur D. Little Company in a report in 1973 for the Council on Environmental Quality. Part of the research was carried out under subcontract by the Academy of Natural Sciences of Philadelphia. The physical and biological issues that were assessed were (1) wetland drainage; (2) clear-cutting of hardwood trees; (3) cutoff of oxbows and meanders; (4) changes in water-table levels and stream recharge; (5) downstream effects; (6) erosion and sedimentation; and (7) channel maintenance.

One of the main criticisms of the Little (1973) report was that none of the projects that had received widespread public attention, namely the Alcovy, Cameron Creole, Starkweather, Chicod Creek, Obion Forked Deer and Cache Rivers, were included in the study. As a consequence of this, the President's Council on Environmental Quality received a flood of complaints in the following year from federal agencies and environmental groups because it appeared to be in favour of continued stream improvement (Gillette, 1972a,b).

The principal consequences of channelization are summarized in Figure 4. and include a reduction in the complexity of habitat by elimination of pools, riffles and non-uniformities in channel geometry; increased water temperatures due to removal of bankside and instream vegetation; bed and bank erosion; and downstream flooding and sedimentation. The combined effects of these changes produce a wide range of biological impacts, principally

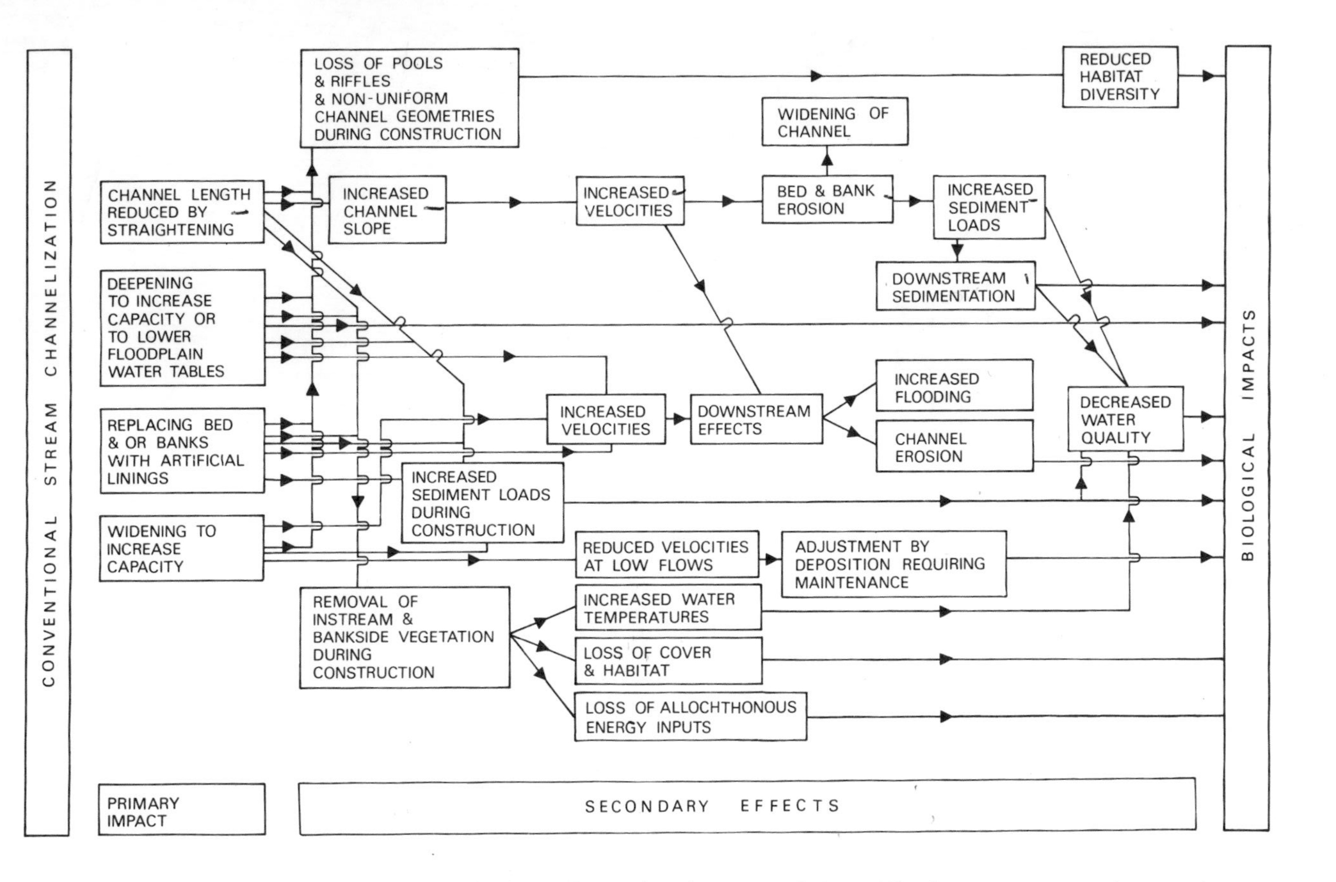

Figure 4. Principal effects of major types of channelization.

upon benthic invertebrates, fish and aquatic vegetation. In addition, lowering of the water-table in an adjacent floodplain, as a direct consequence of channelization, can produce impacts on natural vegetation and wildlife.

The intention of this book is to summarize the developments in scientific research on the physical and biological impacts of channelization, to indicate future avenues for research, and to recommend ways in which river channel management can be improved during the next decade. The book encompasses five main components: 'Introduction and problem' outlines the need for river channelization and the historical background (Chapter I) before reviewing the methods and limitations of conventional engineering design (Chapter II). Following the realization that rivers have been extensively modified by such techniques, new or revised legislation has recently been developed with the intention of minimizing the environmental impacts. A number of new management methods and procedures have been produced to meet the requirements of this legislation (Chapter III; 'Legislative Framework'). The undesirable physical and biological consequences of conventional channelization methods are detailed in Chapters IV, V and VI ('Effects of Channelization') and alternative techniques and recommendations which aim to work with a river rather than against it are outlined in Chapters VII and VIII ('Recommendations'). The prospects for river management are then reviewed in Chapter IX ('Postscript and Prospects').

CHAPTER II

Conventional River Engineering

The main objective of this chapter is to show what is entailed in conventional river engineering methods, including resectioning, realignment, diversion channels, embankments, bank protection and training works, flood walls, culverts and maintenance. Procedures available for the design of stable channels are also discussed and the limitations outlined. In several countries channelization has traditionally been justified on the basis of benefit–cost analysis which fails to account for the environmental costs. Solutions to flooding problems which do not entail directly modifying the channel, but have been used less frequently, include flood storage reservoirs located in the basin upstream, flood proofing of buildings and effective floodplain zoning.

Conventional Engineering Methods

Several methods of channelization exist for the purposes of flood relief, agricultural drainage, erosion control or navigation. River engineering works invariably require considerable modification of the environment. Examples of the more conventional types of river channel engineering are depicted in Figure 5. Flood channels are usually wider and deeper that pre-existing natural channels to contain the design discharge and frequently trapezoidal sections are constructed in non-cohesive sediments to achieve bank stability. Rectangular sections can also be built but with the vertical sides and bed lined with durable material such as concrete. Channels may also be deepened to lower the water-table in an adjacent floodplain, thereby potentially increasing agricultural productivity. During excavation channels may be regraded and the natural pool–riffle sequence and large-scale roughness elements removed. Erosion control can take several forms but in high energy environments attempts have been made to restrain braided channels by the extensive use of bank revetments. Meandering channels are straightened, thereby increasing the slope, and speeding floods downstream.

A large proportion of the works undertaken on rivers throughout the world have involved more than one method, these works being described as either 'comprehensive' or 'composite' (Johnson, 1966). Each individual river problem is treated as a case study by an engineer and the most practical

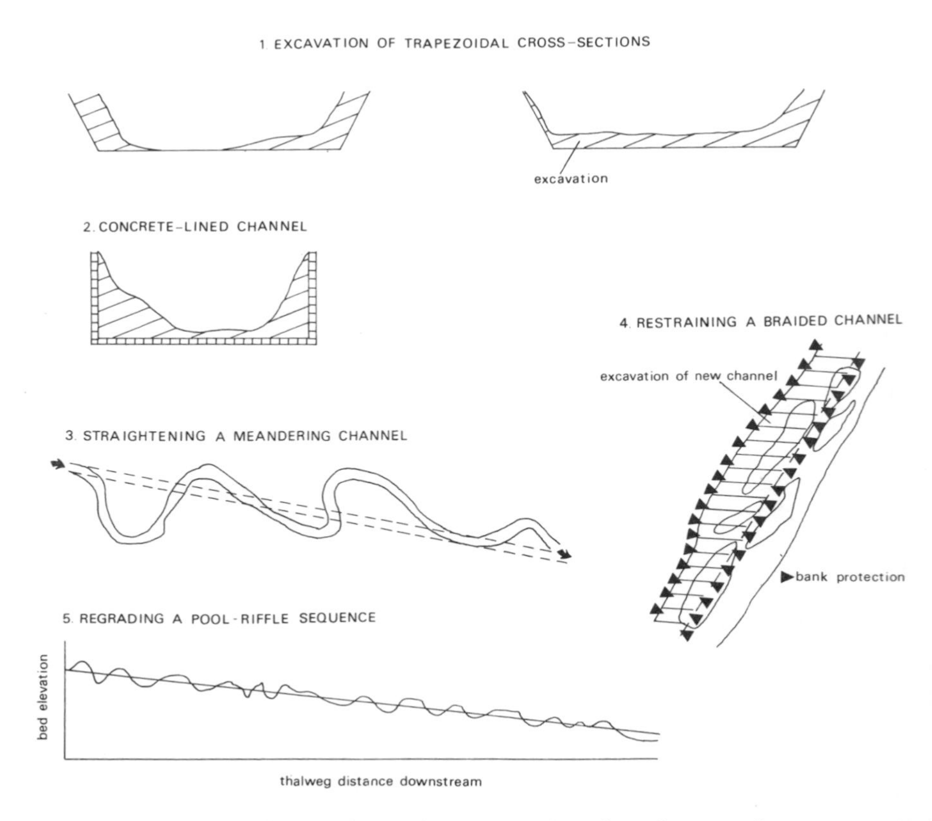

Figure 5. Conventional river channel engineering practices.

and economic solution is usually sought for a particular river. Numerous textbooks covering the design and construction of stable channels have been published this century and include for the United States of America: Elliot, 1919; Etchverry, 1931; Ayers, 1939; Pickles, 1941; Rouse, 1950; Chow, 1959; Leliavsky, 1966; Karaki *et al*., 1974; Vanoni, 1975; for Britain: Minikin, 1920; Adkin, 1933; Richards 1950; Institute of Water Engineers, 1961; Raudkin, 1976; for Ireland: Lynn, 1970; France: Aubert, 1949; The Netherlands: NEDECO, 1965; Jansen *et al*., 1979; India: Malhotra, 1951; Joglekar, 1971; Garde and Ranga Raju, 1977; and for Australia: Strom, 1962. Engineering journals are also replete with examples of schemes involving a variety of methods (e.g. Szilagy, 1932; Grant, 1948; Doran, 1959; Duckworth and Seed, 1969; Jackson and Bailey, 1979; Watson, 1981). Other projects have combined channel works with alternative strategies such as flood control basins or washlands in the headwaters (e.g. Spieker, 1970).

Resectioning by Widening and Deepening

This involves enlargement of a channel by widening and/or deepening to increase the conveying channel cross-section so that water which would previously have spread onto the floodplain is contained (Figure 6). Widening or deepening of a channel permits a given quantity of water to flow through at a lower level than in the unimproved channel. For flood control purposes the size of the modified channel is determined by the flood discharge which is to be contained. Resectioning of a river is usually combined with regrading of the bed, and associated works may include the underpinning of structures or their protection by the construction of inverts and the rebuilding of bridges. Regrading of the bed and widening may also be necessary to improve water levels for storm-water outfalls from urban developments or drainage outfalls in fields and as a means of lowering the water-table in an adjacent floodplain to improve agricultural efficiency. In theory a channel should have a cross-sectional area which provides the maximum efficiency

Figure 6. Resectioned small stream in the English Lake District. Work undertaken by the North West Water Authority in 1982.

in discharge with the minimum of excavation, but in practice this is not always possible due to factors such as bank instability, porosity of the bed and the proximity of structures which restrict the width or depth to which the channel can be constructed. Channels with unlined earth banks are often designed with trapezoidal cross-sections to provide side slopes for stability. Since the rectangle has vertical sides it is commonly used for channels built of stable materials such as concrete. A triangular section is often used for small ditches and roadside gutters (Chow, 1959).

Realignment or Straightening

Several scales of channel realignment can be distinguished, ranging from an 'improved' alignment introduced by dredging, to the more conventional forms of cutting-off bends in rivers. The objective may be to design an adequate channel to convey flood flows. A certain degree of realignment can be introduced into a channel as a result of maintenance dredging, being achieved by the removal of shoals which may have accumulated as point bars on the inside of bends. The technique is intended to reduce the flood level in a reach by increasing the velocity of flow. A particular practice used on gravel-bed rivers in New Zealand is 'rock raking' (Table VII), whereby the heavier material in shoals is loosened and removed (Acheson, 1968). In some circumstances bridge protection is achieved by moving boulders up to 4.1 tonnes in weight.

Table VII. Channel development by rock raking as carried out by the Ministry of Works, Christchurch, New Zealand (summarized from Acheson, 1968)

(1) a ditch 1.5 m deep was excavated by a bulldozer down the fairway margins and unsorted material pushed outwards to form a base for the 3 : 1 rock bank

(2) all stone greater than 25 cm diameter was then raked into the ditch to protect against undermining in the early stages

(3) all exposed stone on the bed was lifted and raked to continue the rock-bank development

(4) the channel was then left for floods to remove all the unwanted fines and expose further stone before the process of rock raking was repeated

(5) many over-sized stones which appeared after degrading were blasted to prevent channel obstructions

(6) provision was made for extensive planting of native trees and shrubs

The results of the rock raking procedure are not spectacular but allow a steady channel improvement. After 2–3 years a channel of good alignment and section, with uniform gradient and paved with rock that will scour during normal floods, is achieved.

A more significant form of realignment involves the shortening of a river by means of a cutoff, often at the scale of an individual meander (Figure 7 (a) and (b)). In South Wales (UK) it has been shown that if the sharp bends are eliminated on gravel-bed rivers then the gradient becomes excessive and erosion results from the increased velocities (McLeod, 1970). Low flow weirs may therefore be introduced to reduce the gradient. Programmes of cutoffs have been carried out on a number of rivers including the Lower Mississippi (Winkley, 1976), the Middle Yangtse (Ching-Shen *et al.*, 1977) and the Yellow River (Huang Chung, 1972). In extreme cases straightening may extend for several hundred kilometres (e.g. Campbell *et al.*, 1972). Cutoffs are used to reduce flood height by increasing the gradient and therefore the velocity. They are also used to improve manoeuvrability during navigation (e.g. Pickles, 1941; Matthes, 1949).

Diversion Channels

Relief channels have been constructed which have the purpose of diverting flood flows only away from an area to be protected, the existing channel carrying the normal flows. These are favoured in urban areas where it is not possible to widen the existing channel due to development. In New Zealand 'floodways' are used to provide a means of diverting floodwater across the neck of a meander or a series of meanders (Acheson, 1968). The inlet to a floodway may be controlled or uncontrolled. Das (1976) described a cut made on the River Bhargavi in India which was used to bypass floodflow. A diversion is only effective in reducing the flood stage if the distance between the point of diversion and point of return is sufficiently large to overcome backwater effects.

Finally, cutoff channels can be used as an entirely separate system for the purpose of diversion of all flow away from an area. A classic example is the Great Ouse Flood Protection Scheme in England, which directs the flow of the Rivers Lark, Little Ouse and Wissey along a 43 km long cutoff channel from Mildenhall to Denver Sluice (Nixon, 1966).

Embankments or Levees

Embankments are also known as flood banks, levees, bunds or stopbanks. Their purpose is to artificially increase the capacity of a channel so that high flows which would normally have spread onto an adjacent floodplain are now confined. They are one of the oldest forms of flood protection, used in either rural or urban areas provided that there is sufficient space for construction. Some of the great rivers of the world have extensive embankment systems such as those that extend for more than 1000 km alongside the Nile, 700 km along the Hwang He and 1400 km on the Red River in Vietnam (United Nations, 1951; Ward, 1978). They are key components in flood control systems

Figure 7. Realignment by dragline of the River Wylye near Stapleford, Wiltshire, England during Winter 1982. The course of the river was changed from position (a) to (b) to facilitate road widening.

along the lower courses of large rivers such as the Mississippi, Missouri and Sacramento Rivers in the United States and are intended to protect major towns and cities which have become established on wide floodplains. Older embankments often require extensive maintenance, including strengthening with willow fascines and steel sheet piles, and maintenance dredging to remove silt from the channel that would formerly have been an overbank deposit at times of high flow. Examples of the design of embankments include Marsland (1966), Jaffrey (1967) and Ibad-Zade (1973).

Embankments are built to contain a design discharge and the entire floodplain may be protected when two banks are located sufficiently close together, although this is extremely expensive because the banks need to be very high. In other circumstances the embankments are placed just outside the meander belt of a migrating river to avoid erosion, and if a high discharge capacity is required for a given stage then the embankments are placed far apart. Both the topography and human infrastructure on a floodplain influence the alignment of embankments. The elevation is primarily selected according to the design flood discharge and its accepted probability of exceedance, the design flood being routed through a section and flood stages calculated at desired positions along the channel. An embankment cannot be built too high because of the increased danger to the population if the bank were to be overtopped or breached during an extreme flow event. To allow for subsidence following construction a freeboard may be incorporated above the design level but in general the slopes must not be over-steep, otherwise failure may occur. Trapezoidal sections have been used, typically with 1 : 2 side slopes, although in rural areas it is recommended that a bank top width of 3 metres is used, with a slope of 1 : 7. Berms may be incorporated between the channel and the foot of the structure, thus allowing improved access and a higher discharge capacity for a given stage. Top widths should exceed 2 m and this often has to be wide enough to allow maintenance traffic. Embankments are normally constructed from material excavated either from the channel or from a borrow pit in the floodplain, but can also be built from imported materials. Suitable construction materials include rubble and clay, shingle, sands and clay.

New banks require a site investigation to determine information on the soil type and properties such as shear strength. This should be obtained by making boreholes at regular intervals. Low-strength soils will limit the height of the embankment. Embankment stability should be checked using slip circle analysis in cohesive soils, including checks under floods and conditions of high groundwater-table. The effects of seepage should be determined, especially if the bank is to be built from silt. A filter (e.g. geotextile) could be used to offset the effects of seepage. To prevent fissuring of clays in dry weather it may be necessary to plant a good grass cover. Table VIII summarizes the design considerations used for flood banks in New Zealand by the Ministry of Works.

Table VIII. Design considerations for stopbanks in New Zealand (after Acheson, 1968)

(1) Establish the justification for stopbanks. This is determined by the width of the valley floor, its present condition and potential; the width of the flood channel required to take the design flood and the relationship of benefits and costs.

(2) Do not locate too close to an eroding river bank and avoid unstable ground. Width between banks should not be confined such that flood levels are raised to an undesirably high level. On braided rivers the location of banks on terraces often has considerable advantages.

(3) Where a high degree of protection is required banks are designed to the 25–100 years flood standard with 0.6–1.0 m freeboard depending on the characteristics of the river and the exposure of the bank to direct attack or wave action. Where a low to moderate standard of protection is required then the banks may be designed either to withstand overtopping or floodways/spillways may be constructed to carry flows at high flood stages.

(4) Check foundations, especially where the stopbank crosses peat swamps or an old river channel.

(5) Bank section depends on flood characteristics of the river, the type of material and construction methods. Preferred section for low to medium banks is 2 m top width and 2 : 1 side slopes. Provision of clay blanket is desirable.

(6) Key the bank into the foundations; satisfactory compaction of material is essential.

(7) Grazing of banks is necessary to limit growth of vegetation and controlled grazing is desirable.

Bank Protection Methods and River Training Works

Channel stabilization and training works may be undertaken to protect against abrasion and bank slip, particularly at locations where undermining of a bank would result in an impediment to flow, thereby causing flooding (Petersen, 1964; Carey, 1966; United Nations, 1953; US Army Corps of Engineers 1966, 1969; Brooks, 1976). Groynes, dikes or spurs are structures which are built transverse to the river flow and extend from the banks into the channel. Their purpose is either to guide or deflect the axis of flow, create a desired channel width, promote scour, or build up the river banks by trapping the sediment load and inducing deposition (Hedman, 1965; US Army Corps of Engineers, 1969). In New Zealand bank protection work is carried out after channel clearance, particularly where the river channel is braided. Stabilization may be required where excessive shoaling causes bank erosion (McLeod, 1970), where flood banks are threatened by a migrating river or because of the increased wash from boats (Goodman and Staines, 1969). The methods of channel stabilization have been reviewed by a number of authors (e.g. Johnson, 1966 and Charlton, 1980 in the UK; Richardson *et al.*, 1975 in the United States). A wide variety of methods practised in New Zealand is summarized in Table IX.

Economic considerations dictate the use of cheaper and lighter types of bank protection in New Zealand (Table IX) and these have a higher failure rate. Acheson (1968) stresses the need for correct siting of the works and recommends that such works fit in with a general training plan for the whole river or section of river. Authorities in New Zealand encourge the planning of comprehensive schemes of work rather than piecemeal protection for this reason.

Groynes can be either impermeable or permeable, depending on whether it is desirable to trap the bed material or allow water to pass through the materials, thereby inducing deposition of suspended material between the groynes. The width, slope, shape and spacing of groynes and location of the axes in relation to the current vary with the purpose for which the training works are being used and with the area to which they are being applied. Training fences comprise a line of piles driven parallel to the bank, with cross-fences between the longitudinal line of piles and the bank. Their function is to induce accretion by allowing a river to deposit behind the fence and build up the bank (Charlton, 1980). Permeable groynes are often built of timber or lines of stakes interwoven with willows, and impermeable groynes are constructed of concrete, earth-fill, boulders or gabions.

Revetments provide direct protection to the channel by armouring the banks and protecting the underlying soil layers against erosion. In constructing a revetment, irregularities are first removed and banks graded to acceptable slopes so that the structure will not be damaged due to improper support. The shear stress, and in particular the component along the slope of the weight of the revetment, is taken into consideration, and in theory the slope must be less than the critical angle of repose of the material (US Army Corps of Engineers, 1969). Slope values for alluvial material have been suggested as 1 : 6 below water and 1 : 4 above. A general recommendation is that revetments should be extended below the toe of the underwater slope to prevent failure through progressive scouring.

Depending on the materials used in construction, revetments may either be pervious or impervious. If the groundwater-table follows the water level during falling river level then an impermeable revetment built of concrete, stone or plastic sheet, would serve to obstruct this adjustment and excessive pressures would develop behind the structure. Pervious revetments act as filters but retain bank particles, and riprap is preferable. Other materials used for pervious revetments are of varying durability and include willow piling and fascines, tripods, gabions and synthetic materials such as glass-fibre (Danels *et al.*, 1960; Snell, 1968). Gabions and mattresses are rock-filled wire devices which have been used for controlling erosion and stabilizing soils for centuries (Keutner, 1935; Simons *et al.*, 1984). They have been used extensively in Italy, Germany and Austria and have been applied in the USA

Table IX. Methods of bank protection used in New Zealand (after Acheson, 1968)

	Procedure	Description
(a)	Stake planting	planting of willow or poplar
(b)	Shingle and stake	material bulldozed from the bed against low eroding banks and light willow poles or brush placed against the material
(c)	Mattress work	bank graded and covered with willow or poplar fascines or loose brush anchored with live stakes or netted stone 'bolsters' can be used to weigh down the fascines. Long line willow or poplar poles can be laid along the bank instead of fascines ('Hayman method')
(d)	Pile and fascine revetment	timber or old rail piles are driven in front of an eroding bank; fascines, brush or branches of willow are then placed between the piles and the bank
(e)	Live pile and stake fence	as above but live poles planted behind the line of piles and held in position with wires; space behind the piles may be filled with material dredged from the channel
(f)	Anchored willow and poplar work	large willow or poplar branches are anchored along an eroding river bank; may be planted in a trench; brush end facing downstream and held in place with a wire. Heavier construction to cope with severe bank erosion (e.g. heavy duty wire weighted with stones)
(g)	Rock raking	using coarse material from bed
(h)	Netted stone gabions and mattress	stone-filled gabions
(i)	Spur groynes	series of short boulder gabion groynes, built normal to the bank or pointing slightly upstream
(j)	Riprap	placement of quarried stone; size of material depends on local conditions
(k)	Concrete block; concrete or stone; masonry	used in urban areas at bridge sites
(l)	Sheet pile	timber or steel, backed with rubble, boulders or concrete
(m)	Heavy pile, cable and netting	heavy construction used in mountain environment. Cables attached to two rows of rails in chevron pattern; netting then attached by wire
(n)	Hedgehogs or sputnics Concrete tetrahedrons	concrete block with rails set into them. Heavy blocks which do not roll easily; placed in clusters

for over 60 years. Modern versions consist of a rectangular compartment made from thick steel wire which can be galvanized or coated wth PVC (Agostini *et al*., 1981, 1985). One of the main advantages of gabions is that the wire mesh permits the rock-filled basket to change shape without failure due to unstable ground or scour from moving water.

However specific methods have been developed for particular regions, depending on the local physical conditions and the preferences and experiences of individual engineers. Based on his experiences on small streams in the south-eastern United States, Lester (1946) identified three inexpensive categories of erosion control. These are mechanical control, vegetative control and maintenance. Mechanical control is required as a preliminary step towards the establishment of vegetation. The intention is that in the long-term mechanical structures such as timber or rock jetties will decay and disappear. They can be installed with farm equipment and labour. Old steel wire cable is used to anchor a mass of small woodland trees held together with smooth wire. The trees are lashed to a boom which projects from the bank, the tops of the trees facing downstream. If rocks are locally abundant then these may be used in conjunction with a simple woven-wire stocking. The aim is to direct flow away from the bank (Figure 8) and erode a point

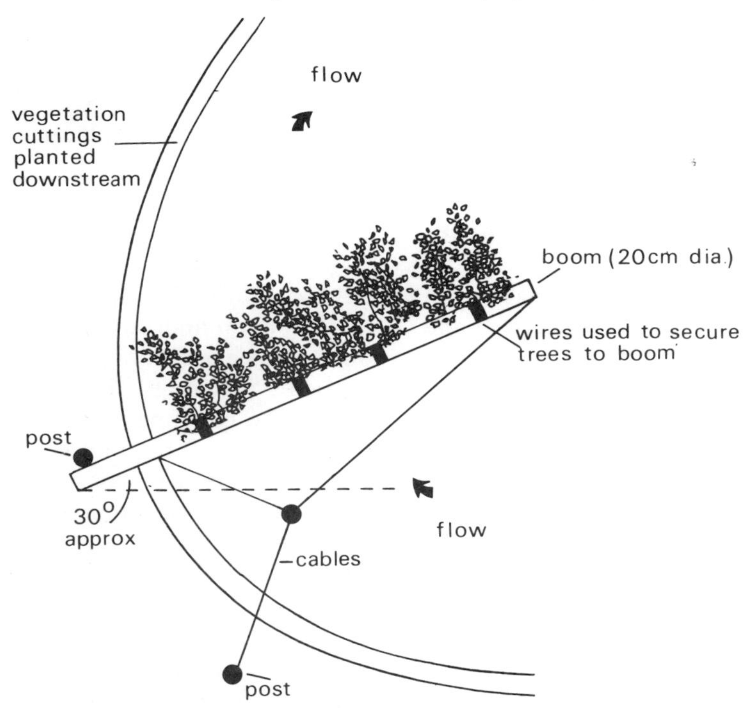

Figure 8. Protective measures of bank protection applied in the south-eastern United States. (Based on Lester, 1946. Reproduced by permission of American Society of Agricultural Engineers.)

bar on the opposite side. Sand and silt deposited downstream from the jetty may allow vegetation to become established. A series of jetties may be required around one bend to provide satisfactory protection and jetties should only project 30–40% of the stream width at flood flow to prevent excessive scour. A 30–45° angle of projection is suggested.

Secondly, vegetative control should be the ultimate aim in stabilizing eroding stream banks but can be used in conjunction with mechanical control. Plantings of shrubs and willows should be made at the bottom of the bank close to the water's edge to prevent undercutting. It is recommended that cuttings are taken and planted in holes made with a posthole borer in sand or silt. Planting of trees or shrubs along the bank can also prevent encroachment of agriculture. The position of the bank can be preserved by using a mat of protective vegetation which includes live willow poles. The eroding bank is typically graded to a 45° angle and willow poles laid on the bank about 0.3 to 0.6 m apart. This is done when the willows are dormant. The poles are covered by brush to a thickness of 15 cm and secured by woven wire. This can then be anchored to posts driven into the bank top. Heavy stone weights are used to anchor at the low-water mark. The willows root in spring and a dense mat is formed. Finally, it is crucial to maintain protective vegetation annually. If trees and shrubs are allowed to grow too large they may restrict streamflow.

The application of these techniques depends mainly on the physical conditions, including the rate of channel migration, the stream gradient, the presence of sand bars, and the presence or absence of vegetation.

Flood Walls and Lined Channels

Fully lined channels have been constructed throughout the world, particularly in urban areas such as the cities of London and Los Angeles, where space for other forms of channelization such as earth floodbanks is limited and where access for maintenance is restricted (Figure 9). The size of these channels depends on the flood discharge to be contained. Lined channels are frequently rectangular in shape, with a horizontal bed and vertical sides built from concrete (e.g. Erichsen, 1970). These may be constructed from reinforced or mass concrete, steel sheet pile or trench sheeting. Concrete diaphragm walls may be necessary where the soils are permeable. Steel sheets may be faced with brick or stone to improve the visual appearance.

Lined channels reduce the roughness dramatically, thereby increasing the velocity and decreasing the discharge for a given stage. A reduction in Manning's 'n' coefficient from 0.025 to 0.015 would increase a channel's carrying capacity by about 80%. Since the bed and banks are protected from erosion the purpose of such channels is to be maintenance free. The gradient is designed so that no deposition occurs. Furthermore the environment is unsuited to the growth of aquatic plants.

Culverts

In the course of urban development many urban watercourses have been culverted with masonry arches (e.g. Figure 10; Medrington, 1965), whilst on smaller streams large-diameter concrete pipes have often been used. Barton (1962) gives an account of how the streams in the City of London became lost beneath the streets. There are several disadvantages with these structures including problems of access for maintenance and the possibility of the design flood being exceeded. The culverting of streams in urban areas of New Zealand, subject to flash flooding, is not recommended because of the risk to life and property (Acheson, 1968).

Maintenance

This involves the dredging, cutting and removal of obstructions. It is obvious from flow equations such as the Manning equation that a reduction in the roughness coefficient at a particular slope in a channel will lower the stage for a given discharge. This reduction can be achieved by removing fallen trees, overhanging branches, and other obstacles from a channel.

A simple form of dredging requires material to be broken up, loosened and left to be transported out of a channel by the current, whilst in other circumstances sediment may be pumped ashore or discharged into barges before being dumped at selected locations. Mechanical dredgers are used to remove the material from the bed in pipes (Joglekar, 1971) whilst hydraulic excavators or draglines attached to land-based vehicles are used on smaller watercourses to remove silt and gravel shoals. The frequency of dredging varies with the environment and the rate of sedimentation. For example clay streams in East Anglia (UK) experience much silting and require dredging every 5 to 10 years (Haslam, 1978).

The annual growth of aquatic plants is one of the greatest problems in the maintenance of channel capacity in highly productive rivers such as chalk streams in both Britain and North America (Haslam, 1978). Prolific growth of vegetation can impede drainage of lowland areas. For the River Frome, a chalk stream in Dorset (UK), it has been estimated that the amount of weed removed during a cut is between 1.3 and 2.6 tonnes per hectare (Westlake, 1968b). To a limited extent plants physically reduce the capacity of a channel but particularly increase the roughness as well as the accumulation of silt, and in certain areas they have to be controlled several times a year. Submerged, emergent or marginal bank plants are managed by a variety of methods, including cutting either manually or by paddle-driven flat-bottomed boats with reciprocating swallow-tail cutting knives (Figure 11), chemical control by the use of herbicides, or by grazing with fish such as carp (*Cyprinus* spp.) (Chancellor, 1958; Robinson, 1969, 1971; Robinson and Leeming, 1969; Robson, 1973, 1977; Robson and Fearon, 1976; Miles, 1976; George, 1976; Brooker, 1976; Zon, 1977; Barrett, 1978; Ministry of

Figure 9. Typical concrete lined channel: River Towe at Towcester, England in 1981.

Figure 10. Culverted watercourse near Southampton, England in 1981.

Agriculture, Fisheries and Food, 1979). The control of aquatic plants is an important part of fishery management on chalk streams (Soulsby, 1974). Efficient control facilitates rod and line fishing and the spawning of salmon and trout, and encourages the presence of flies necessary for sport fishing.

Figure 11. Weed cutting machine on the Hampshire Avon, England during summer 1982.

The term 'clearing and snagging' encompasses removal of fallen trees and debris jams from the channel as well as the harvesting of timber from the banks and floodplain to increase the hydraulic capacity and prevent hazards to bridges or navigation. As a means of flood control on small streams in North America, the conventional practice has been to remove all obstructions from the channel (snagging) and to clear all significant vegetation within a specified width on both sides of the channel (clearing) (Shields and Nunnally, 1984). In Britain much 'pioneer' tree clearance was undertaken in the 1930s since branches trailing in the water created flood problems whilst whole trees blocked river channels, collecting further material and causing localized scour. Early channel clearance work in New Zealand involved the complete removal of willows by mechanical methods such as traction engines and bulldozers. Follow-up work was often undertaken to spray young growth and stumps. However Acheson (1968) recommended that in many circumstances complete removal is not advisable because willows form satisfactory bank protection.

Design of Stable Channels

In designing channel works a fundamental requirement is to ensure that the proposed channel will remain stable (Lane, 1955a). This will ensure the minimum burden of maintenance. In a lined channel the velocities required to cause erosion are relatively high, usually in excess of 3 m sec^{-1} and therefore the problem is one of constructional economics. However in erodible alluvial channels design is made extremely difficult by the processes of erosion and deposition. Alluvial rivers have a maximum capacity to transport sediment of a particular size range and if this capacity is exceeded then excess sediment will be deposited. Conversely if less than the required amount of sediment is supplied then the river will erode bed and banks composed of alluvial materials, provided that the forces exerted by the flowing water exceed the forces resisting sediment motion. The forces exerted by the flow arise from fluid flow and particle characteristics; the resisting forces are principally caused by cohesion and friction amongst particles. In the design of stable channels engineers have attempted to balance these two forces using concepts and formulae derived initially from empirical studies and subsequently from transport theory. Two principally used methods of channel design, together with their limitations, are outlined here: (a) regime theory, a wholly empirical approach, originally developed by irrigation engineers in India and Pakistan; and (b) maximum permissible tractive force. For further detail the reader should refer to one of the many textbooks on hydraulic engineering (e.g. Blench, 1957; Chow, 1959; Garde and Ranga Raju, 1977).

Regime Equations

The empirical approach to channel design is known as 'regime theory' and is based on the measurements of irrigation canals in India and Pakistan which were determined to be stable or 'in regime'. The equations related hydraulic and geometric characteristics of rivers and could be applied to design stable channels under similar physical conditions. Application to channels with dissimilar bed and bank sediments and sediment discharge can produce disastrous results. Lindley (1919), working in India, showed that the width, depth and slope of uncohesive sediments were related to the discharge, the amount and type of bed sediments being moved and the nature of materials forming the channel. Subsequently much work in India related the variables.

Many equations have been developed since the publication of the classic work of Kennedy in 1895 and it is not intended to review them in detail here. As a result of a greater availability of data and an improved understanding of the factors controlling channel morphology, more recent equations have greater validity and have been developed for a number of countries (e.g. Blench, 1952; Simons and Albertson, 1960; Charlton *et al.*, 1978; Hey and Thorne, 1986). Furthermore statistical analyses have been applied more recently. Research has refined and extended the regime equations using results from flume studies and sediment transport theory. The equations can be classified on the basis of bed material size into cohesive bed, sand-bed and gravel-bed equations. For each of these types of channel the width, depth and slope can be determined. Clearly considerable care is necessary in ensuring that regime equations are applied to the same geotechnical conditions as those for which they were derived because both the intercepts and exponents vary. Regime theory studies have met with limited success and no set of general equations has been produced which have wider application to natural streams. A recent review of these is given by Richards, 1982, amongst others.

The design procedure is exemplified by the system used by Lacey in 1930 for straight sand-bed channels. This is a relatively simple procedure, involving the following steps, in which the coefficients have been converted from fps units to SI units:

(1) Obtain a value for the 'silt factor' (f_L) from:

$$f_L = 1.59\sqrt{D_{50}}$$

where D_{50} = median diameter of sediment

(2) Determine the wetted perimeter (p) from:

$$p = 4.84\,Q^{0.5}$$

where Q = water discharge

(3) Calculate the bed slope (s):

$$s = 0.0003\,f_L{}^{5/3}\,Q^{-1/6}$$

(4) Determine the hydraulic radius (R):

$$R = 0.47\,Q^{1/3}\,f_L{}^{-1/3}$$

(5) Determine the velocity (v):

$$v = 0.44\,Q^{1/6}\,f_L{}^{-1/3}$$

(6) The wetted area can be derived from:

$$A = Q/v$$

where A = channel cross-sectional area.

The best hydraulic trapezoidal section, that is the channel having the least wetted perimeter for a given area and therefore the maximum conveyance, can then be determined from:

$$A = bd + zd \qquad \text{and} \qquad p = b + 2\sqrt{1 + z^2 d}$$

where b = bottom width and d = depth
z = bank slope (expressed as the ratio of the horizontal and vertical distances)

Tractive Force Theory

This has been widely used in North America for stable channel design in unconsolidated materials. The concept of tractive force was first introduced by duBoys (1879) last century, where the unit tractive force (τ_0) is:

$$\tau_0 = \omega R s$$

where ω is the specific weight of water
R is the hydraulic radius
s is the bed slope

However the approach was not considered widely in the United States until Lane in 1955a proposed a set of criteria for designing stable channels. To ensure stability the design procedure determines the minimum tractive shear values required to initiate movement (the critical tractive shear values) on the bed and on the sides.

The design of channels in non-cohesive sediments using tractive force is based on four principal assumptions:

1. The channel side slope at or above the water surface is equal to the angle of internal friction of the sediment. This angle (ϕ can be obtained if the grain size and shape of the sediment is known).
2. At incipient motion it is assumed that traction is opposed by the component of the submerged weight (W') of the grain acting normal to the bed, multiplied by the tangent of the angle of internal friction.
3. It is assumed that in the middle of the channel motion is initiated only by the shear stress exerted by the flow. Here the side slope θ is zero.
4. On the side slopes, where the angle is greater than zero, there is an additional downslope component of the submerged weight of the grains.

The design procedure involves the following steps:

(1) Calculate the coefficient K from the equation:

$$K = \cos\sqrt{1 - \frac{\tan^2\theta}{\tan^2\phi}}$$

where ϕ is the angle of repose and θ is the angle of the channel side slope.

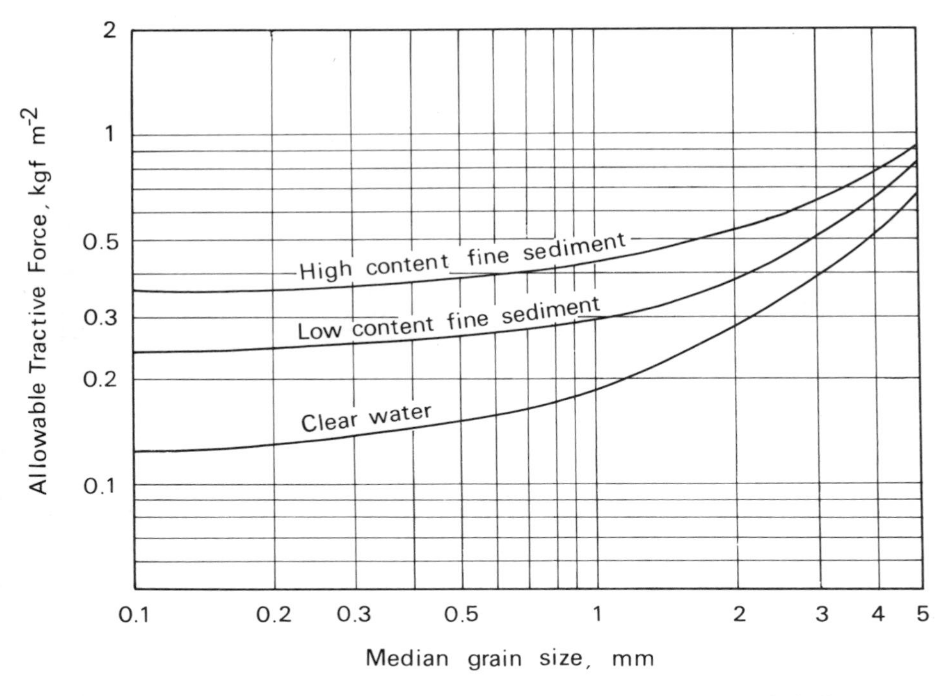

Figure 12. Determination of critical tractive force. (Adapted from Soil Conservation Service, 1977a. Reproduced by permission of the Soil Conservation Service.)

(2) Using the median grain size of the sediment determine the critical tractive force from Figure 12.

(3) Calculate the permissible shear on the side slopes (τ_s) from the equation:

$$\tau_s = K\,\tau_c$$

(4) The maximum *permissible* tractive force on the channel bed (τ_b) is given by:

$$\tau_b = \omega\, sd$$

where ω is the specific weight of water
s is the slope
d is the depth

The tractive force acting on the banks (τ_s) is less than that acting on the bed and is a function of the width–depth ratio. The maximum tractive force ratio for the side slope (τ_s/τ_b) is obtained from Figure 13.

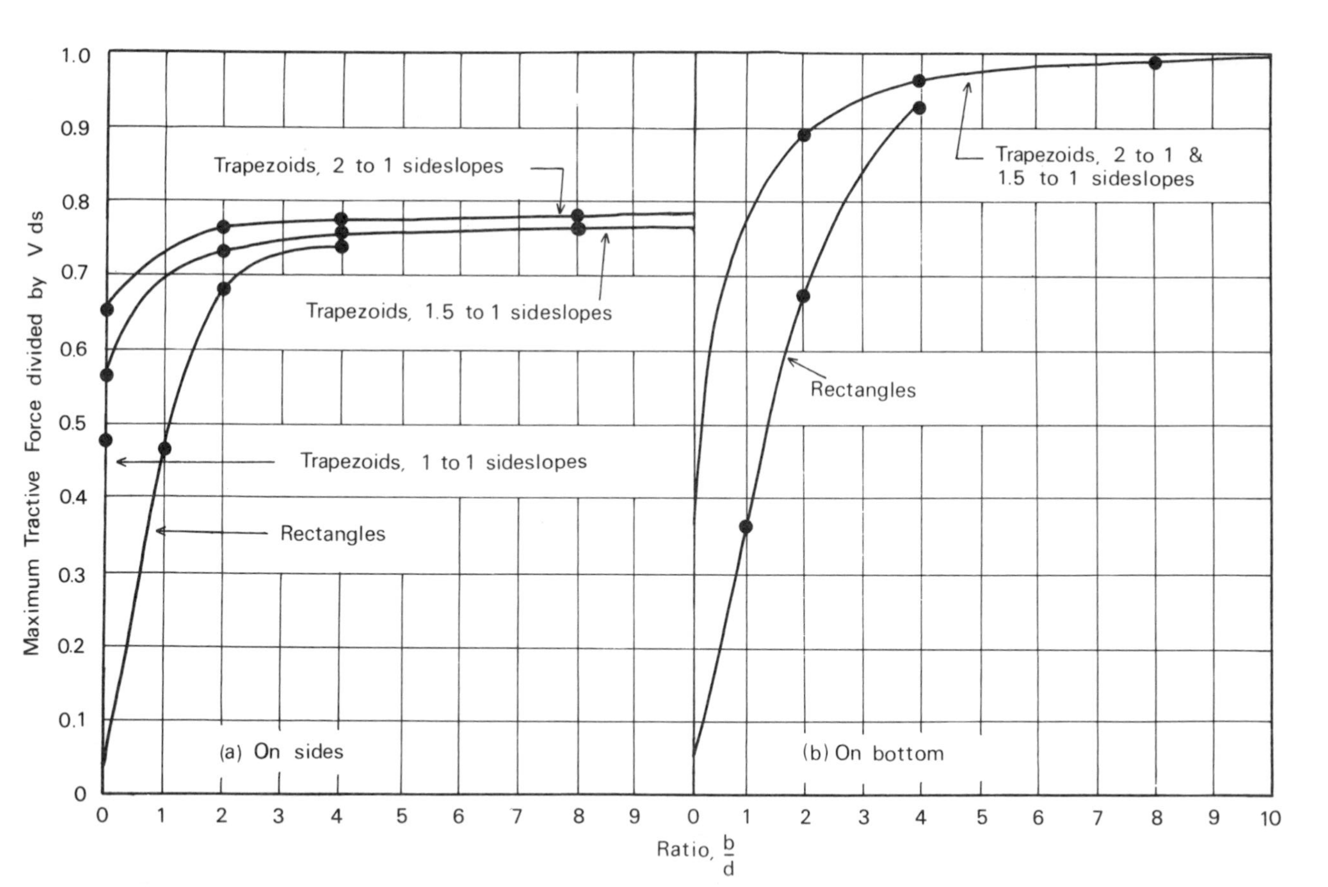

Figure 13. Variation of τ_s/τ_b in a trapezoidal cross-section (after Lane, 1955a. Reproduced by permission of American Society of Civil Engineers.)

The critical tractive force (τ_c) is given by:

$$\tau_c = \frac{\tau_s}{\tau_b} \omega sR$$

where R (hydraulic radius) is set equal to d in the basic tractive force equation

(5) Set τ_s equal to τ_c and solve the equation for d

(6) The cross-sectional area can then be obtained from the Manning equation:

$$v = k_1 R^{2/3} S^{1/2} n - 1$$

where k_1 is taken to be 1 in SI units. Then,

$$A = Q/v$$

(7) The wetted perimeter (p) is then determined from:

$$p = A/R$$

(8) The dimensions of the cross-section can then be determined using the formulae:

$$A = bd + zd^2 \qquad \text{and} \qquad p = b + 2\sqrt{1 + z^2 d}$$

Trapezoidal channels are often constructed because they are more akin to natural channels and easier to excavate than other shapes such as cosine-function sections. However the sharp corners in trapezoidal sections cause secondary currents and a non-uniform distribution of shear stress.

Limitations of Design Equations

The regime and tractive force approaches have been criticized for a number of reasons. There is apparently no agreement in the various regime equations on either the choice of width or depth parameters. The hydraulic radius or the mean depth have been used as the depth parameter, whilst the wetted perimeter or average width have been used to characterize the width. Regime equations do not include sediment load as an independent variable in channel design. In most cases no attention is given to the sediment transport rate during design. One of the major limitations of the regime approach is that the constants in the equation are not known over a wide range of conditions. Regime equations are also strictly valid for one discharge condition only, which in natural streams is taken to be the dominant discharge.

By contrast the resistance laws or sediment transport laws used in tractive force theory are applicable to streams with varying discharges and are therefore more general. However because of the incomplete understanding of flow resistance and sediment transport mechanics, the tractive force method may give inaccurate designs. The tractive force method has worked well for channels carrying clear water in coarse unconsolidated material. More research is required for channels carrying sediment-laden water.

The tractive force approach also assumes steady, uniform flow. However complex secondary eddies produced by decelerating flow may cause localized scour in channels designed for uniform flow (Simons and Senturk, 1977). Sinuous channels present special problems which cannot be resolved through tractive force computations.

Applying the methods often results in over-wide, shallow channels. Such cross-sections are impractical in urban areas where space is restricted, they are often expensive to construct and severely affect the biological and aesthetic qualities of watercourses (e.g. Nunnally and Keller, 1979).

Both methods only serve as guides, and do not replace experience and sound engineering judgement. Many river engineering problems unfortunately lie outside areas where research or routine observations are presently concentrated and these have been solved by a combination of intuition and past experience (Kellerhalls *et al*., 1976). However channel design always considers the river to exhibit a static equilibrium over the period for which the design is intended.

Future Developments

There is a requirement to develop more rational, theoretically based solutions to engineering design problems, which rely more on fundamental hydraulics and mechanics and less on 'rule of thumb' procedures. Hey (1986) advocates that no distinction should be made between theoretical and natural approaches to design, and that methods should be developed using the laws of motion and continuity and energy principles. It is also suggested that theoretically sound equations can be developed as the understanding of natural channel processes increases, and this will broaden the means available to the river engineer.

Hey (1974) advocated that river channels possess five degrees of freedom which may respond to changes of water and sediment discharge, bed and bank sediments and valley slope (Table X). Hey (1978) further concluded that it is possible to define bankfull hydraulic geometry by the simultaneous solution of nine process equations in which the dependent variables are bankfull hydraulic radius, channel slope, bankfull dune wavelength, bankfull dune height, bankfull wetted perimeter, maximum bankfull depth, channel sinuosity and meander arc length. The variables controlling this response are the discharge, sediment load, the nature of the bed and bank materials, valley slope and bank vegetation. Theoretically

Table X. Degrees of freedom (after Hey, 1974)

Changes in:	may cause adjustments of:
water discharge	width
sediment discharge	depth
bed and bank material	slope
valley slope	velocity
	planform

the simultaneous solution of these equations will enable the regime dimensions of the channel to be determined from values of the controlling variables. However at present this cannot be achieved because insufficient information is available about all the process equations. In particular feedback mechanisms operate during periods of instability to control the channel geometry and Hey (1986) believes that a mathematical model will eventually be developed which incorporates these feedback mechanisms and process equations. This model would simulate scour and fill and predict the response of natural or man-induced changes.

Benefit–Cost Analysis

Justification for government-funded channelization projects in many countries has roots in benefit–cost ratios that incorporate tangible or measurable values but do not consider the possible adverse environmental effects as part of the anticipated costs (Committee on Government Operations, 1973a). Benefit–cost analysis is a technique for evaluating and comparing the benefits and costs of a scheme, and is based on the assumption that the 'best' solution is represented either by the lowest-cost alternative or by that scheme which maximizes the desired objective (O'Riordan and More, 1969; Ministry of Agriculture, Fisheries and Food, 1985). Benefits and costs which have traditionally been considered are listed in Table XI.

The benefits can be tangible and direct (prevention or reduction of direct physical damage; increased intensity of land use), indirect (prevention or reduction of indirect damages) or intangible (prevention of death, injury, sickness and stress). In many instances the merits of a scheme have been judged only on the tangible and direct benefits.

Even following a scheme there will be some degree of 'residual' damage, for example when an embankment overtops. Benefits can therefore be expressed as follows:

$$\text{benefits} = \begin{array}{c}\text{damages which would result}\\ \text{from taking no action}\end{array} - \text{residual damages}$$

Costs are the total expenditure of implementing a particular response to a flood hazard. Cost estimates should include all tangible costs, both direct and

Table XI. Typical costs and benefits used in the conventional economic appraisal of schemes

A.	Costs	(i)	engineering (including capital costs and subsequent maintenance)
		(ii)	study, design and supervision (e.g. hydraulic model, benefit–cost analysis and topographic survey)
		(iii)	compensation (e.g. loss of land)
B.	Benefits	(i)	urban (prevention of damage to property; indirect losses to community and businesses avoided)
		(ii)	agricultural (crop damages averted; increased crop production)
		(iii)	road traffic (disruption avoided)
		(iv)	reduction of interruption of services (electricity, gas, water, sewage, telephones)
		(v)	intangible benefits (e.g. prevention of loss of life; reduced sickness and stress)

indirect, which are incurred from the start of a preliminary investigation of a scheme to completion.

Construction costs are usually incurred at the outset, whilst maintenance costs, together with benefits, are distributed throughout the post-construction period. Fluctuations in monetary value throughout the economic life of a scheme also present problems of assessment. To overcome these difficulties, benefits and costs are based on prices prevailing at the time of analysis and are normally expressed as annual averages over the life of the scheme. Benefits and costs are compared:

$$\text{benefit–cost ratio} = \frac{\text{benefits}}{\text{costs}}$$

This ratio can be used to determine the viability of a scheme and to compare alternative solutions.

For a more detailed understanding of the economic appraisal of flood alleviation and arterial drainage schemes the reader should refer to one of the many handbooks which are available (e.g. Penning-Rowsell and Chatterton, 1977; Parker *et al.*, 1987; and Brandon, 1987).

Reliance on Structural Solutions

The most frequently used approach of Federal authorities in the USA to solving flood problems between 1936 and 1966 has been the structural solution involving channelization (Muckleston, 1976). There are a number of other conventional solutions to flood problems, which do not require direct modification to the channel, and these can be categorized as either corrective or preventive (Table XII; James, 1965; US Water Resources Council, 1976). The 'American Flood Control Controversy' debated the relative merits of flood protection, by channelization in areas subject to flooding, and flood abatement, by managing the basin upstream through small dams or land-use treatment to reduce runoff (Leopold and Maddock, 1954). Although these alternative solutions have been increasingly implemented during the past 20 years, the majority of countries still rely principally on river channelization.

Table XII. Alternative types of adjustment on floodplains

Other corrective	Preventive
Reservoirs—flood storage areas	Floodplain zoning
Watershed treatment	Building codes
Flood proofing	Open spaces
Flood forecasting	Flood insurance
Evacuation	Warning signs
Urban redevelopment	Tax adjustments

It is the reliance on conventional channelization, particularly for flood control, navigation and agricultural drainage, which has recently generated so much public controversy in numerous countries. Projects have often been appraised and justified on tangible economic values such as prevention or reduction of flooding, or increased agricultural productivity, and relatively little attention has been given to the environmental ramifications, which may have been less easy to quantify. Increased public concern is reflected in the new or revised legislation and methodologies which have been devised to protect the environment, and these are described in the next chapter prior to discussing the environmental impacts in Chapters IV, V and VI.

Environmental Legislation

CHAPTER III

Legislative Framework

INTRODUCTION

The engineering practices described in the previous chapter have been at the centre of the public controversy which has arisen in various countries during the past two decades, but particularly in the USA (see Chapter I, p.20). Much of the legislation described in Chapter I was directed principally at achieving the engineering objectives of flood control, agricultural improvement and navigation improvement and failed to adequately consider the environment. Increased environmental awareness has led to new or revised legislation intended to protect the environment from the repercussions of a particular action. Before considering the physical and biological impacts of river engineering and alternative solutions, it is necessary to establish the legislative framework. The major objectives of this chapter are therefore to show how environmental legislation has evolved, and then to describe examples of the methodologies and procedures which have been applied in evaluating the consequences of channelization. Although methodologies such as project appraisal and environmental impact assessment are widely practised, others such as the evaluation of aesthetic qualities are not universally applied.

Developments Leading to Revised Legislation and Procedures

Many citizens and environmental organizations in the USA became increasingly concerned about the adverse consequences of channelization, and also more and more critical of the increasing use of channelization by Federal agencies since the 1950s. One American citizen, Martin Heuvelmans, wrote a book entitled *The River Killers* in 1974, which described how the public works of the Corps of Engineers had 'systematically ruined the nation's rivers' and recommended that the Corps should be abolished. Frustration with the inadequacy of environmental evaluations of channelization projects by Federal agencies prompted citizens to turn to the courts to obtain injunctions on several major projects. Practical handbooks have also been written in the USA with the objective of encouraging ordinary people to organize effective community river conservation programmes (Diamant *et al.*, 1984).

On 14 December 1972 the US Court of Appeals for the the 8th Circuit ruled that the Corps of Engineers', environmental impact statement of December 1970 concerning the Cache River–Bayou DeView Channelization Project was 'vague, too general and too conclusionary'. The Court felt that the Corps had not made a full environmental consideration under the National Environmental Policy Act (NEPA) of 1969 and required the submission of a revised statement and a review of the Corp's decision to proceed with the project (Committee on Government Operations, 1973b, p.27).

Another major case involved the Corp's channelization of 176 km of the Obion River System and 170 km of the Forked Deer River System in Tennessee, a largely rural area. In December 1972 the Federal District Court prohibited further channelization because the Corp's Environmental Impact Statement had not complied with the requests of NEPA (Committee on Government Operations, 1973b, p.28). In particular there was inadequate discussion of project alternatives such as floodplain planning and zoning, floodplain insurance, upstream structures or deferring channelization work. Furthermore the project had not taken into account the Department of Agriculture's views as to the effects of channelization on land use and the views of the State Highway Department on project maintenance costs. Inadequate consideration had been given to the environmental impacts, especially on wetlands and water quality.

The Example of Chicod Creek, North Carolina

The Chicod Creek lawsuit may be the landmark case over channelization in the USA. The outcome modified the court's interpretation of the National Environmental Policy Act and altered the Soil Conservation Service (SCS) compliance with the law in other projects.

A work plan involving 106 km of stream excavations was drawn up by the Soil Conservation Service in August 1971 to remedy severe crop losses due to flooding in the Chicod Creek, Pitt and Beaufort Counties, North Carolina (Coffey, 1982). Flooding was primarily the result of blockages by fallen trees and progressive sediment accumulations arising from deforestation in the headwaters in the early 1800s. The plan included measures to mitigate the loss of fish and wildlife habitat. However in November 1971 the Natural Resources Defense Council and other local and national organizations filed a lawsuit opposing the project on the grounds that a satisfactory environmental impact statement (EIS) had not been produced. The court ruled in favour of the plaintiffs and the lawsuit proceeded for a total of 6 years. During that time the SCS prepared and later completely revised the EIS, and it was not until September 1977 that a compromise agreement was reached. Under this agreement the SCS agreed to undertake before and after studies of environmental impacts on water quality, groundwater, fish and wildlife habitat and studies of anadromous fish populations. Modifications to the project included clearing and snagging only in wooded swamp rather than excavation; and that no construction should occur between February 1 and June 30 to avoid damage to the spawning runs of fish. Sediment traps were excavated at critical points in the stream channel; a 4.5 m grass strip was established on both sides of the channel; excavation was from one bank only; all spoil was seeded; in wooded areas clumps of trees were preserved at intervals of 60–90 m; and there was some variation in the slopes of channel banks.

SCS studies have been undertaken for 5 years following completion of works in 1981 to assess impacts (Wingate and Weaver, 1977). This case has not only changed the construction techniques now used but has changed the planning philosophy. Improved and earlier input by fish and wildlife interests in the planning stages has avoided controversy and delay. The SCS has also produced guidelines which take into account environmental values, whilst also meeting drainage and flood protection objectives.

Amberley Wild Brooks (UK)—Public Inquiry

In the UK a period of intense questioning of policies relating to agricultural drainage schemes followed a public inquiry in 1978 into a government proposal to drain the Amberley Wild Brooks, a shallow wetland area of 365 hectares, forming part of the River Arun floodplain (Penning-Rowsell and Chatterton, 1977; Parker and Penning-Rowsell, 1980). Although this is a tidal area, flooding of the brooks occurs principally from the surrounding hills, embankments and a series of sluices protecting the area from inundation by the river. The proposed solution was to improve agricultural productivity by channel widening and to control water levels by pumping the brooks into the river. Although not solely a channel improvement, this is an historic case because it represents the first public inquiry into a drainage scheme in England and Wales.

Opposition to the scheme was based on the loss of conservation and amenity value of an important wetland. Eighteen hectares of the area was a Site of Special Scientific Interest (SSSI). The Inspector criticized the insufficient consultation on the part of the water authority, who had proposed the scheme, and the conservation agencies. It was recommended that the scheme not be undertaken because of the unique nature conservation interest of the site. The Minister accepted the recommendations but rejected the call at the inquiry for clearer guidance on how to balance the requirements of nature conservation and agriculture, suggesting that these should be established for individual sites.

Environmental Legislation

It is evident that the extent to which legislation has developed has varied significantly from one country to another, depending on the date at which legislation was enacted (Table XIII). In the United States there is a long history of legislation dating from the Rivers and Harbors Act of 1899 and Table XIII refers to the Fish and Wildlife Coordination Act of 1958, the

Table XIII. Legislation and guidlines for channel management

Country	Legislation and guidelines	Implications of legislation for channelization works
United States of America	Rivers and Harbors Act 1899	Regulates the excavation and filling in any manner to alter or modify the course, location, condition or capacity of any navigable water
	Fish and Wildlife Coordination Act 1958	Requires that wildlife be considered in the design and implementation of water resource development programmes and the mitigation of development impacts
	National Environmental Policy Act 1969	Requires preparation and review of environmental impact statements for *major Federal action significantly affecting the quality of the human environment*
	Wild and Scenic Rivers Act 1968	Designates selected rivers of the nation, *which with their immediate environments, possess outstandingly remarkable scenic, recreational, geologic, historic, cultural or other similar values, shall be preserved in free-flowing condition*
Canada	Canada Water Act 1970	Provides *for the management of water resources of Canada including research and the planning and implementation of programs relating to the conservation, development and utilization of water resources*
	Clean Environment Act 1975 (New Brunswick)	Watercourse Alteration Regulation regulates *any change made intentionally at, near or to a watercourse or waterflow in a watercourse, both temporary and permanent*
New Zealand	Water and Soil Conservation Act 1967	Requires that every board (Regional Water Board) have regard to recreational needs and the safeguarding of scenic and natural features, fisheries and wildlife

Denmark	Watercourse Act 1949, ammended 63, 65, 69, 73	Gives drainage priority over all other uses, but regulates dredging and the timing of weed control
	Environmental Protection Act 1973	Requires bodies *to safeguard environmental qualities essential to the ... maintenance of the diversity of plants and animals*
	Watercourse Act 1982	Requires that works are to be planned and undertaken with regard to the stream quality (this includes the physical form of the stream, including the pools and riffles)
		Supplement to the 1982 Act states the legal procedure for restoring watercourses
Great Britain	Water Act 1973 section 22	Water authorities must *have regard to the desirability of preserving the natural beauty, of conserving the flora and fauna and geological or physiographical features of special interest*
	Wildlife and Countryside Act 1981	Extends the above requirements. Water Authorities and other bodies concerned with land drainage *shall ... so exercise their functions ... as to further conservation and enhancement of natural beauty and the conservation of flora, fauna and geological and physiographical features of special interest*

From Brookes and Gregory (1988). Reproduced by permission of John Wiley & Sons Ltd.

National Environmental Policy Act of 1969 and the Wild and Scenic Rivers Act of 1968. The latter two involved requirements for managing the rivers of the nation and particularly in relation to environmental impact statements.

General environmental awareness in the US in the 1950s and 1960s culminated in the National Environmental Policy Act. This Act, and others, including the Fish and Wildlife Coordination Act, are recent steps which recognize environmental concerns connected with channelization projects. The Flood Control Act of 1936 and Watershed Protection and Flood Prevention Act of 1954 were intended for the implementation of river channelization works primarily for flood protection, and did not adequately consider the environmental issues. The NEPA Act requires an environmental impact statement prior to the implementation of federally financed projects, whilst the Fish and Wildlife Coordination Act demands that consultation should be made with the US Fish and Wildlife Service and state agencies responsible for the management of wildlife at the planning stage to consider the effect on wildlife.

The National Environmental Policy Act (NEPA) of 1969 had the purpose:

> To declare a national policy which will encourage productive and enjoyable harmony between man and his environment; to promote efforts which will prevent or eliminate damage to the environment and biosphere and stimulate the health and welfare of man; to enrich the understanding of the ecological systems and natural resources important to the Nation; and to establish a Council on Environmental Quality.

Section 102 (c) known as the Environmental Impact Statement (EIS) clause stated:

> Include in every recommendation or report on proposals for legislation and other major Federal actions significantly affecting the quality of the human environment, a detailed statement by the responsible official on
>
> (i) The environmental impact of the proposed action,
> (ii) Any adverse environmental effects which cannot be avoided should the proposal be implemented,
> (iii) Alternatives to the proposed action,
> (iv) The relationship between short-term uses of man's environment and the maintenance and enhancement of long-term productivity, and
> (v) Any irreversible or irretrievable commitments of resources which would be involved in the proposed action should it be implemented.

The importance and necessity for filing a comprehensive EIS only evolved during the 1970s. A Statement is required for major developments on public land so that environmental values are now taken into account as well as economic and technical considerations. It ensures that a full examination of the environmental consequences of all possible alternative solutions to a particular river problem are examined. The Act requires: (1) the use of a systematic interdisciplinary approach in the planning and decision making of projects that may have an impact on the environment; (2) identification and development of methods and procedures for environmental assessment in consultation with the Council on Environmental Quality; (3) the production of an Environmental Impact Statement for major developments, summarizing unavoidable adverse impacts, alternatives to the proposed action, relationships between short-term and long-term productivity, irreversible and irretrievable commitments of resources and the comments of other organizations which either have jurisdiction by law or special expertise; and (4) utilization of ecological information in planning and developing resource-oriented projects.

Environmental impact studies were first developed in the United States but during the 1970s aroused considerable interest in other industrial nations. Australia, Canada, France and Ireland adopted similar procedures, whilst in other countries the method, although not formally adopted, may have given planners a new outlook (Lambrechts, 1983). The danger with impact studies is that they can be so detailed that they miss the point: they can be confused and indigestible (Lewin, 1976a). Generally speaking, impact studies have little effect in law. Courts may ensure that such studies have been carried out, but the final decision is made by the public authorities. The major benefit of impact studies is that they provide valuable information that can be used in the decision-making process.

Within the United States, individual States have passed acts protecting watercourses. For example, Montana formulated a 'Natural Streambed and Land Preservation Act' in 1975, when it was realized that many kilometres of streams had been bulldozed, with the result that brown trout (*Salmo trutta*) populations were considerably reduced (Bianchi and Marcoux, 1975).

In New Zealand and Canada there have been other acts leading to different degrees of legislative control on watercourses. In the case of drainage authorities in England and Wales the requirements are not as stringent as those for the United States. Under Section 22 of the Water Act of 1973, the provision that authorities should 'have regard' to the needs of conservation was of limited value and it was a significant advance when, in 1981, this was amended by the Wildlife and Countryside Act to give authorities the duty to 'further' conservation. In Britain the imminent requirement for environmental impact statements, arising from European Community legislation, may lead to a greater significance being attached to these developments.

Denmark

The trend towards more environmentally sensitive legislation is demonstrated in Denmark. The notion of modifying channels for the purpose of agricultural drainage has been implicit in Danish legislation dating back to the Watercourse Law of 29 July 1846 (Andersen, 1977; Madsen, 1983; Brookes, 1987c). The Watercourse Act of 11 April 1949 was amended by

Acts in 1963, 1965, 1969 and 1973. However permission to modify channels was often granted without consulting experts on technology or biology. The Environmental Protection Act (Miljøbeskyttelsesloven) of 13 June 1973 was designed to control pollution of rivers, air and soil. The principal objective was to 'safeguard physical environmental qualities essential to the ... maintenance of the diversity of plants and animals'. The aim was to achieve a large number of habitats and greater diversity of species, and to protect watercourses for the future. However it was not until the Danish Watercourse Act of 9 June 1982 that land drainage and water quality became jointly administered. For the first time the Act gave consideration to the morphology of the channel and river works are now planned and undertaken with regard to the physical form, including pools and riffles (Brookes, 1987b,c). Alternative management techniques are being sought for streams, where originally maintenance resulted in the destruction of gravel spawning areas or the complete removal of vegetation annually. Furthermore the Watercourse Law supplement of September 1983 has allowed for the restoration of watercourses using techniques which are more sympathetic to the natural environment.

The Conservation of Nature Act (Naturfredningsloven) of 18 June 1969, amended by Acts of 7 June and 26 June 1975, had the objective of protecting the nature and landscape values of Danish streams of particular interest for scientific or educational reasons. These watercourses were selected by the conservation departments of the regional authorities. Ditches and very heavily polluted streams were not protected by this Act.

Revised Methods of Management

Project Appraisal

Until the early 1970s in the United States, and the early 1980s in the UK, the appraisal of flood alleviation and arterial drainage schemes remained bound with the measurements of engineering costs and financial benefits (Little, 1973; Brandon, 1987). For any commitment of resources to be economically viable the benefits had to be shown to exceed the costs by the use of techniques which have already been described in Chapter II (p.47). Such techniques do not allow significant indirect environmental values and costs to be evaluated systematically with traditional gains and losses. There have been few trade-offs between tangible benefits and intangible physical and environmental consequences.

During the past decade in the UK a series of handbooks detailing the conservation aspects of rivers have been produced by organizations such as the Nature Conservancy Council (Newbold *et al.*, 1983), the Water Space Amenity Commission (1980a,b) and the Royal Society for the Protection of Birds and Royal Society for Nature Conservation (Lewis and Williams, 1984) and by individual water authorities (e.g. Purseglove, 1983).

The increased environmental awareness has shown such feasibility studies to be limiting and has led to the development of alternative appraisal methods. For example, in the UK, the Ministry of Agriculture, Fisheries and Food (1985) suggested that appraisals should state the objectives and methods of proposed schemes, identify the quantifiable benefits and costs, describe in full the non-quantifiable impacts and justify the scheme using both economic and non-economic criteria. Schemes submitted for government aid should have a standardized table attached for project appraisal. Techniques of project appraisal have also been devised in Bavaria (Schmidtke, 1987).

US Soil Conservation Service Procedure

The US National Environmental Policy Act of 1969 required that all Federal agencies should develop methodologies which allow the consideration of wildlife resources in project planning and decision-making. Several methods of assessment are available, including the Environmental Assessment Methodology of the Soil Conservation Service.

The methodology allows wildlife, amenity and recreation to be considered at the planning stage of a river engineering scheme (Soil Conservation Service, 1977c). In particular all possible alternative solutions to a particular problem are considered, including the 'do nothing' alternative. The potential environmental consequences of each alternative, including the existing situation, are then documented in a concise report which is then used in the decision-making process. The method incorporates all the procedures outlined in Table XIV.

Table XIV. Soil Conservation Service Environmental Assessment Methodology

(1) *Initial phase*	A multidisciplinary team formulates the aims of a proposed project, alternative solutions and the information required to study the alternatives. A preliminary field survey is undertaken, together with collation of available information. Summary Report is produced, on which a decision can be made on whether to proceed or not to more detailed studies.
(2) *Detailed phase*	Strategy is developed which describes the existing situation, predicts the future of each alternative solution and calculates the potential environmental impacts. Summary Report produced and a check is made to determine if all issues have been covered sufficiently. If not, then a supplemental phase is initiated.
(3) *Supplemental Phase*	Further investigations may be required if all issues were not clarified at the Detailed phase.
(4) *Final phase*	The multidisciplinary team produce a detailed technical summary of the likely impact of each of the alternative solutions. A graphical comparison of the alternatives is incorporated. A non-technical executive summary is also produced.

Design Procedure Adopted by the US Army Corps of Engineers

Table XV details the process of planning and designing flood control projects as adopted recently by the US Army Corps of Engineers (Nunnally and Shields, 1985). This is a hierarchical process that commences with a definition of the project objectives, proceeds to the selection and evaluation of alternative procedures for achieving these objectives and concludes with the detailed design of the channel and environmental features. Nunnally and Shields (1985) recommend that environmental concerns and opportunities for enhancement should be considered at an early stage, otherwise they may be overlooked.

The selection of environmental features should depend not only on the environmental objectives but also on basin and stream conditions that may affect the successful operation and maintenance of environmental features. Once selected alternatives are evaluated according to planning procedures (Office of Chief Engineers, 1982a,b).

The detailed planning of projects commences with the collection and analysis of data describing project conditions and provides information needed in the design process (Table XV; item IV A). The design procedures found in standard handbooks can then be followed (e.g. Soil Conservation Service, 1977a).

The Environmental Laboratory of the Waterways Experiment Station in Vicksburg, Mississippi, is currently developing a computer program 'ENDOW' (ENvironmental Design Of Waterways). This is an expert system for the selection of environmental features for streambank protection and flood control channels. The program is user friendly and can be run on a PC with the DOS operating system.

Bavaria: West Germany

In West Germany the responsibility for the maintenance and improvement of larger watercourses is borne by the individual states (Binder *et al.*, 1983). In Bavaria major land drainage proposals are dealt with by the Bayerischen Landsamts für Wasserwirtschaft, and minor ones by an organization called Verbande. All schemes are passed to the biotechnical engineer's department at the planning stage and their proposals for mitigation and enhancement measures for conservation and landscape are included in the scheme. This involves an interdisciplinary team of engineers, biologists and landscape architects. Successful cooperation depends on mutual trust between engineers and environmental parties and a willingness to involve these issues at an early stage in the planning of a project. This is now the accepted procedure in Bavaria. To achieve the objectives of biotechnical engineering and safeguard environmental works, the State buys all the land between the floodbanks and 5 metres beyond. In addition the State also purchases wooded oxbow lakes which are opened up for recreational purposes.

Table XV. General procedure for the design of flood control projects adopted by the US Army Corps of Engineers (after Nunnally and Shields, 1985)

I. Establish project objectives.
   A. Flood damage reduction; desired level of protection
   B. Environmental (water quality, recreation, fish and wildlife)
II. Choose alternatives for achieving project objectives.
   A. Non-structural
   B. Structural (reservoirs, levees, flood channels)
III. Evaluate alternatives and select general plan.

IV. Detailed project design for flood control channels.
   A. Data collection and analysis of existing conditions
      1. Watershed (climate, topography, soils, geology, sediment yield, land use, cover, existing and recent changes, hydrology)
      2. Stream and floodplain
         (a) hydrology (flood frequency series, storage data, flow duration curves)
         (b) hydraulics (determine *n* values; assess amount and size distribution of bedload and suspended load)
         (c) geomorphology (survey cross-section, slope; establish relationship of cross-sectional geometry to discharge; measure pool–riffle spacing and meander geometry and relate to discharge and channel width; evaluate bed and bank stability; measure size distribution of bed and bank materials; measure cohesiveness of banks; identify rock controls in bed or bank)
         (d) stratigraphy (determine stratigraphic sequence and describe units)
         (e) existing structures (types, location, design, morphological effects)
         (f) ice (thickness, dates of freeze and thaw, damage, flow patterns and blockages)
         (g) ecology (map and describe riparian vegetation, terrestrial and aquatic ecology)
         (h) water quality (turbidity, chemical, temperature)
         (i) aesthetics (identify, describe and photograph)
         (j) cultural and recreational resources (e.g. archaeology)
   B. Flood control and channel design
      1. Determine the exact location and alignment of channels
      2. Hydraulic Design
         (a) rapid flow—use lined channels: choice of environmental features is therefore severely limited
         (b) tranquil flow—select best combination of cross-section and alignments and construction techniques to meet flood control and environmental objectives
            — select additional features to meet environmental objectives
            — determine downstream water surface elevation and the water surface control line, including freeboard
            — select Manning's '*n*' values for each reach
            — size channel
            — check channel stability and if unstable alter cross-section or provide protection
      3. Review the design for maintenance consideration—adjust if necessary
      4. Review the design for aesthetics—adjust if necessary
   C. Design environmental enhancement features that are not part of the channel proper.
   D. Analyse costs; if too high then modify design, starting at step IV B.

Thames Water Authority (UK)

There are considerable benefits to be gained by adopting an interdisciplinary approach to river management and by focusing upon management in the context of the drainage basin as a whole. Within the UK problems of flood control and land drainage have traditionally been solved by an engineering design (Figure 14). However environmental and planning issues have made the implementation of river schemes a contentious and increasingly difficult operation in recent years. Thames Water has therefore recently developed a new interdisciplinary approach to river management which is unique to the UK Water Industry (Brookes and Gregory, 1988). Through this holistic approach, which is indicated by route 2 (Figure 14), potential schemes are appraised prior to design in terms of environmental, political and economic factors, including impact assessments. It is believed that detailed appraisal creates a better basis for the decision-making process and ensures a useful programme and structure for scheme development. A four-stage appraisal process has been adopted (Figure 15), whereby the nature of the problem is defined, baseline surveys are then undertaken which involve the collection of all the necessary data; a range of alternative options, including the 'do-nothing' state, are then evaluated, before a final option is selected.

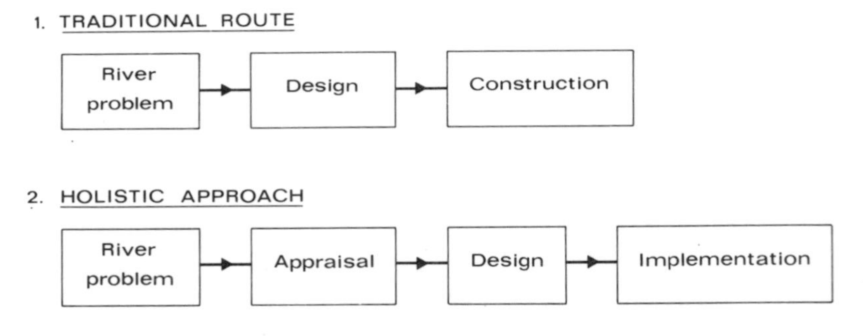

Figure 14. Approaches to channel management (From Brookes and Gregory, 1988. Reproduced by permission of John Wiley & Sons Ltd.)

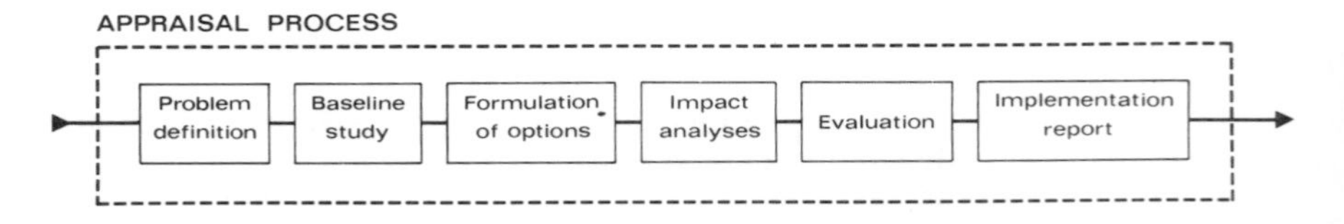

Figure 15. Stages in the appraisal process (From Brookes and Gregory, 1988. Reproduced by permission of John Wiley & Sons Ltd.)

The appraisal team within Thames Water has produced a manual for the guidance of river managers, and this manual necessarily reflects interdisciplinary expertise (Thames Water Authority, 1988b). The first chapter is devoted to geomorphology and contains advice on geomorphic appraisal, including changes which occur at the scale of an individual reach and within the river basin. The remaining chapters include scheme engineering, hydraulics and hydraulic modelling, hydrology, agricultural benefit evaluations, benefit–cost analysis, aquatic biology, fisheries and water quality. Attention is also given to post-project appraisal which aims to evaluate the impacts of existing engineering schemes. The handbook has drawn heavily on North American and European experience and techniques in the absence of sufficient information specific to the UK. The lack of an adequate database has prompted studies for the Thames basin. Such interdisciplinary approaches can then provide information to allow public policy decisions to be made in the most informed way.

Computational hydraulic models are now established as a method of designing flood alleviation works and have been used for a number of different rivers in the Thames basin (Gardiner *et al*., 1987). Such models are capable of representing the physical processes in the prototype, and enable the engineer to consider a large number of alternatives in searching for the most cost-effective and environmentally acceptable solution.

METHODS OF ENVIRONMENTAL ASSESSMENT

Environmental Impact Assessment

Methods for the assessment of environmental impacts have been devised as a means of identifying and in some cases quantifying the nature and magnitude of the impacts resulting from projects which have either already been implemented or are proposed (e.g. Leopold *et al*., 1971; Munn, 1979). There are three principal categories of methods for identifying environmental effects and impacts (Warner and Preston, 1973; Warner *et al*., 1974):

1. *Checklists* of a wide range of effects and impacts intended to stimulate the analyst to think broadly about the possible consequences of contemplated actions.
2. *Matrices* typically list human actions in addition to a list of impact indicators. For example, the Leopold matrix (Leopold, 1969) lists 88 environmental characteristics. By relating the two in a matrix cause and effect relationships can be identified.
3. *Flow diagrams* can be used to visualize the relationships between an action, the environmental effects and the impacts.

There are three methods for the interpretation of impacts. Firstly, the impacts can be displayed in a checklist or matrix. If the set is relatively small then a visual assessment of the aggregate impact may be obtained. A second method involves ranking alternatives within groups of potential impact. This enables the determination of alternatives that have the least adverse, or most beneficial, impact. However, no attempt is made to assign weights to the potential impacts and hence the total impacts of alternatives cannot be compared. A third group of methods have therefore attempted to assign numerical weights, although the scales on which individual impacts are measured must be in comparable units. To achieve comparability a normalization technique can be used. For example in the Battelle system (Dee *et al*., 1972) environmental quality ranges from 0 (very bad) to 1 (very good) and all impacts can be simply transferred arithmetically to this scale. Perhaps the most controversial issue is the assignment of numerical weights to obtain an aggregate index for comparing alternatives, but this can be used to quantify value judgements. The major criticism is that a range of impacts are aggregated into a single index, on which a decision is then made.

Environmental impact assessment also demands an Executive Summary or Statement and every effort should be made to avoid incomprehensibility or ambiguity (see Munn, 1979).

Aberdeen University developed a method of EIA for the Department of the Environment (UK) for assessing the impact of engineering works on several British rivers (Department of the Environment, 1977; North West Water Authority, 1978; Thames Water Authority, 1988a,b; Countryside Commission, 1988). Environmental impact statements have also been applied to drainage projects in Ireland (McCarthy and Glass, 1982). Figure 16 is a checklist of the potential environmental impacts of regrading a 1.3 km length of the Frays River adjacent to an industrial area in West Drayton, to the west of London. This is a stable, well-shaded channel with a gravel bed and a diversity of habitats. The proposal involved deepening the channel by the removal of up to 700 mm of bed material and impacts were identified for a range of environmental disciplines, including aquatic biology, fisheries, angling, archaeology, landscape, land use and terrestrial wildlife. Impacts can be described for both the short term, during construction, and for the longer term following implementation of the works (Figure 16).

The relative significance of individual environmental impacts can be described by various characteristics (Figure 16). *Strategic impacts* (S) are significant changes in the existing environmental character of an area whilst *Local impacts* (L) refer to localized changes. *Adverse* (A) and *Beneficial* (B) impacts are self explanatory. *Long term* (Lt) and *short term* (St) impacts do not have specific timescales attached, since the duration varies from one type of impact to another. *Reversible impacts* (R) are those which can be overcome, either by modifying the nature of the engineering works or through the processes of natural adjustment which follow initial development. *Irreversible impacts* (I) cannot be overcome. Furthermore the impacts can be either *Di-*

rect (D) or *Indirect* (Id) and can be of an *intermittent* (Int) or *continuous* (cont) nature.

Although the probable interactions between engineering proposals and the environmental character of an area can be summarized in a matrix, it does not, however, assign values to the relative importance of one type of impact against another.

INSTREAM FLOW METHODOLOGY

Several methods are available for assessing the existing wildlife resources of streams and for predicting the future resource if any of the alternative solutions are selected for action.

Instream Flow Incremental Methodology (IFIM)

This approach was originally developed in the United States as a means of assessing how much water could be extracted from a river at various times of the year, but without adversely affecting the fishery resource (Bovee, 1978; Bovee and Hardin, 1983; Stall and Herricks, 1982).

The approach is based on the concept that a particular species can be correlated with particular habitat requirements such as water quality, velocity, depth, substrate, temperature and cover. If these requirements are known then an assessment of the habitat suitability for a particular species can be made by measuring the quality of the available habitat. This approach has the advantage that it is easier to measure the existing habitat conditions than the fish population itself. It is also easier to predict the habitat conditions that will be produced by a proposed water extraction project.

The method can be adapted to predict the effects of river channelization on fish populations and for comparing the effectiveness of different habitat improvement devices. In theory it is also possible to extend the technique to other members of the aquatic community. Table XVI outlines the basic methodology. The objectives of the study determine the application of the data. A time series of the available habitat over a year can be produced using historical data on the likely discharge each month. The limiting conditions for particular life stages can then be identified for dry or wet months.

However, the method could be used to determine the effects of proposed river engineering works on any available habitat (Bovee and Hardin, 1983). Channel works often alter the morphology of a channel to produce new depths, velocities and cover which can be predicted and related to available habitat. The potential decrease in abundance of species resulting from a decrease in usable habitat can be estimated. Conversely, the potential effects of habitat improvement works can be quantified.

ENVIRONMENTAL RESPONSE TO PROPOSALS FOR WORKS ON THE FRAYS RIVER
(a) MATRIX SHOWING IMPACTS DURING CONSTRUCTION

	A/B	S/L	Lt/St	Int/ Cont	D/Id	I/R	Comments
Agriculture	—	—	—	—	—	—	No impact
Amenity	A	S	St	Cont	D	I	Construction activity occurring in large number of private gardens.
Angling	A	L	St	Cont	D	R	Disruption to channel environment.
Aquatic biology	—	—	—	—	—	—	No impact.
Archaeology	A/B	—	—	—	—	—	Archaeological watching brief.
Fisheries	A	L	St	Cont	D	R	Disruption to channel environment.
Landscape	A	L	St	Cont	D	I	Potential major problems with disposal of spoil.
Planning	A	S	St	Cont	D	I	Access and works would cause much disruption for a large number of residents.
Recreation	—	—	—	—	—	—	All works contained within private land with no public access.
Water quality	—	—	—	—	—	—	No impact.
Wildlife	—	—	—	—	—	—	No impact.

ENVIRONMENTAL RESPONSE TO PROPOSALS FOR WORKS ON THE FRAYS RIVER
(b) MATRIX FOR SHOWING IMPACTS AFTER IMPLEMENTATION

	A/B	S/L	Lt/St	Int/ Cont	D/Id	I/R	Comments
Agriculture	—	—	—	—	—	—	No impact
Amenity	A	L	Lt	Cont	D	I	Grave concern over private amenity as hundreds of gardens back on to the river channel.
Angling	A	L	Lt	Cont	D	I	Loss of fast water fisheries, need angling access on high banks.
Aquatic biology	A	L	Lt	Cont	D	I	Difficult to recreate this stable fast flowing reach.
Archaeology	—	—	—	—	—	—	No impact.
Fisheries	A	L	Lt	Cont	D	I	Difficult to recreate this stable fast flowing reach. Important spawning area.
Landscape	A	L	Lt	Cont	D	I	Likely to be adverse even with sensitive working methods, materials and designs. Likely to result in substantial bankside tree loss and may necessitate bank stabilization works.
Planning	A	L	Lt	Cont	D	I	Loss of residents' amenity likely to generate considerable local reaction. Need to ensure no impact on Frays island Nature Reserve.
Recreation	—	—	—	—	—	—	All works contained within private land with no public access.
Water Quality	—	—	—	—	—	—	No impact.
Wildlife	A	L	Lt	Cont	D	I	Difficult to recreate this stable fast flowing reach. Inevitable loss of many mature trees.

Key: Categorization of impact: A/B, adverse/beneficial; S/L, strategic/local; Lt/St, long-term/short-term; Int/Cont, intermittent/continuous; D/Id, direct/indirect; I/R, irreversible/reversible.

Figure 16. Example of the application of a matrix approach to the planning of river engineering works: Lower Colne Study (Reproduced by permission of Thames Water Authority.)

Table XVI. Instream Flow Incremental Methodology

1. Characteristics of the stream are assessed (flow, sediment, water chemistry, temperature).
   Need to determine whether the system is stable or responding to change.

2. Stream is divided into reaches which have constant morphology, chemistry and temperature.
   Shallow areas such as riffles which may present difficulties for fish migration should be noted.

3. Representative sampling reaches are taken and sampling is made through a range of discharge conditions.
   Transects are then taken so that each variation in morphology (pool, riffle, undercut bank, debris jam) is represented. The elevation of the water surface, depth, velocity, substrate and cover are measured for each transect.

4. The measurements are taken to be representative of the habitat conditions within the 'cell' bounded laterally by the spot sampling points across the transect and longitudinally by the point midway between adjacent transects.

5. The measured habitat conditions within each cell are then compared with data on the habitat requirements for each species/life stage. A selection of species is taken, which can include those which are sensitive to change or those which are economically valuable.

6. Models, based on probability of use (POU) curves, are available to assess the habitat requirements of many species. The suitability of each habitat attribute is assessed between 0 (unsuitable) and 1 (optimum). Combinations of attributes are obtained by multiplying POU values, thereby giving an overall POU for each cell.

7. The product of the POU value and surface area of the cell gives a weighted usable area (WUA). The sum of the WUA value for all the cells provides a measure of the available habitat for each species/life stage under prevailing discharge conditions.

8. Calculations can be made over a range of discharges to produce a relationship between discharge and the available habitat for each species/life stage. This forms the output from the IFIM simulation programme.

Habitat Evaluation Procedure (HEP)

This procedure was developed by the US Fish and Wildlife Service in conjunction with other agencies and research organizations, so that terrestrial and aquatic wildlife resources can be considered as part of an environmental impact assessment.

The methodology is based on the concept that wildlife is dependent on the habitat for food, shelter and reproduction. The habitat within a proposed development area is grouped into a reasonable number of categories (e.g. deciduous woodland, herbaceous wetland, natural grassland). The quality of the habitat types can then be assessed in relation to the requirements of species, and the least valuable areas can then be selected for development. Furthermore, if the quality of the habitat after development can be predicted it is possible to select the least damaging technique, to quantify the resultant loss of wildlife should construction take place, and to determine the extent of any wildlife mitigation works.

The procedure produces a numeric value expressed in 'habitat units' which can be related to the quantity and quality of wildlife habitat within a given area. Habitat suitability models are used to determine these values. Several models are already available (e.g. Raleigh, 1982; Schroeder, 1982). Table XVII outlines the habitat evaluation procedure (HEP) methodology as used

Table XVII. Habitat evaluation procedure for rivers

(a)	Clearly state the objectives of a scheme	— Problem (flooding, erosion, drainage) — Alternative solutions (flood storage; channel works) — Main concern (impact on environment)
(b)	Assess area affected by each alternative	— Examples: channel, floodplain and tributaries
(c)	Assess the cover types with the affected area	— Using aerial photographs or maps, sub-divide total area into cover types
(d)	Select evaluation species	— Species sensitive to change or economically valuable species
(e)	Develop habitat suitability models for the evaluation species	— Using the known requirements of a species for food, cover and reproduction, linear relationships can be developed between physical and chemical variables and carrying capacity. If no model is available for a species then one must be be developed.
(f)	Assess suitability of the cover area relevant to each species	— Sample selected areas and measure the model variables. Between 3 and 15 sites is recommended for each species and cover type. Averages of the measurements for each variable are then taken and compared with graphical relationships. The suitability of a habitat to support a species can be assessed. — An overall habitat suitability index (HSI) is obtained by averaging the suitability of individual variables.
(g)	Calculate the overall suitability of the habitat within the affected area for each evaluation species	— The product of the area of the cover type and the HSI for each species gives a measure of the potential resource, expressed in habitat units (HU)
(h)	Sum the habitat units for all the evaluation species	— This gives an overall value for the wildlife resource

for rivers (US Fish and Wildlife Service, 1976; 1980a,b,c; Schamberger and Krohn, 1982).

The habitat evaluation procedure can be used to compare two or more sites which are being considered for development, and the site which will have least damage on the wildlife resource can be determined. However, the procedure can be extended to assess the likely impacts of channel works on future wildlife potential where the likely changes in cover type and quality can be assessed. For example channelization may drain a wetland, thereby changing the cover type, and may also dramatically alter the instream habitat. Alternative solutions can then be compared on the basis of net losses of wildlife resources. The methodology can also be used to evaluate the need for compensation or mitigation works.

Biotic Indices

The distribution of macroinvertebrates may also be determined by dissolved substances in streams such as oxygen, salinity, acidity and hardness (Hynes, 1970). Organic inputs, derived from sewage, industrial or agricultural sources, also have a major impact on invertebrates. Many rivers throughout the western world have been affected by sewage pollution, which has caused the disappearance of many species, even though levels have been within the permitted safety limits. Species can be listed in terms of susceptibility to pollution:

1. Stoneflies, mayflies and caddis-flies may be very sensitive to organic pollution because of their requirement for a high oxygen concentration.
2. Shrimps and mussels are only moderately susceptible.
3. Chironomid midges, worms and maggots are extremely tolerant.

As a result of this variety of response of macroinvertebrates to pollution, several biotic indices have been developed to assess the quality of water, particularly in North America (e.g. Winget, 1984). The most commonly used in the UK water industry at present is the Department of the Environment/National Water Council Biological Monitoring Working Party Water Quality Score (BMWP). Scores are allocated to individual families, and for each site the total score for individual families is obtained. The higher the score then the cleaner the water. This system of scoring does tend to reflect habitat diversity in addition to water quality. Typical scores range from 0 in highly polluted rivers to 200 in relatively unpolluted chalk streams in southern England with an abundance of species.

The Freshwater Biological Association commenced a 'Rivers Communities Project' in October 1977 with the aid of water authorities and River Purification Boards in Scotland. The aim of the project has been to provide the water industry and scientific bodies with a greater knowledge and understanding of the variation in macroinvertebrate communities found in British rivers. With experience, the macroinvertebrates at a site can be used to confirm that the site is unpolluted, or alternatively that the site is subject to one or more of a wide range of environmental stresses. The two major objectives are:

1. to develop a biological classification of unpolluted running-water sites based on the macroinvertebrate fauna;
2. to examine the relationships between the physical and chemical features of sites and their invertebrate communities.

A procedure has been developed for predicting macroinvertebrates to be expected at a site from a knowledge of a small number of environmental features. The environmental variables include width, depth, surface velocity, substrate characteristics, slope and chemical data. This is based on samples from 370 sites on 61 river systems. TWINSPAN (TWo-way INdicator SPecies ANalysis) was used to classify the 370 sites on the basis of their invertebrate fauna into 30 end groups. The classification reveals the major differences between northern and southern Britain, which can be attributed to geological, topographical and climatic variations which influence the physical and chemical characteristics of rivers. Progressive changes down individual river systems were also identified.

The classification enables water authorities to catalogue the range of unpolluted rivers as a basis for future management decision making. The system offers an objective method for predicting the fauna to be expected at a site from environmental features on the assumption that it is unpolluted. In the future it may be possible to predict changes arising from various forms of environmental stress, including channel modifications.

AESTHETIC EVALUATION

The evaluation of aesthetic quality is a subjective issue. Visual and other sensual experiences associated with contrasts of slow, shaded versus sunlit water; deep, slow water in pools versus fast, shallow water on riffles and marginal vegetation versus trees along banks are important aesthetic values (Eiserman *et al.*, 1975; Keller, 1976). Unfortunately, channelization which involves channel straightening, removal of vegetation and other debris and replacement of overhanging banks by bank stabilization is detrimental to the aesthetic quality of a watercourse.

However, methods are available which allow a comparison of the aesthetic values of channelized and natural reaches and which permit evaluation of the quality of natural reaches which are planned for modification (Dunne and Leopold, 1978). Aesthetic degradation may occur when a natural stream is changed towards a channel with greater symmetry and less diversity of vegetation (Little, 1973). Clearly the judgement of the extent of a reduction

Table XVIII. Inventory of physical. biotic and human use criteria used to define 'river uniqueness'

	Criteria	Categories				
		1	2	3	4	5
Physical features						
1	Channel width (m)	<3	3 to 9	9 to 30	30 to 90	>90
2	Low flow discharge (cumecs)	<0.3	0.3 to 1.5	1.5 to 3.0	3.0 to 6.0	>6.0
3	Average discharge (cumecs)	<0.3	0.3 to 3.0	3.0 to 15	15 to 30	>30
4	Basin area (km^2)	<25	25 to 250	250 to 1250	1250 to 2500	>2500
5	Channel pattern	sinuous (pool and riffles)	meandering (pool and riffles)	sinuous (without riffles)	meandering (without pool and riffles)	braided
6	Ratio valley width to height	<5	5 to 12.5	12.5 to 25	25 to 50	>50
7	Bed material	alluvium (A) 100%	A(75)R(25)	A(50)R(50)	A(25)R(75)	Rock (R)100%
8	Bank and valley material	uncon. (U) 100%	U(75)R(25)	U(50)R(50)	U(25)R(75)	Rock (R)100%
9	Bedslope	<0.0005	0.0005 to 0.001	0.001 to 0.005	0.005 to 0.01	>0.01
10	Width of valley flat (m)	<30	30 to 150	150 to 300	300 to 1500	>1500
11	Erosion of banks	stable	–	slumping	–	eroding
12	Valley slope($x°$)	0 to 10	10 to 30	30 to 50	50 to 70	70 to 90
13	Sinuosity	<1.25	1.25–1.5	1.5–1.75	1.75–2.0	>2.0
14	No. of tributaries	none	1 to 3	4 to 5	6 to 7	>7
Biological and water quality						
15	Water colour	clear and colourless	–	green tints	–	brown
16	Floating material	none	vegetation	foamy	oily	variety
17	Algae	none	bed and bank partly covered	–	bed and banks mostly covered	everything covered

18	Landplants—floodplain	open	wooded with brush	wooded	cultivated	mixture cultivated and other
19	Landplants—hillslope	open	wooded with brush	wooded	cultivated	mixture cultivated and other
20	Water plants	absent	—	—	—	abundant
Human use and interest						
21	Trash per 30 m	<2	2 to 5	6 to 10	11 to 50	>50
22	Variability of trash	equally distributed	–	–	–	predominantly in localized areas
23	Artifical control	free and natural	partially controlled	partially channelized	completely channelized	dammed
24	Utilities, bridges, roads	none	<4	5 to 10	11 to 20	>20
25	Urbanization	no buildings	cabins trailers campsites few farmhouses	farmhouses	mixture 2, 3 and urban	predominantly urban
26	Historical features	none	1	2	3	>3
27	Local scene	pleasing	—	—	—	nauseating
28	View confinement	open	—	—	—	closed by hills, cliffs
29	Rapid and falls	none	—	—	—	abundant
30	Land use	agriculture	recreation	urbanization	recreation and urban	agriculture and urban
31	Misfits	none	1	2	3	>3

From Melhorn *et al*. (1975). Reproduced by permission of the Purdue University Water Resources Research Center in aesthetic quality depends on the individual. For example, beneficiaries of flood schemes have a greater tolerance for aesthetic degradation, whilst those who place heavy emphasis on aesthetics are usually not affected locally in a direct social or economic way (Little, 1973). This distinction is important because the defence of aesthetic qualities does not have strong advocacy in project formulation and design.

The methods of aesthetic evaluation are based either on measurable physical attributes, or on subjective personal responses. Interview techniques involve a random sample of people who are asked about their preferences and reactions. Another method of evaluation is to expose randomly selected groups to visual stimuli, usually colour slides or drawings. This technique can be combined with an interview.

Leopold and Marchand (1968), Leopold (1969) and Melhorn *et al.* (1975) provided a method for qualitative comparison of some aesthetic factors among rivers. No sophisticated tools are required and determination of parameters can be made quickly in the field or from maps/aerial photographs. The approach is a numerical description of a riverscape based on a range of factors which describe (1) physical characteristics; (2) biology and water quality; (3) human use and interest (Table XVIII). Each factor for every stream studied is assigned a category rating ranging from 1 to 5 on an arbitrary scale. The rating depends on measurement or evaluation at observation points. A 'uniqueness ratio' for each stream is then derived by taking the reciprocal of the number of stream sites showing the same category rating. The addition of 'uniqueness ratios' for all factors provides a 'total uniqueness ratio'. This allows streams to be ranked in hierarchical order on the basis of uniqueness, although the approach does not distinguish between relative good or bad categories of uniqueness.

Leopold (1969) derived a 'scale of river character' for 12 reaches in the Hell's Canyon on the Snake River in Idaho, where a dam was proposed. Figure 17 shows that the axes of the criteria are orientated at 45°, implying an equal weighting to each. Semi-graphical procedures were also developed to show rating scale of 'valley character' and 'scenic outlook'. Melhorn *et al.* (1975) extended the Leopold approach by developing the LAND system (Landscape Aesthetics Numerically Defined). This was designed as a relatively objective evaluation. This system discounted the unpleasant categories to generate an aesthetic river index.

Little (1973) distinguished between short- and long-term aesthetics of channelization. During and immediately after construction, the majority of channels are an eyesore: banks are not colonized by vegetation; the surrounding land has been affected by heavy machinery and disturbance to the total stream results in highly turbid water. Appearance improves as the banks become seeded and the extent of erosion is minimized. However, the degree of subsequent maintenance determines the extent of aesthetic improvement.

Non-structural alternatives have no effect on aesthetics, whilst embankments set back from the channel have a minimal effect unless material for the

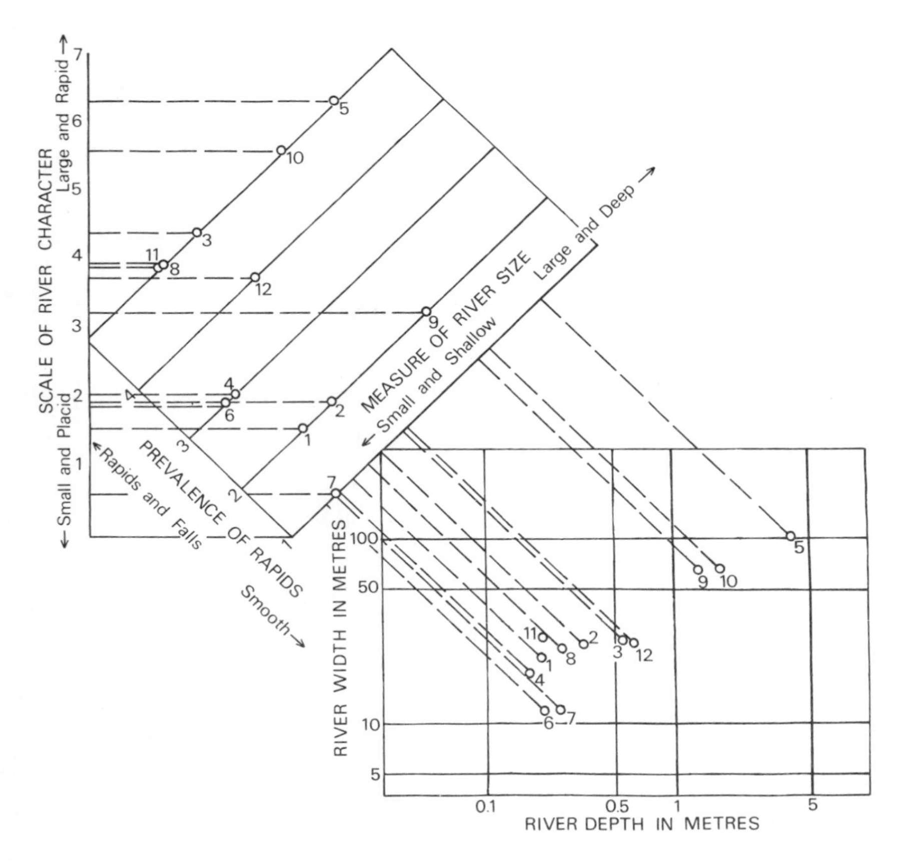

Figure 17. Analysis of river character. Twelve reaches in Idaho have been plotted, of which number 5 is Hell's Canyon on the Snake River (From Leopold, 1969. Reproduced by permission of the United States Geological Survey.)

banks is excavated from the channel itself. Physical change is minimized by clearing and snagging of obstructions and by following the natural alignment of the stream channel, rather than by excavating and by realigning. Earth channels are less degrading than paved or riprap channel; excavation from one bank is aesthetically better.

By referring to particular examples this chapter has demonstrated the types of environmental framework within which the negative impacts of river channelization can be evaluated. It is the objective of the following discussion to describe in detail the environmental repercussions of channelization (Chapters IV, V, VI) and alternative environmentally sensitive designs (Chapters VII, VIII) which can also be considered within these frameworks.

Effects of Channelization

CHAPTER IV

Physical Effects

> Natural streams are essentially open hydraulic systems in equilibrium. The system state variables are those same dependent variables that describe discharge: channel width, depth and slope; boundary roughness; and the size and concentration of the sediment load ... the effects of change in any one of these interdependent variables must be compensated for by changes in the others.
>
> (Nunnally, 1978a)

INTRODUCTION

To achieve the engineering objectives of river channelization it may be necessary to substantially alter the existing channel. Channelization involves changing one or more of the interdependent hydraulic variables of slope, width, depth, roughness or size of the sediment load. For example, to accommodate a flood discharge the capacity of a channel may be increased by widening or deepening, whilst at a site of highway construction it may be necessary to shorten the course of a river channel, with a corresponding increase of slope. The conventional methods of river channelization are described further in Chapter II.

Through altering one or more of the interdependent hydraulic variables the existing equilibrium will be disrupted, and to compensate for this there will be natural changes in the remaining hydraulic variables in an attempt to attain a new state of equilibrium. For example, a straightened river may immediately react to the increased slope by increasing the sediment discharge through bank erosion. Eventually the channel may widen through erosion, with a corresponding reduction of velocity, and the adjusted cross-section will be more efficient in dissipating the energy.

The planning and design of river channelization projects is a difficult task since rivers are naturally dynamic. Particularly when the design is based on inadequate data there may be instability in the improved channel reach and upstream and downstream of the reach. Each section may undergo erosion or deposition. Unless modified channels which are adjusting are regularly maintained then the hydraulic efficiency may be decreased.

Unfortunately many river problems have been solved by a combination of intuition and past experience, which includes the empirical design rules and equations of the 'regime theory' described in Chapter II (Jansen *et al.*, 1979).

The river engineer may attempt to design a channel which does not undergo undesirable and costly adjustments but his job is further complicated by the wide variation in river size, shape and rate of activity throughout the world and within individual regions. It is suggested that even in situations where quantifiable methods of computation and the necessary data are available, the engineer should check his results against geomorphological evidence such as the differences in the course of a river as highlighted by comparing two or more maps of different dates (e.g. Hooke, 1977) and should ascertain whether the channel cross-section is adjusting naturally or to some form of upstream land use changes such as urbanization (Wolman, 1967; Hammer, 1972; Gregory, 1977b) or mining (Richards, 1979) or reservoir impoundment (Petts, 1979).

This chapter is concerned with the documented examples of morphological adjustments which have occurred within reaches which have been realigned, enlarged, lined, embanked, diverted, restrained or affected by clearing and snagging. An understanding of the physical impacts of channelization and subsequent adjustments is essential to interpreting the biological effects described in Chapter V. Physical and biological impacts of channelization also extend to the natural reaches downstream and these are outlined in Chapter VI.

However, as a basis for interpreting the range of environmental repercussions and for developing alternative strategies, it is necessary to understand the variety of form and behaviour of river channels and this is considered first.

Morphological Classification

Throughout the world a wide variety of river channel types have been channelized, each type reacting differently to the engineering measures imposed. Several classifications which define the wide variation in river channel morphology are now available to engineers (e.g. Simons and Senturk, 1977). Kellerhalls *et al*. (1976) devised a comprehensive classification of river processes encompassing a wide range of channel types. The widely used conventional classification of planform as meandering, straight or braided is too simple because these categories are not mutually exclusive. Kellerhalls *et al*. (1976) proposed a classification under three principal headings: (a) channel pattern; (b) islands; (c) channel bars and major bedforms (Figure 18). Channel pattern (a) includes (i) straight channels with very little curvature—these are extremely rare in nature; (ii) sinuous channels which have a slight curvature with a belt width less than approximately two channel widths; (iii) irregular—where there is no repeatable pattern; (iv) irregular meanders—where a repeated pattern is vaguely present in the channel plan; (v) regular meanders—which have a clearly repeated pattern. The angle between the channel and the general valley trend is less than 90°. Freely meandering gravel-bed channels with a high slope are often regular; so too are

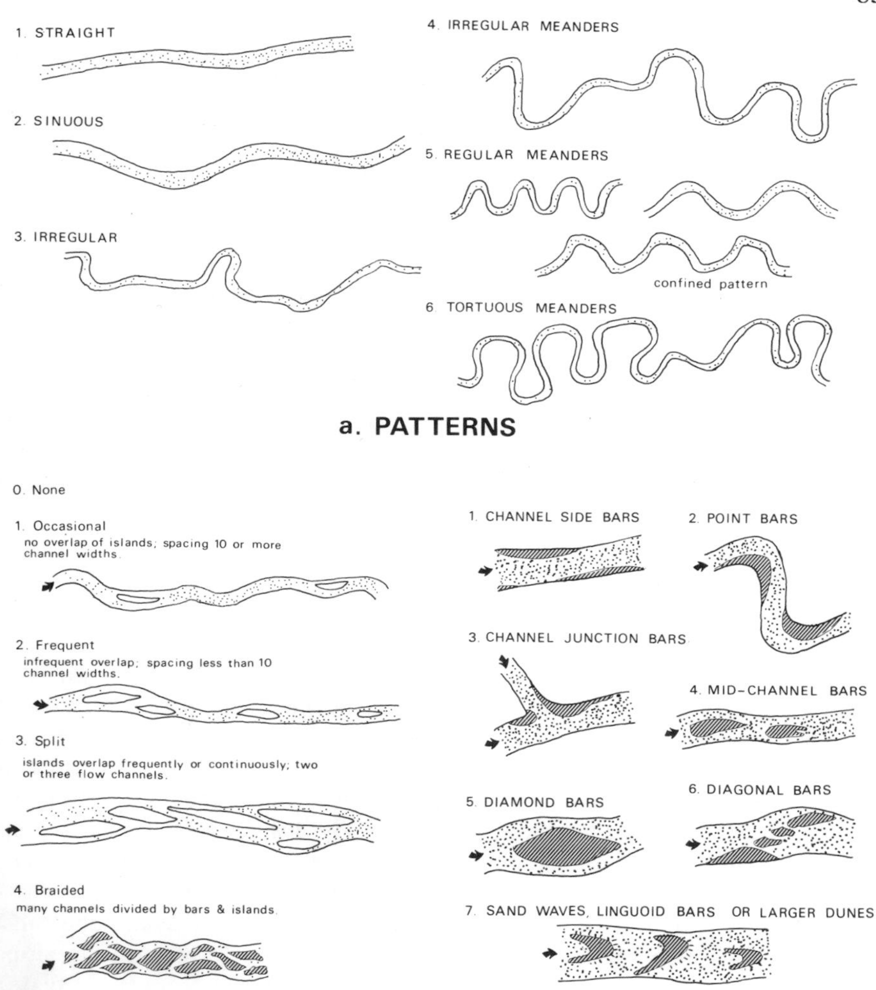

Figure 18. Classification of river processes. (From Kellerhalls *et al.*, 1976. Reproduced by permission of American Society of Civil Engineers.)

confined meanders; (vi) tortuous meandering which is a repeated pattern characterized by angles greater than 90° between the channel axis and the valley trend.

Islands (b) are relatively stable, vegetated and generally extend to the floodplain level (Figure 18). Under the Kellerhalls *et al*. (1976) classification it is quite possible to have an irregular meandering channel which is 'split' by islands. In braided channels the bars are generally covered by water in high floods. The channel bars (c) classification encompasses those reaches with side bars (1) which may cause the flow to follow a more meandering path;

point bars (2) which form on the inside of well-developed bends; channel junction bars (3) which develop at tributary junctions; mid-channel bars (4), found particularly in gravel bed rivers; diamond bars (5) which are an extreme development of mid-channel bars; diagonal bars (6) found only in gravel-bed rivers; (7) sand waves, linguoid bars or large dunes, commonly found in active sand-bed channels. They have a dune profile, with a gentle upstream face and a downstream face at the angle of repose. Channels with no bars are also included.

The classification shows that rivers present a wide spectrum of intermediate forms between the familiar classic braided and meandering types which reflects the wide variety of flow distribution, bed material size, sediment transport and channel stability. Channels can also be codified on the basis of lateral channel activity. Morphological adjustments are perhaps most dramatic when freely meandering gravel-bed channels with a high slope are straightened, or where braided river systems are confined to a single-thread channel.

Physical Consequences

There are many instances where the application of conventional designs has resulted in adverse adjustments because inadequate consideration has been given to the natural form and process of the river channel. Pre-existing equilibria can be disturbed and accelerated processes may be initiated. The majority of studies have compared maps, aerial and ground photography or engineering drawings to assess the extent of channel change. A few types of channelization, such as dikes, are purposely designed to induce adjustments of channel morphology. However, in the majority of cases adjustments arising from channelization are undesirable, requiring extensive and costly maintenance, which in turn inhibits biological recovery. Perhaps the most dramatic adjustments within engineered reaches occur in response to slope changes caused by channel shortening or regrading, or to over-excavated bottom widths.

Morphological Effects of Realignment

> 'The cutting of the meanders of rivers has been a terrain for some classical engineering trials. The reason is perhaps that a meandering river, upon preliminary inspection, looks like a very inefficient system, in need of serious correction. That the contrary is the case, is obviously not obvious'
>
> (Ryckborst, 1980)

Perhaps one of the earliest and most vivid descriptions of the effects of a man-made cutoff was that made on the River Dorback in Scotland by Lauder in 1830. Immediately after the cut had been executed, the onlookers:

> heard the deadened roar of the river, as it poured over the clayey bank, in a fall of fifteen feet, carrying everything before it... Huge stones were continually rolling down... the banks of the cut, being undermined, rapidly gave way.

Lauder felt that this description provided practical information on how to deal artificially with a river. The consequences of realigning channels are exemplified by the Lower Mississippi, described in Chapter I (p. 18).

A Model for Change in Straightened Stream Channels

Lane (1947) demonstrated the effects of cutoffs in both non-erodible and erodible channels. Non-erodible channels do not adjust. However long-term changes occur after straightening of erodible channels. For the East and West Prairie Rivers in Alberta, Parker and Andres (1976) found that straightening a meandering stream increased the slope by providing a shorter channel path (Figure 19). This increase of slope enabled the transport of more sediment than was supplied at the upstream end of the channelized reach and the difference was obtained from the bed, causing degradation which progressed upstream as a nickpoint. An excess of load was then supplied to the downstream part of the channelized reach and because the flatter natural reach downstream could not transport this sediment it was deposited on the bed. The excess may be deposited in gradually decreasing quantities with distance downstream. Degradation within the straightened reach may also cause bank collapse. The range of adjustments which might occur in response to straightening are summarized in Table XIX, and include local effects within the engineered reach such as a steeper slope, higher velocities, increased transport and channel degradation. The downstream effects are covered more thoroughly in Chapter VI.

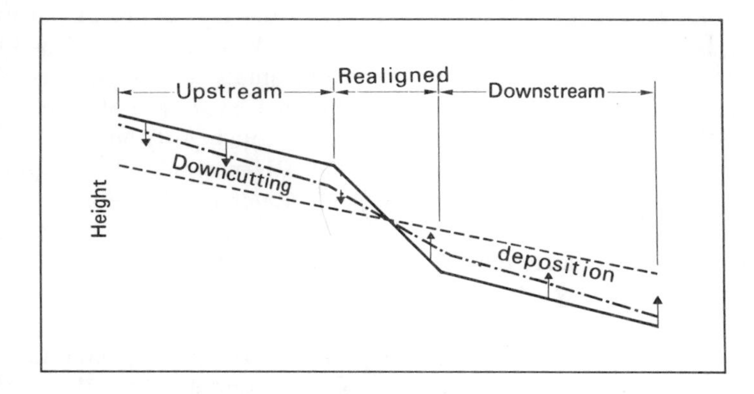

Figure 19. Degradation in straightened alluvial channels. (From Parker and Andres, 1976. Reproduced by permission of American Society of Civil Engineers.)

Table XIX. Effect of straightening a reach by cutoffs (after Simons & Senturk, 1977)

Local effects	Upstream effects	Downstream effects
1. Steeper slope	1. See local effects	1. Deposition downstream of straightened channel
2. Higher velocity		2. Increased flood stage
3. Increased transport		3. Loss of channel capacity
4. Degradation and possible headcutting		
5. Banks unstable		
6. River may braid		
7. Degradation in tributary		

Case Studies

Entrenchment of the Willow Drainage Ditch, Harrison County, Iowa, since 1919–20 was reconstructed by comparison of the original profile of the ditch, historical records, and a survey of the ditch in 1958 (Daniels, 1960). The channel adjusted by channel scour and headward movement of a series of nickpoints (Figure 20). The gradient of the original river averaged 1.0 m per km in the lower reaches and 1.4 m per km in a reach further upstream. The average slope of the drainage ditch which replaced these two reaches was 1.5 m per km and 1.7 m per km respectively. The Willow River was straightened over a distance of approximately 42.1 km, representing a slope increase of 0.2 to 0.4 m per km. The ditch had a trapezoidal section with 1:1 side slopes. A maximum increase in channel size of 440% between 1919–20 and 1958 was observed in the upper reaches (Figure 20). At the Monona–Harrison County line the ditch had increased from an original depth of 3.4 m to a depth of 13 m. The 1920 top width of 9 m had increased to 33.5–36.6 m. Nickpoints were observed to move upstream rapidly during periods of high flow. Passage of a nickpoint caused the channel banks to collapse through slumping. The channel also adjusted through scouring of the bed.

The effects of straightening the Tarkio River which flows southwest through Iowa and Missouri to the Missouri River (drainage area: 1400 sq km) have been described by Piest *et al*. (1977). A new channel, completed in 1920, effectively shortened the course by 25.7 km at its confluence with the Missouri River. By comparing cross-sections for 1846, 1932 and 1975 accelerated channel erosion was shown to have occurred since straightening, with a two- to fourfold increase in cross-sectional area at specific locations. Drastic channel incision and enlargement was observed along the Blackwater River in Johnson County, Missouri, as a result of straightening 60 years previously (Emerson, 1971). The shortened course was 24.6 km less than

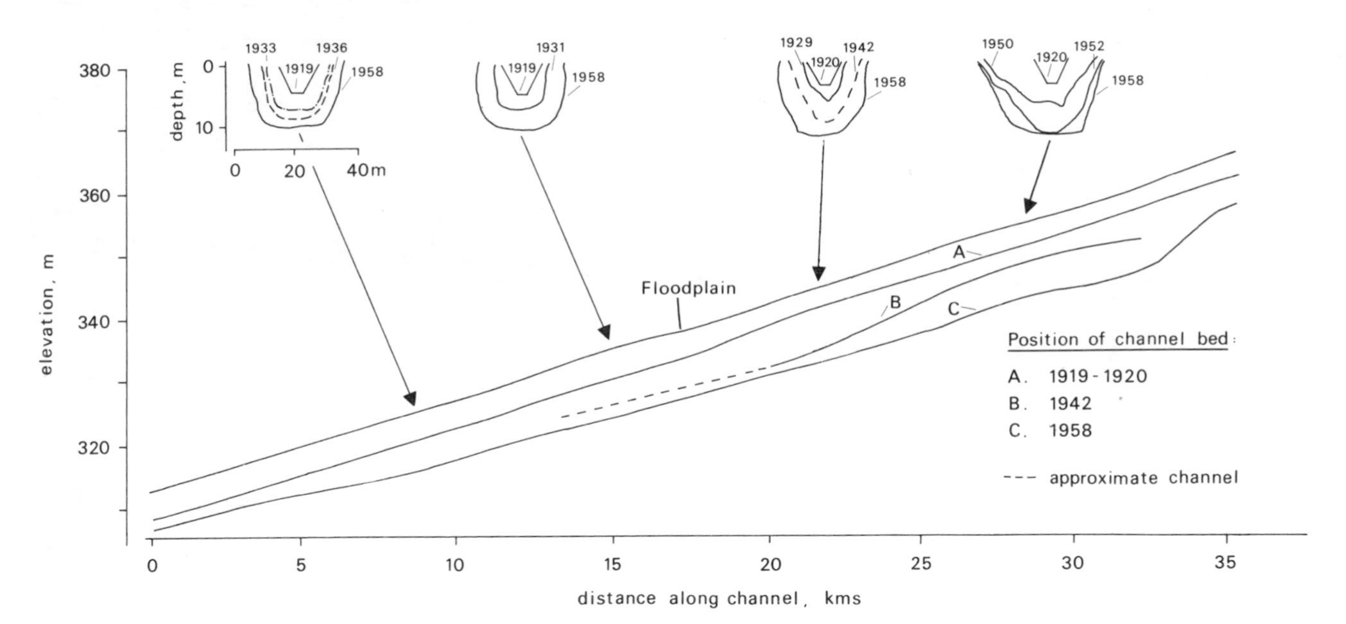

Figure 20. Morphological adjustment of the Willow Drainage Ditch, Harrison County, Iowa. (Based on Daniels, 1960. Reproduced by permission of the *American Journal of Science*.)

the original and the gradient was nearly doubled. The present channel had increased from a cross-sectional area of 125 m^2 when newly dredged to a size ranging from 525 to 1589 m^2. The maximum value represents an increase in area of 1173% over 60 years.

The Tippah River, Pigeon Roost Creek and Oaklimiter Creek in northern Mississippi have exhibited varying degrees of adjustment since straightening in 1965, 1968 and 1965 respectively (Schumm *et al.*, 1984). The responses of the three channels over periods ranging from 9 to 15 years are compared in Table XX.

Table XX. Summary of responses of three channelized streams in northern Mississippi (after Schumm *et al.*, 1984)

River	Response	Change of cross-section area	Comments
Oaklimiter Creek	1965–80	436–995%	maximum activity in middle reaches; not yet recovered
Tippah River	1965–80	45–401%	maximum adjustment in middle reaches
Pigeon Roost Creek	1968–77	not quantified	middle reaches

The slope of the Oaklimiter Creek was increased by an order of magnitude and the channel subsequently degraded, with a series of nickpoints. The cross-section area increased by 436% in the lower reaches of the channel, where sediment had been deposited, and by 995% in the very active middle reaches. A similar evolutionary sequence was observed for the Tippah channel and the Pigeon Roost Creek, least change occurring in the headwater reaches and maximum adjustment in the active middle reaches. Channel changes were again smallest where sediment deposition had occurred in the lower reaches.

The Peabody river in New Hampshire was shortened by approximately 260 m and immediately after construction the channel began to adjust through erosion and scour (Yearke, 1971). However, major changes occurred within the first year and adjustments were of decreasing significance in subsequent years (Figure 21). The original channel had a fall of 10 m per km and the relocated channel was steepened to 15 m per km. The channel adjusted itself to 14 m per km after two years and to 13 m per km seven years after construction.

Channelization of the Big Pine Creek in Benton County, Indiana (drainage area: 42 sq. km) between 1930 and 1932 reduced the mainstream length by 30% from 16.3 to 11.6 km and increased average gradients by 33% (0.0012 to 0.0016) (Barnard, 1977). During the six years immediately after construction the channel adjusted dramatically, particularly in reaches with the

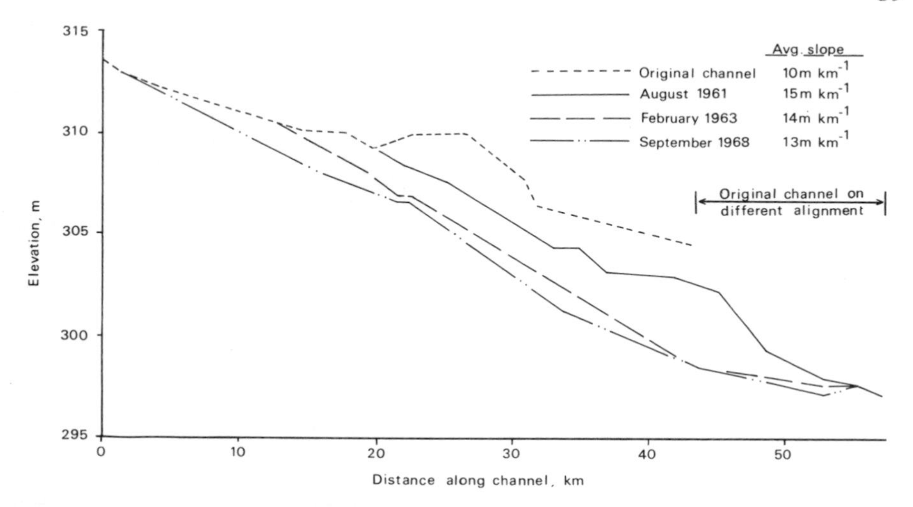

Figure 21. Morphological adjustments along a relocated segment of the Peabody River, New Hampshire (1961–1968). (From Yearke, 1971. Reproduced by permission of the American Society of Civil Engineers.)

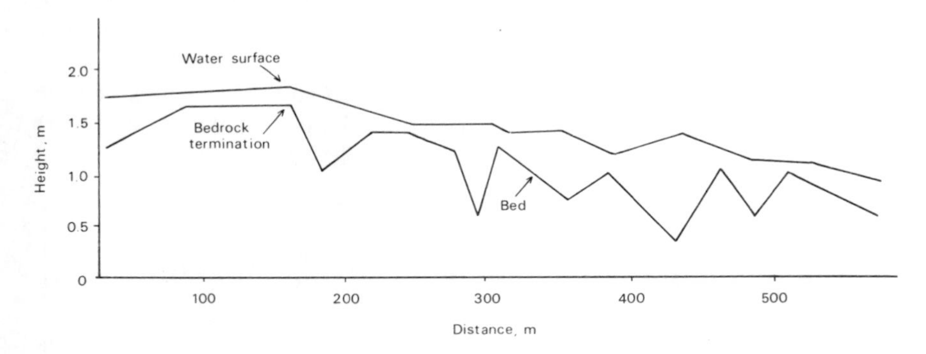

Figure 22. Advanced stage of pool–riffle development 40 years after channelization of the Big Pine Creek, Benton County, Indiana. (From Barnard and Melhorn, 1982. Reproduced by permission of George Allen & Unwin.)

highest gradients. The soils are glacial till and outwash deposits. However, the upstream reach became stabilized by a rock control, with minor changes. In more rapidly eroding reaches adjustment included incision, widening and the development of higher sinuosity. In these reaches considerable bank slumping had occurred with deposition on the inside of bends to form point bars and advanced pool–riffle development (Figure 22).

Brice (1981) assessed the stability of 103 stream channels in different regions of the United States that were realigned for purposes of road or bridge construction, mostly during the period 1960–70. The length of natural channel relocated ranges from 70 to 4200 m (8 to 550 channel widths).

Table XXI. Types of instability of relocated channels in the USA, tabulated according to stability of channel (after Brice, 1981)

Stability class	Number of sites	Kind of instability at a specific site		
		Bank erosion	Degradation	Aggradation
Rare or absent	36	4	1	–
Local only and not severely cut or slumped	41	41	6	1
Local only but severe	15	15	6	–
Generally cut or slumped	8	6	4	–

Stability of the relocated channel was rated as good at 36 sites, fair to good at 41 sites, fair at 15 sites and poor at 8 sites (Table XXI). Three sites were totally lined with concrete. In comparison with bank stability of the prior channel, bank stability of the relocated channel was about the same at 45 sites (52%), better at 28 sites (32%) and worse at 14 sites (16%).

Minor channel degradation was discerned at 17 sites but serious degradation threatening bridges was apparent at only 3 sites. Banks at 78% of the relocated channels were classed as stable and 28% as unstable.

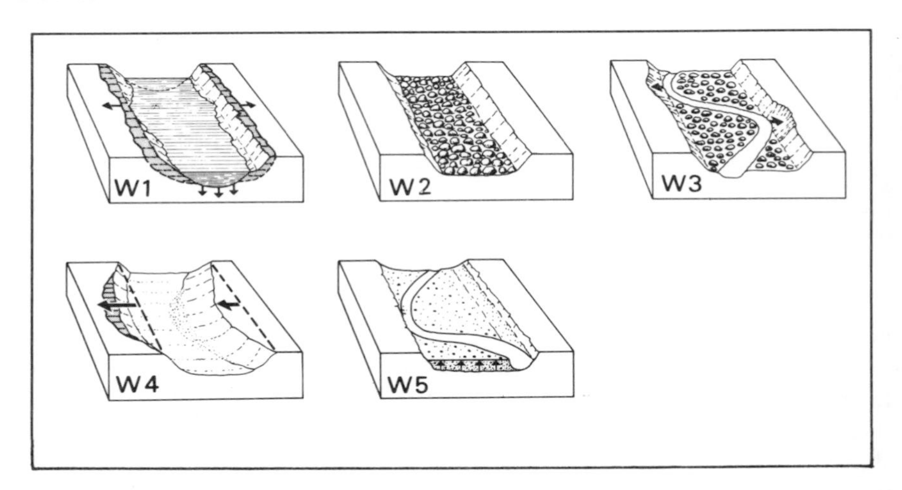

Figure 23. Principal types of adjustment in straightened river channels in Denmark: degradation of the channel bed (W1), armouring (W2), development of a sinuous thalweg (W3), recovery of sinuosity (W4) and development of a sinuous course by deposition (W5). (Based on Brookes, 1987c, Reproduced by permission of John Wiley & Sons. Ltd.)

Measurements at about 300 sites of channel straightening in Denmark revealed five major processes of adjustment in the absence of maintenance (Brookes, 1987c). Many streams which had been straightened but not stabilized by bed-fixation structures had adjusted by bed degradation, with associated bank slumping (Figure 23). However downcutting had been restricted by segregated gravel or cobble layers which had developed by the selective removal of fines. Where the channel had been straightened, but also widened beyond the original width of the natural stream channel, a sinuous thalweg had developed to carry low flows. The sinuous course had formed by erosion at locally high slopes, leaving remnants of the enlarged channel at a higher level. At lower slopes the channel had narrowed by deposition. This process was observed particularly in channels which had been recently modified and may represent the initial step towards full recovery of channel sinuosity. Natural recovery of the original sinuosity was observed at only 13

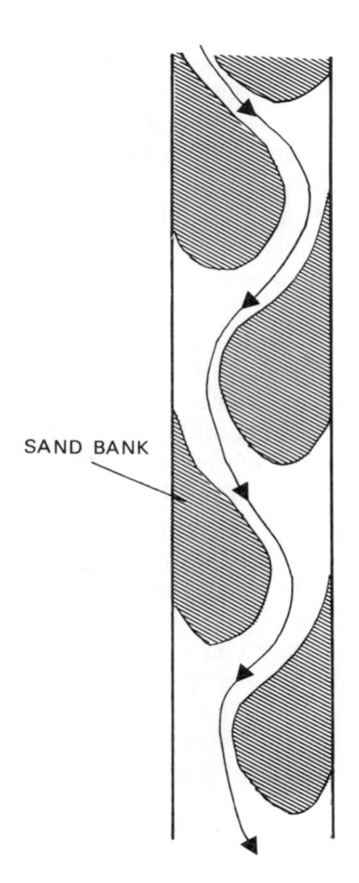

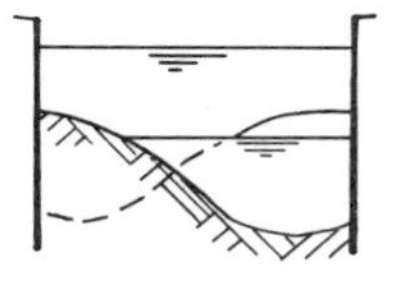

Figure 24. Propagation of sand banks in straight channels. (From Jansen *et al.*, 1979. Reproduced by permission of Professor M. de Vries.)

sites, each having a very high slope. The final type of adjustment occurred in over-widened straightened reaches at very low slopes and with a high sand load.

Sand banks may propagate in artificially straightened channels (Figure 24). Helical flow causes erosion at the upstream ends of the sand banks and sedimentation at their downstream ends. During low-flow these sand banks may be partly emerged (Jansen *et al.*, 1979).

On a much smaller scale adjustments have been observed on a total of 53 New Forest streams in Hampshire (UK) following realignment and enlargement by the Forestry Commission (Tuckfield, 1980). In particular, channels on steeper slopes underwent significant erosion and entrenchment; although stability had been attained within a maximum of ten years.

Meander Growth

Research has shown that at a particular location the size of meanders, as expressed by the wavelength, can be related to values of discharge (Carlston, 1965) and to channel width (Zeller, 1967). There also appears to be a close link with the pool–riffle wavelength. The interaction of discharge and properties of the bank sediment controls the width, which in turn controls the pool–riffle wavelength. Under conditions of equal discharge a channel in sandy sediments is wider and shallower than a channel in silt or clay and has a greater meander wavelength. In the absence of bank protection measures many channels which have been straightened have a tendency to regain their former sinuosity.

Plans were drawn up in 1800 by Joh. Gottr. Tulla for the shortening of the River Rhine between Basel and Strasborg by 14% (Figure 25A). Subsequently, however, the straightened reach had more available energy and the Rhine established a new set of meanders similar to the previous natural meander pattern with the exception of the reinforced outer banks (Figure 25C). The solution was to further restrain the river by protecting all the banks (Figure 25D). However, this caused instability of the bed with periodic scour and fill (Van Bendegom, 1973; Ryckborst, 1980).

Laboratory experiments have shown how meandering channels may develop from straight channels (e.g. Friedkin, 1945; Hussey and Zimmerman, 1953; Martvall and Nilsson, 1972). Noble and Palmquist (1968) studied the Skunk River and Squaw Creek in Story County, near Ames, Iowa, which were artificially straightened in 1900 and which subsequently developed a meandering pattern. They examined a series of aerial photographs taken in 1939, 1953, 1958 and 1966. The rate of meander development was observed to decrease with distance from a bend in a channel. Meander amplitude increased most rapidly in channels with low width–depth ratios and meander wavelength increased least rapidly in channels with high width–depth ratios. Very wide channels cut into sands did not develop meanders because excess energy was expended in moving bedload.

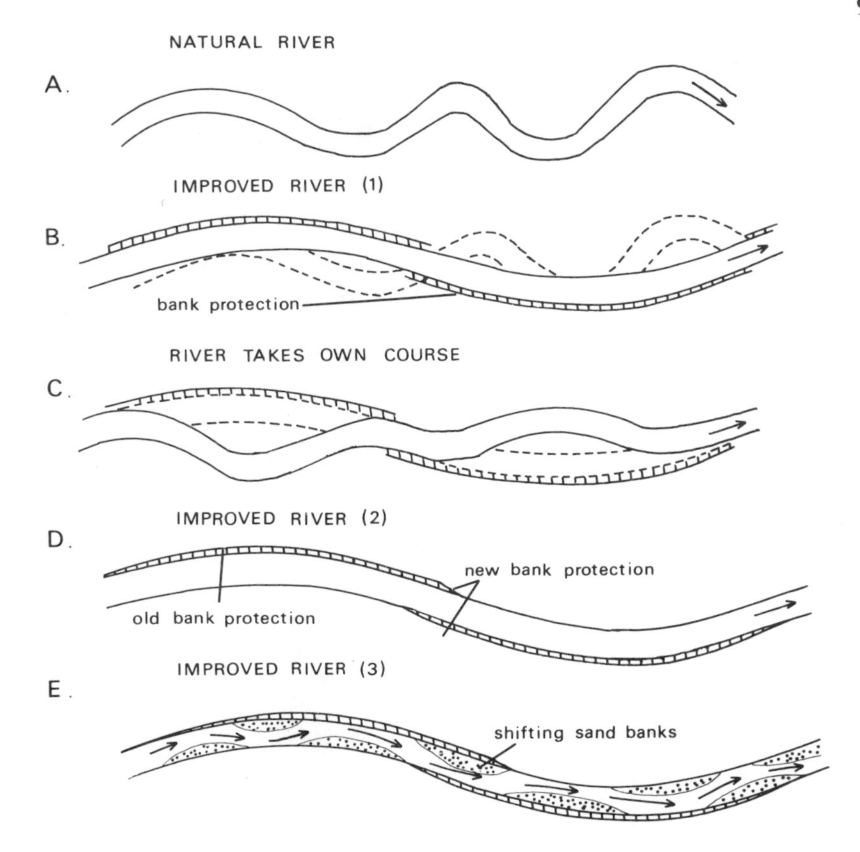

Figure 25. Experience of meander cutoffs on the River Rhine. (From Ryckborst, 1980. Reproduced by permission of Martinus Nijhoff Publishers.)

Meander growth has been observed on a reach of the River Ystwyth, a gravel-bed river in Wales (UK), which had been straightened in 1864 to run parallel with a railway track, and subsequently restraightened in 1969 (Lewin, 1976b). In both cases the operation was carried out without associated bank protection works and stages of meander development were observed during peak flows.

The Big Pine Creek in Indiana was shortened from 15.6 km to 10.9 km (a 30% decrease of channel length) in 1932 to improve drainage. The original meandering channel had a sinuosity of 1.42 compared to a sinuosity of 1.00 for the straightened channel (Barnard and Melhorn, 1982). By using aerial photographs for the years 1938, 1963 and 1971 it was possible to monitor progressive changes over a period of 40 years. Figure 26 relates sinuosity to channel slope for the three dates, and lines have been tentatively drawn to represent the stream at each date. This indicates adjustment from the channelized state (sinuosity = 1.00) towards the original meandering state (sinuosity = 1.42) in the absence of maintenance.

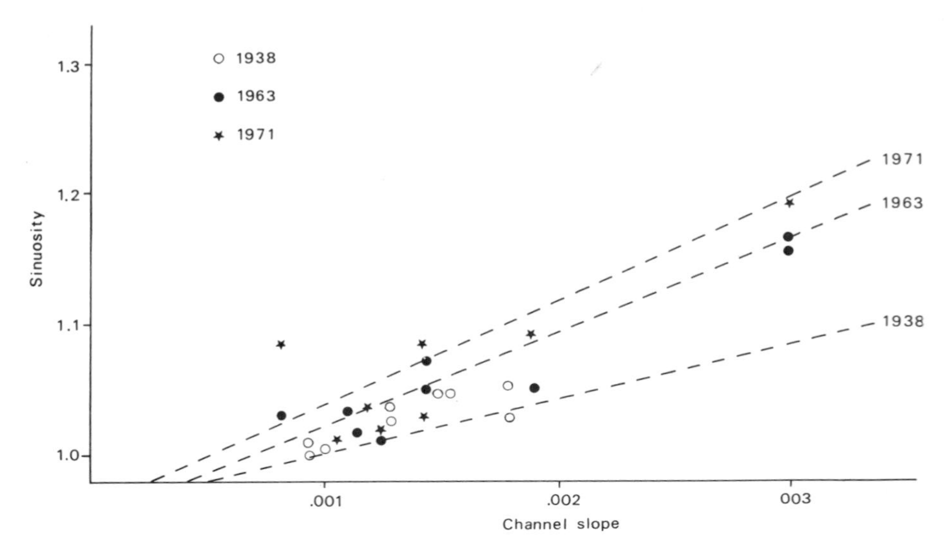

Figure 26. Variations of sinuosity with channel slope for the years 1938, 1963 and 1971, following channelization of the Big Pine Creek, Indiana. (Based on data in Barnard and Melhorn, 1982.)

Effects on Tributaries

Tributaries of the Willow Drainage Ditch in Harrison County, Iowa, underwent downcutting in response to entrenchment of the mainstream (Daniels, 1960). Emerson (1971) described entrenchment of the lower reaches of tributary streams of the Blackwater River in Johnson County, Missouri, sixty years after channelization. For example the cross-sectional area of Honey Creek had increased from 12 m^2 immediately after dredging to 255 m^2. Tributary meanders had become entrenched.

Table XXII summarizes the effects on a tributary stream of lowering the base-level in the main river channel, which can result from straightening.

Table XXII. Lowering of base-level for tributary streams (adapted from Simons and Senturk, 1977)

Local effects	Upstream effects	Downstream effects
1. Headcutting	1. increased velocity	1. increased transport to main channel
2. General scour	2. increased bed load	2. aggradation
3. Local scour	3. unstable channel	3.,increased flood stage
4. Bank instability	4. morphology of changes	4. morphology of channel changes

The local gradient of the tributary stream is significantly increased, thereby inducing headcutting and causing a significant increase in water velocities. This results in bank instability, possibly with major changes in the morphology of the tributary stream and increased local scour. In turn this may have repercussions on the downstream channel.

Controls on Channel Change

Degradation by the upstream migration of nickpoints is the main process by which over-steepened channel gradients are reduced, thereby enabling the channels to evolve to a new condition of dynamic equilibrium. Channel deepening and widening are responsible for the increase in channel cross-sectional area. Schumm and others (1984) attempted to determine the number of headcut events which had occurred in the channelized reaches of Oaklimiter Creek. For a given reach the total increase in depth following construction was calculated. Based on field observation the assumption was made that each headcut event increased the channel depth by 1.2 m. Therefore the number of headcut events was calculated by dividing the total increase in depth by 1.2 m. Figure 27 plots distance along the straightened channel against the number of headcut events. A plot of valley slope against distance is also included and it can be seen that a greater number of headcut events have been associated with steeper valley reaches.

The behaviour of straightened stream channels depends on the character of the bed and bank sediments, their erodibility and stratification. Bray and Cullen (1976) observed that degradation following a cutoff on the Çoverdale River in New Brunswick was controlled by bedrock outcrops, which prevented a nickpoint from migrating upstream and affecting the foundations of a bridge. Computed degradation rates indicated that problems might have occurred if there were no controls. Similarly it is expected that bed degradation will be restricted where a coarse segregated or armoured layer develops. By contrast the dramatic changes observed along the Tarkio River in Iowa and Missouri were attributed to the thick soil mantle of loess and valley alluvium which did not allow the channel to degrade and stabilize (Piest *et al.*, 1977).

The nature of the sediment load conveyed through the channel and the change in character of the sediment load as the channel adjusts may also be important (Schumm *et al.*, 1984). Sediment is initially derived from incision, followed by bank collapse. Subsequently, rejuvenation of the tributaries upstream becomes a source of sediment. A line of trees adjacent to the bank may have the effect of inhibiting adjustment.

Brice (1981) identified three types of factor important to bank stability (Table XXIII), namely site factors existing before modification and which may be relevant to stability after alteration, alteration factors and post-alteration factors.

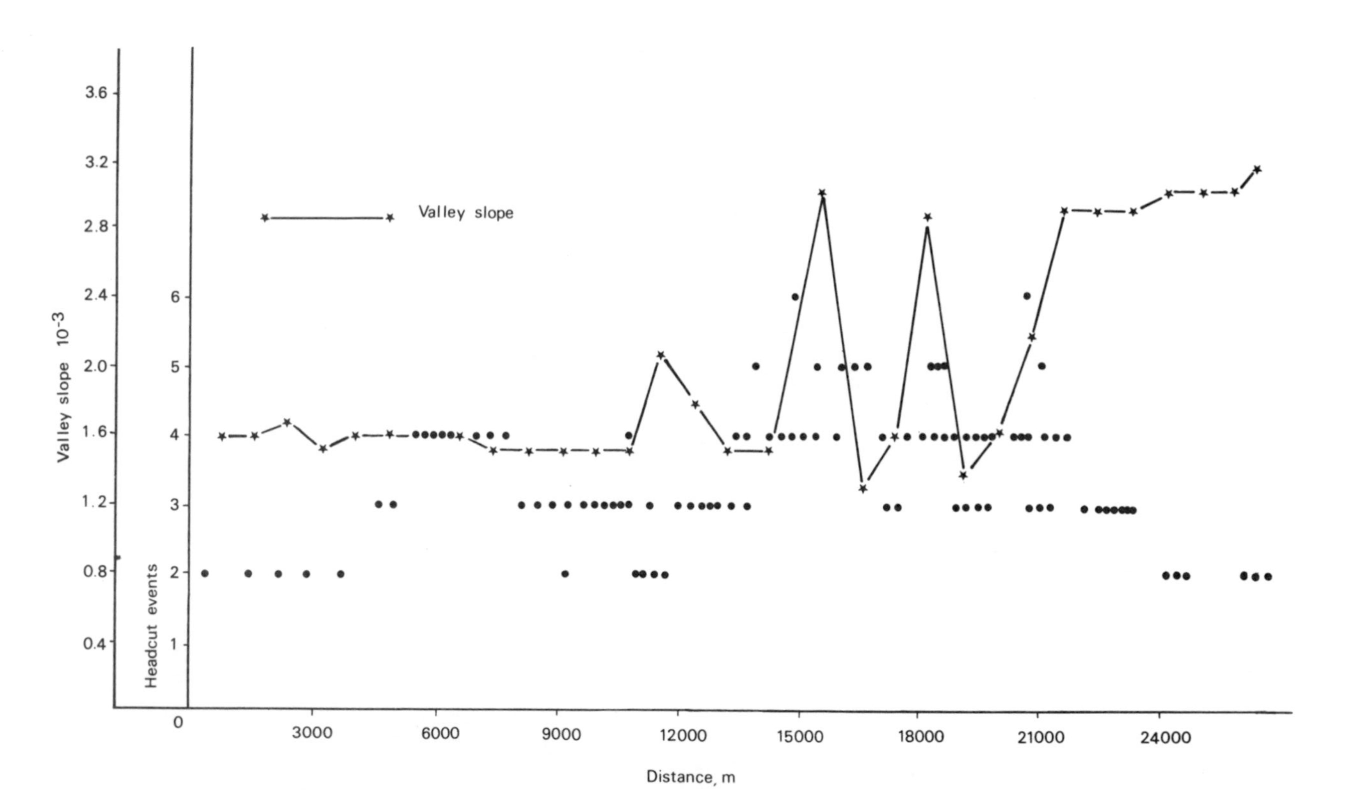

Figure 27. Number of assumed headcut events (dots) related to valley slope with distance along Oaklimiter Creek, northern Mississippi. (Based on Schumm *et al.*, 1984.)

Table XXIII. Factors important to the stability of relocated channels (after Brice, 1981)

1. Site factors	Stream flow habit, drainage area, water discharge, channel width, bank height, sinuosity, stream type, valley relief, channel boundary material, incision of channel, vegetation cover along banks, prior channel stability, works of man
2. Alteration factors	Length of relocation, slope and cross-sections of relocated channels, aspects of channel alignment, measures for erosion control and environmental purposes
3. Post-alteration factors	Length of performance period, streamflow during performance period, post-construction maintenance and addition of countermeasures, growth of vegetation along the channel

Table XXIV. Critical factors contributing to the stability and instability of relocated channels (after Brice, 1981)

Stability		
1.	Growth of vegetation on banks	41 sites
2.	Bank revetment	33 sites
3.	Stability of prior channel	20 sites
4.	Straightness of channel	20 sites
5.	Low channel slope	16 sites
6.	Erosional resistance of bed or bank material	15 sites
7.	Minimal channel shortening	15 sites
8.	Bedrock control	13 sites
9.	Check dam or drop structure	11 sites
10.	Natural or artificial discharge regulation	10 sites
11.	Number of floods in first few years after construction	6 sites
12.	Preservation of original vegetation	3 sites
13.	Dual channel	3 sites
Instability		
1.	Bends in relocated channel	21 sites
2.	Floods of large recurrence interval	17 sites
3.	Erodibility of bed or bank materials	16 sites
4.	High channel side, susceptible to slumping	9 sites
5.	Instability of prior channel	8 sites
6.	Sharp decrease in channel length	8 sites
7.	Failure of revetment	7 sites
8.	Width-change factor too high or too low	6 sites
9.	Cleared field at bankline	5 sites
10.	Flood soon after construction	5 sites
11.	Lack of continuity in vegetal cover along banks	5 sites
12.	Turbulence at check dam or drop structure	4 sites
13.	Flow constriction at bridge	4 sites
14.	Non-linear junction with natural channel	3 sites
15.	Steep channel slope	2 sites

Brice (1981) further listed 13 factors contributing to the stability of relocated channels and 15 factors responsible for instability (Table XXIV). These factors are ranked in order of importance for the 103 relocated channels which Brice studied in North America.

The stability of Danish stream channels, responding in a variety of ways to straightening, has been related to specific stream power (Figure 28). Stream power is an important independent variable which determines adjustments of river morphology, and is highly dependent on the gradient and the discharge. Reaches which have regained their sinuosity following straightening have the highest stream powers, values for individual channels exceeded 100 W m^{-2} (Brookes, 1987c). Other sites exhibiting bed and bank instability have slightly lower stream powers, but lie above a threshold of 35 W m^{-2}. Sites which have adjusted to form a segregated gravel or cobble bed also lie above

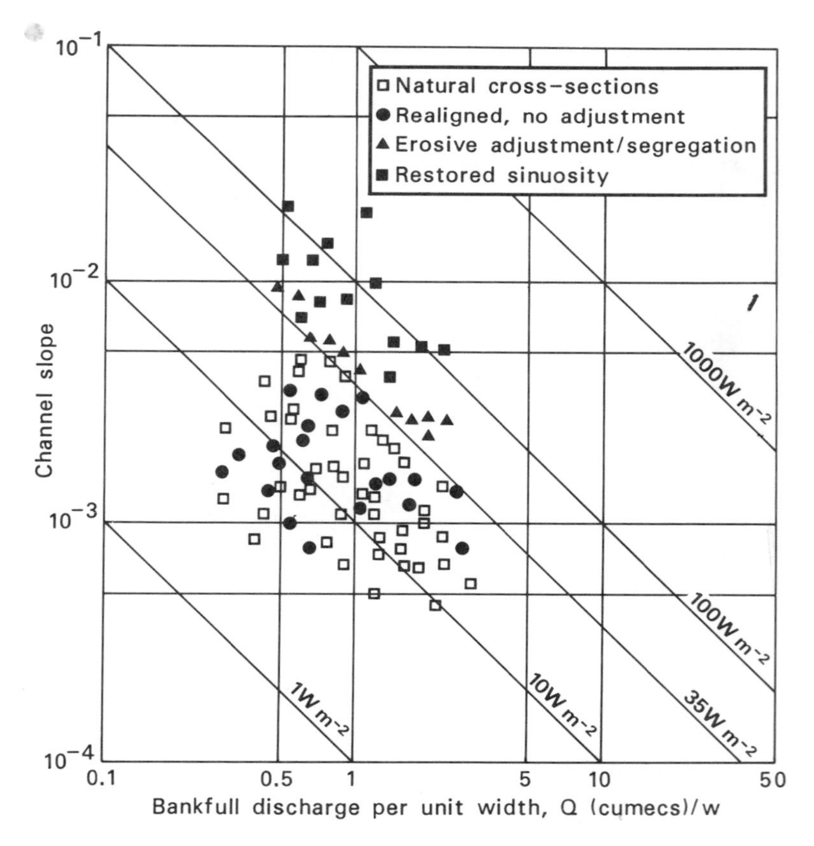

Figure 28. Channel stability of Danish streams related to specific stream power. Each site is plotted as a function of bankfull discharge and channel slope, and lines of equal specific stream power have been superimposed. (From Brookes, 1987c. Reproduced by permission of John Wiley & Sons, Ltd.)

this threshold. By contrast, natural channels and straightened channels not adjusting by erosion, fall below the 35 W m^{-2} line.

For the River Ystwyth in Wales (UK), Lewin (1976b) attributed the reformation of meanders to secondary flow patterns related to bedform modifications in the initial channel (initialization stage). However, once the sinuous channel had developed, the flow pattern and bank erosion were less dependent on initial forms, eventually with lateral accretion becoming a result rather than a cause of bank erosion (development stage). The ultimate pattern of sedimentation proved to be complex, in response to a range of flows of varying regime that occupied varying proportions of the total channel bed.

Relaxation Time and Half-life

The period between the commencement of channel adjustment and the attainment of a new state of equilibrium is referred to as the relaxation time. Major changes of cross-sectional area of the straightened reaches of the Pigeon Roost Creek, Oaklimiter Creek and Tippah River in northern Mississippi occurred during periods of flooding (Schumm *et al*., 1984). The time between disruption of the system and the beginning of change, and the annual rates of change, depend on the stochastic nature of morphogenetic events. Schumm *et al*. (1984) concluded that the relaxation time for the three straightened Mississippi rivers varied between 9 and 15 years.

De Vries (1975) made an analogy between the duration of erosion caused by channel shortening and radioactive decay. The number of years required to lower the bed over 50% of the final amount at an upstream location is given by the half-life of the man-made river disturbance. The half-life was calculated to range from 30 to 1000 years for a distance 200 km upstream of straightened reaches on the River Rhine (Netherlands), Magdalena River (Colombia), Danube River (Hungary), Tana River (Kenya) and the Apure River (Venezuela).

It has been estimated that in the absence of remedial maintenance the Big Pine Creek Ditch will attain the pre-channelization quasi-equilibrium state by the year 2000, or after 165 years of recovery (Barnard, 1977; Barnard and Melhorn, 1982). This extrapolation was based on adjustments which had been measured for a 40-year period, and indicates that recovery rates were initially high and have since been decreasing.

These examples of recovery demonstrate that realignment can be a costly capital operation which requires a long-term (50+ years) commitment of resources for upkeep, improvement and replacement.

Engineering Consequences

Adjustments arising from channelization can have serious implications for structures built adjacent to, or across, the channel. The widening and deepening of streams in the Blackwater River Basin in Missouri as a consequence of straightening has caused many problems (Emerson, 1971). Most bridges in Johnson County have been replaced or lengthened and have had vertical extensions added to the lower supports. In most cases the ends of the present bridges are threatened by bank erosion. One bridge of 27 m width built in 1930 had to be replaced in 1942 and again in 1947 by a bridge of 70 m width, but has since collapsed. Another bridge has been extended from 22.8 m to 124.2 m.

The Lang Lang River in Victoria, Australia, was modified in 1920–23 to be 12 m wide by 2.5 m deep and straightened (Bird, 1980). After flooding, further straightening was undertaken in 1926 by a New Cut, 18 m wide and 2 m deep, most of the excavated material being used to construct a levee bank. Erosion along the Lang Lang River has caused the damage or destruction of seven bridges (Bird, 1980). An estimated 4 million cubic metres of sediment has been eroded from the channel in the past 60 years, mostly being deposited as a broad delta at the mouth of the Lang Lang River. The erosion is generally 7–9 m with a maximum of 15 m. Erosion took place in the New Cut almost as soon as it was excavated. A bridge 12.2 m wide built in 1924 at this section had to be strengthened in 1931 because of scouring of the river bed. Three months later it was expanded 7.3 m to span the rapidly enlarging channel. From the 1940s onwards it required annual repairs until in 1968 a new reinforced concrete bridge 55.5 m wide was built (43.3 m longer than the original one).

Prediction

Attempts have been made to predict the average rate of degradation using bed load equations and flow duration curves, which in the case of five artificial cutoffs in New Brunswick, Canada, ranged from 0 to 0.3 m per year (Bray and Cullen, 1976). However, predicted rates may differ substantially from the observed rates where bedrock or armouring restricts downcutting. For the East and West Prairie Rivers, Canada, Parker and Andres (1976) quantified the expected degradation. This was calculated using simple theory, appropriate simplifying assumptions and some data for the channel prior to straightening.

Forecasting channel stability following channelization is difficult. Without a set of deterministic equations it is not possible to precisely predict the morphological response to alterations of width, depth, slope, roughness or channel planform caused by channelization. However, behaviour can be deduced to a certain extent by observing responses of stream channels that have already been altered. Channels stable prior to channelization are more likely to remain so, however, and initially unstable channels will probably require extensive engineering and maintenance following construction (Keller and Brookes, 1984). Changes in channel slope, bankfull discharge and thus stream power due to channelization, provide an estimate of potential post-construction instability. The plot of channel slope against bankfull discharge for 300 sites of channel straightening in Denmark (Figure 28) is therefore a useful tool for predicting adjustments for proposed works.

It is recommended that in-depth studies be undertaken before any channelization plan is formulated. Chang (1986) described a method for predicting river channel adjustments based upon the qualitative relationships among the variables of water discharge, bed material discharge, slope, sediment size, channel width and depth for sand-bed rivers in equilibrium. In response to changes of certain variables, the directions and magnitudes of adjustments for the others could be determined. The method was applied to a reach of the Mississippi River near Greenville, where a number of cutoffs had been made between 1933 and 1937, and it was determined that the original channel width of 1310 m was sensitive to channel slope along this reach (Chang, 1986). Since slope was increased by the cutoffs, then this should have been accompanied by significant increases of width and width–depth ratio to avoid subsequent channel adjustment. In practice it was desirable to maintain an adequate depth for navigation and the channel was not widened. However, following straightening the channel adjusted naturally by substantial widening and even braiding, with the result that an extensive bank protection programme was subsequently undertaken to maintain the unnatural alignment. Chang (1986) showed that the predicted width of 2800 m for the 1975 channel was greater than the measured value of 2000 m, probably because the bank protection inhibited width development.

Adjustments Within Enlarged Reaches

Adjustments of channel morphology have been noted within reaches which have been widened and/or deepened. The problems of over-wide channels have been exemplified for part of a flood alleviation scheme constructed on the River Tame near Birmingham, in central England (Nixon, 1966). The channel reverted to its original capacity in less than 30 years in the absence of maintenance, and this was due to the enlarged channel being in equilibrium with a particular design flow event, whilst out of equilibrium with the normal range of flows (Figure 29). Widening a channel reduces the unit stream power, thereby decreasing the sediment discharge. The low flows which tend to predominate for most of the time deposit sediment because of the reduced velocities in an over-widened channel, and the deposits may become stabilized to form permanent morphological features. Extensive aggradation in some rivers has caused the development of mid-channel bars and braided flow at low discharges (Nunnally, 1985).

Continuous dredging causes a river bed to degrade until the balance between sediment load supplied to the river reach and the sediment transport capacity is restored (Figure 30). In the long-term the bed degrades to a more gentle slope and greater depth downstream from the point of dredging (Jansen *et al*., 1979). Degradation also occurs upstream, resulting in a

Figure 29. River Tame, England: reversion of enlarged channel to original capacity in less than 30 years between 1930 and 1959. (From Nixon, 1966. Reproduced by permission of Severn-Trent Water Authority.)

lowered river bed with the initial slope. Degradation of a river bed can be restrained by fixation. A series of fixed weirs can be used to absorb energy, reducing the capacity of the flow to transport sediment (Harvey and Watson, 1988). The crest heights and spacing can be determined by an optimization procedure. The banks may also have to be stabilized to avoid accelerated bank erosion. Deepening the main channel also lowers tributary base-levels and increases tributary slopes and headcuts may develop. Rapid tributary erosion may result in aggradation within the main channel.

Cross-sections were resurveyed for a 2100 metre reach of the River Cherwell near Banbury in Oxfordshire, England, (drainage area 246 km^2), which was widened and deepened in 1967 (Brookes, 1983). In the absence of regular maintenance between 1967 and 1981 the channel had adjusted by the deposition of sand and silt to form benches, to a maximum depth of 2 m, which had become stabilized by bankside species of grasses and shrubs (Figure 31). The margins of these benches close to the water's edge were composed of silt, stabilized by reeds. Deposits of silt were also noted below normal water level, but although the channel had been narrowed by these benches, the actual designed channel capacity was reduced by a maximum of only 9%.

Griggs and Paris (1982) reported that within 10 years of completion of a US Army Corps of Engineers flood channel on the San Lorenzo River at Santa Cruz in California, 350,000 m^3 of sediment had been deposited. This reduced the capacity from the designed 100-year flood to a 25–30 year flood. The project had involved deepening the channel by some 0.9 to 2.1 m below the original channel bottom, thus functioning as a sediment trap for the large volumes of sand and silt derived from the urbanizing watershed.

An analogy to resectioning can be found in the literature relating to the impacts of gravel mining. The general effect of removing sand and gravel from a stream bed pit is to cause a local lowering of the bed, resulting in overall downcutting of the channel upstream from the mining operation (Forshage and Carter, 1973; Bull and Scott, 1974; Sato, 1971, 1975; Prudhomme, 1975; Lagasse *et al.*, 1980). When aggregate is excavated from many locations along a stream the removal of sand and gravel generally exceeds the rate of replenishment. The consequent lowering of the bed increases the potential for undermining bridge piers during major floods. Whilst deepening tends to increase the channel capacity, the banks are likely to be more susceptible to erosion. Perhaps the most dramatic impacts occur in inactive channels, where the sediments are not replenished at the same rate at which they are removed.

Simons and Lagasse (1975) reported how the extraction of gravel and other materials from the Lower Mississippi River had a serious impact on the morphology and stability of the channel by removing significant quantities of the coarser sediments. This coarser fraction, especially gravel, has a tendency to armour the bed through hydraulic sorting and therefore inhibits scour and stabilizes banks. Simons and Lagasse (1976) also demonstrated how maintenance dredging on the Lower Mississippi does not significantly alter the river morphology. However, the combined use of dredging, dikes and the disposal of dredged material in the dike fields can induce major changes in the cross-sectional characteristics of a river. The redistribution of sediments from the main channel to the dike fields has decreased the width–depth ratio markedly in many sections of the Upper Mississippi River. Modification of the channels interrupts the natural downstream movement of bedload

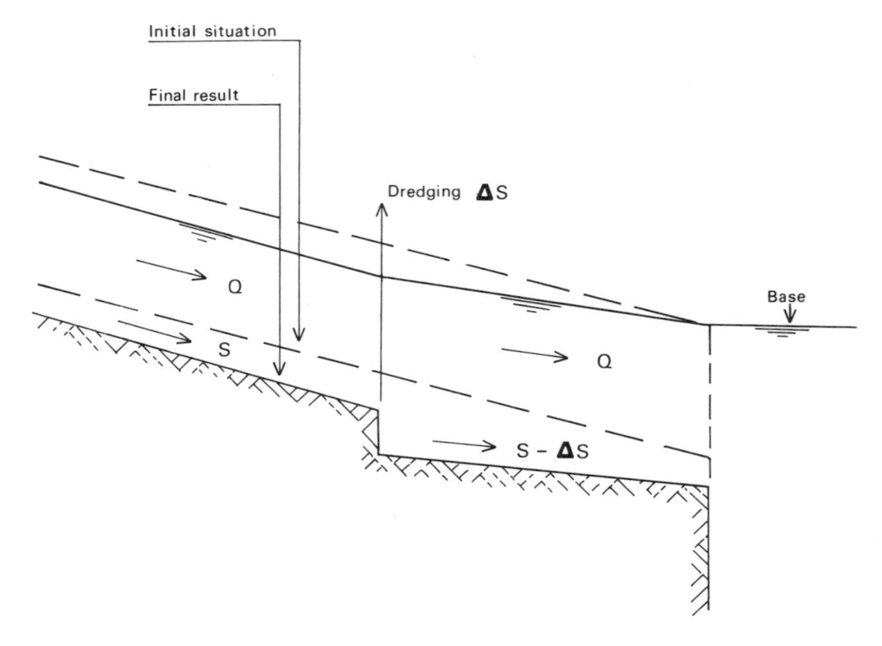

Figure 30. Degradation due to continuous dredging. (From Jansen *et al.*, 1979. Reproduced by permission of Professor M. De Vries.)

Figure 31. Formation of berm in a widened reach of the River Cherwell, England, between 1967 and 1981.

sediments both directly by dredging and indirectly by reducing the width–depth ratio and decreasing the channel's capacity to transport bedload.

Trapezoidal cross-sections are often unsatisfactory for gravel-bed rivers in which bedforms can rapidly re-establish. Alternate bars in a resectioned reach of the Afon Trannon in Wales (UK) caused renewed channel sinuosity and instability (Newson, 1986). Similarly, within 3 years of creating a uniform bulldozed channel on the Tracy Creek, Binghamton, New York, rapid changes occurred (Morisawa and Vemuri, 1975). The flat bed changed as pools and asymmetrical shoals formed. The creek regained a winding thalweg with pools and riffles, spaced at intervals of between 80 and 100 m corresponding to a bankfull width of 25–30 m.

Lined Channels

Adjustments occurring specifically within concrete-lined channels are less well documented, although there is often a requirement to remove sediment which has accumulated above the artificial substrate (Morisawa, 1985). Adjustments frequently occur where the channel has been deepened, widened or straightened prior to protection. Where the banks of an actively migrating channel have been protected by riprap, gabions or similar protection materials, then adjustments during flood events may destroy these structures. The Little Choconut Creek, near Binghamton, New York was riprapped during the construction of a highway (Morisawa and Vemuri, 1975; Morisawa, 1985). The catchment is steep and partly urbanized and flood flows in 1975 caused the banks behind the riprap to be washed away. Quarry stone riprap and gabions placed to protect a resectioned reach of the gravel-bed Afon Trannon in Wales (UK) in 1979 subsequently failed (Newson, 1986). Alternate bars formed in the trapezoidal section, promoting renewed channel sinuosity at low to medium flows. The gabions were then extensively undermined and boulder riprap outflanked, particularly in reaches with composite banks, and during moderate floods. The use of the bed sediments to form bank protection structures, the lack of adequate foundations to avoid basal scour, and the placement of structures near to strong secondary currents, all contributed to the failure (Newson, 1986).

An unpredicted consequence occurred on the River Irk at Chadderton, England, when a cutoff developed over a period of 12 hours in 1964 (Johnson and Paynter, 1967). This change was probably initiated by the deflection of flood flow from a concrete wall built to protect local housing from flood damage, which caused erosion of the glacial sands and gravels forming the bank.

The Effects of Embanking

Adjustments may arise from embanking a channel because larger flows are confined than previously and the greater velocities associated with these flows may cause degradation of the bed and/or banks. Maddock (1976) suggested that because of this additional stress placed in the channel, levee systems often graduate to bank protection systems.

DIVERSIONS

In Alkali Creek in Wyoming, flow returning from a diversion channel caused degradation. The channel degraded to an armoured bed of large gravel and cobbles and subsequent accelerated bank erosion required stabilization measures (Vanoni, 1975).

DIKES

Kuprianov (1977) classified hydraulic structures according to whether they affect channel processes either locally, or over longer reaches (active structures), or whether they do not induce changes in water and sediment discharges (passive structures). Structures causing changes of river bed foundations are: (1) bank structures (groynes, spur dikes, lateral protections) which are attached to one of the banks. By reducing the width of the cross-section, lateral erosion occurs which changes the slope and depth of the reach. Local scour always occurs at the foot of these structures and the works may increase the channel roughness; (2) other structures used to protect a bank may not significantly reduce the width and only produce local scour; and (3) bottom fixing structures including weirs, dams, etc. The principal effect of these works is to cause deposition and raising of bed levels upstream and erosion downstream.

Hydraulic structures may change the local slope of an alluvial river channel and redistribute velocities. Velocities are locally higher near the structure and turbulence may develop. This in turn causes local erosion and sedimentation until the channel attains a new equilibrium slope.

Dikes have been used extensively on the Lower Mississippi to help maintain navigation channels, principally since 1960. By constricting the flow area, velocities are locally increased and shoals are scoured and secondary channels and chutes are closed such that all flow is confined to the main channel. By using photographs and hydrographic surveys taken between 1962 and 1976, Nunnally and Beverly (1983) demonstrated the morphological changes in diked and undiked reaches. The total surface area of the river between river miles 320 and 954 remained relatively constant between 1962 and 1976. However, this area was classified on the basis of main channel used for navigation, secondary channels which carry flow all year, sloughs or slack-water areas with a single inlet or outlet, chutes or narrow channels with relatively little flow, and pools which were found on sand banks. Within the diked areas secondary channel area decreased by 42.1 km^2, a decrease of 38.6%, but this was offset by increases in sloughs (53.2%), chutes (44.8%) and pools (2423.8%) (Figure 32). Pools, sloughs and chutes are all

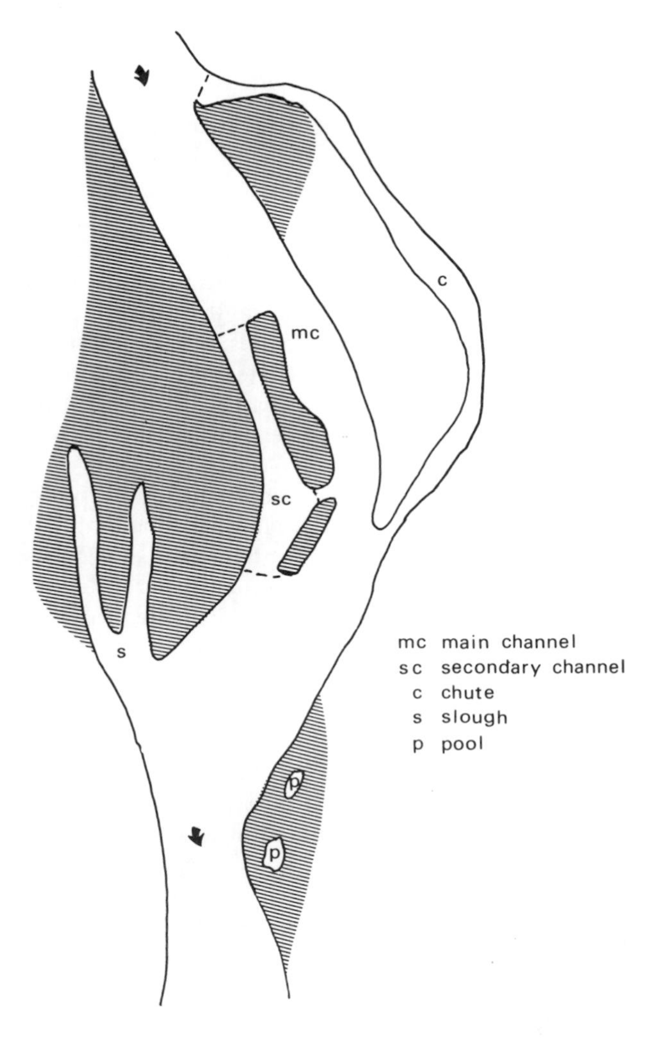

Figure 32. Aquatic habitats of the Lower Mississippi (adapted from Nunnally and Beverly, 1983. Reproduced by permission of the American Society of Civil Engineers.)

considered to be valuable types of aquatic habitat; secondary channels are considerably less valuable.

Dike fields on the Lower Mississippi do not completely fill with sediment and thereby destroy valuable aquatic habitat. Nunnally and Beverly (1983) also suggested that these may be permanent features of the river, a new equilibrium being attained rapidly following construction of the dike fields in response to low sediment load, low slope and highly variable discharge. However, a definitive answer will depend on a better understanding of the mechanics of dike field sedimentation and its relation to dike location, dike field design, flow hydrology and sediment flow.

Restraining Braided Rivers

The effects of channelization of braided rivers are less well documented. In 1817 Tulla initiated the channelization of the braided Alsatian section of the Rhine and is quoted as saying that 'As a rule, no stream or river needs more than one bed' (Ryckborst, 1980).

Undertaking improvements on braided channels requires great care (Jansen *et al*., 1979; Jaeggi, 1984). There are two principal methods of confining a river to a single channel. Firstly, a single channel can be dredged to a depth and width of the unbifurcated reaches, with suitable alignment and eventually stabilized by bank fixation and later closing the remaining channel(s). Alternatively the remaining channels can be closed off, leaving the main channel to scour. This will lead to temporary deposition downstream and aggradation upstream.

Rock Creek is a braided channel near Roberts in southern Montana, characterized by a steep slope (0.013), coarse-grained load and high velocity flows (drainage area: 1480 km^2). The river was channelized between 1957 and 1958 by bulldozing river bedload against the channel banks to form rubble dikes over a length of 27 km (Ritter, 1979). This was intended to prevent the river from occupying high-stage channels and causing the flooding of woods, pasture and buildings, which had occurred during the 100-year flood in 1957. Subsequently, during high flows more water was confined to the main channel. In the first 17 years following channelization there was some lateral erosion and a 10-year flood in 1975 caused more significant change, particularly along unprotected banks. Diking was therefore an improper procedure, serving as a catalyst for greater environmental problems.

Approximately 4 km north of Roberts, Rock Creek changes from a braided to a single-thread actively meandering reach in a downstream direction. The same channelization technique was applied (Ritter, 1979). However, the use of bedload to form dikes is a clear indication of a failure to understand the dynamics of the reach because at high flows the river has sufficient energy to entrain this material.

Effects of Clearing and Snagging

There are few studies which describe the before-and-after effects of clearing and snagging on specific streams (Shields and Nunnally, 1984). Trees, snags and log jams have a significant impact on channel morphology. Trees have been shown to retard bank erosion, whilst fallen trees and log jams may trigger bank erosion and bed erosion, particularly in small streams. In meandering channels it has been shown that log jams frequently result in local channel widening, deposition and mid-channel bars downstream of the obstruction (Keller and Swanson, 1979; Keller and Tally, 1979). Vegetation influences channel width, depth and slope (Zimmerman *et al*., 1967). Removal of debris and bankside vegetation increases the hydraulic efficiency, increases current velocity adjacent to the bank and reduces bank resistance to erosion. Documented changes in Manning's '*n*' following the removal of vegetation, snags, log jams and mid-channel bars are summarized in Table XXV.

Table XXV. Examples of changes in Manning's '*n*' following clearing and snagging and after regrowth

Source	Location	Condition	Manning's '*n*'
Wilson (1968)	15 m wide,	clean	0.022
(1973)	3.6 m deep	after one growing season	
	channelized stream	(summer foliage)	0.045
	near Jackson,	after 6 years (summer foliage)	0.070
	Mississippi	after 8 years (winter foliage)	0.070
Pickles (1931)	4.5–16.8 m wide	clear weeds and	0.032
	drainage ditches in	willow to 1.2 m	
	central Illinois	height	0.050
Burkham (1976)	Gila river during	dense growth of mesquite	
	a flood	and salt-cedar after	0.080
		eradication	0.024
Ramser (1929)	natural river	sinuous, sandy,	
		choked with snags	0.150

From Shields and Nunnally (1984). Reproduced by permission of the American Society of Civil Engineers.

Clearing and snagging have been shown to cause bank erosion and widening of channels (e.g. Strauser and Long, 1976). It is possible to estimate the effects of clearing and snagging by using an analytical approach such as that outlined by Chow (1959). Three of eight factors identified are influenced by clearing and snagging, namely vegetation, channel irregularity (bar removal) and flow distribution (log and debris jams; snags). A formula for composite resistance has been produced by Petryk and Bosmajian (1975), which incorporates the density of vegetation. To estimate the hydraulic effects of clearing and snagging, changes in the resistance factor need to be estimated using an analytical method, and then the estimate is refined by undertaking work and quantifying the effects (Shields and Nunnally, 1984). Large obstructions, such as log jams or mid-channel bars, may be incorporated in a backwater programme rather than being treated as roughness elements.

Current Research

Assessment of the morphological effects of river works has been undertaken by Hydraulics Research Ltd (UK) funded by the Ministry of Agriculture, Fisheries and Food in conjunction with the UK water authorities (Hydraulics Research, 1987). For 41 schemes selected in a preliminary assessment it was concluded that a significant proportion of these works resulted in morphological problems. In the majority of cases these problems were minor, but in a few cases a long-term and expensive maintenance commitment was required to retain the design standard. In the most extreme case the reduction of channel capacity between maintenance works was of such concern that further works were considered in an attempt to reduce the level of maintenance.

Sediment has to be regularly removed from channels which have been substantially enlarged by widening and/or deepening. The problems included large-scale sedimentation in a 500 m reach of the River Usk in Brecon, Powys, which was widened and deepened in 1980 as part of a comprehensive flood alleviation scheme. Between 5000 and 8000 tonnes of gravel has to be removed at least once per year. The majority of schemes reduced flow velocities by enlarging the channel capacity and therefore large-scale erosion was found to be rare. Along the River Derwent in Matlock, Derbyshire, which was deepened by up to 2 m in 1980/81, bank instability was noted along a 1.25 km reach. This subsequently required extensive bank protection (Hydraulics Research, 1987).

The objective of this chapter has been to review the range of morphological adjustments which have occurred following various types of channelization in a number of different environments. There is clearly scope for research to further determine the factors which control the nature and extent of adjustments. Where bedforms, naturally shaped cross-sections and meanders reform following channelization, then this may facilitate biological recovery. However, morphological adjustments are frequently controlled through further intervention or regular maintenance, and this can offset biological recovery.

CHAPTER V

Biological Impacts

Introduction

Stream channelization is an extreme physical disturbance that disturbs not only the ecosystem which occupies the stream bed but also the substrate. Figure 33 compares the morphology and hydrology of a natural stream with a typical channelized watercourse. Channelization can alter the original dimensions and shape of a channel, the slope and the channel pattern, changing a heterogeneous system into a homogeneous one (McClellan, 1974; Neuhold, 1981). Bank cover is eliminated, pools are lost, flow approaches a laminar character, and the substrate approaches homogeneity throughout the channel. Boulders and debris may be removed to increase the hydraulic efficiency of the channel. The result, in ecosystem terms, is that habitat diversity and niche potential are reduced and, depending upon where in the 'continuum' channelization is incurred, the quality and functions of the species occupying the system are changed (Vannote *et al.*, 1980; Neuhold, 1981).

Cummins (1974) examined the concept of the stream ecosystem as a continuum, wherein the ecosystem is driven by allochthonous organic and nutrient input at its uppermost reaches and is supplemented by autochthonous production in its middle reaches—all of which serve to feed a heterotrophic filtering system at its lowermost reaches. Margalef (1963) observed an increase in structural complexity from the upper to the lower reaches. Thus a disturbance in a stream ecosystem at any point along the continuum will affect the linear order of that continuum and along with the type of disturbance will influence the character of change and the character and rate of recovery.

If channelization disturbed a first-order stream, allochthonous organic particle input would be reduced or eliminated with the elimination of bank cover, the channel would be opened to direct solar input, and the essentials for autochthonous communities would have developed. Production relative to respiration (P/R) will increase and structural diversity will decrease. If the disturbance occurred in the autotrophic pattern of the continuum much the same effects would be experienced. Similar effects would be experienced in the heterotrophic position of the continuum.

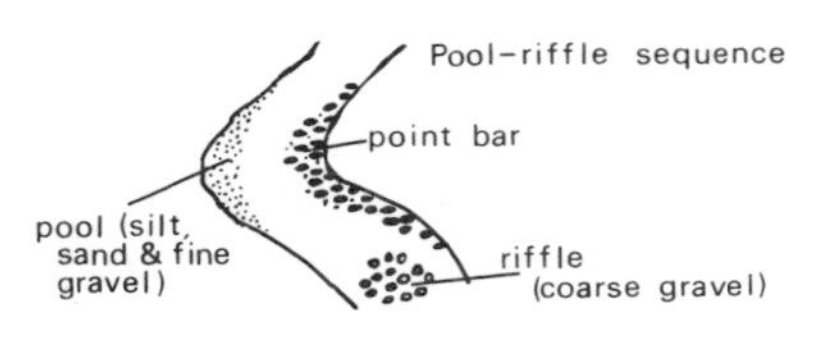

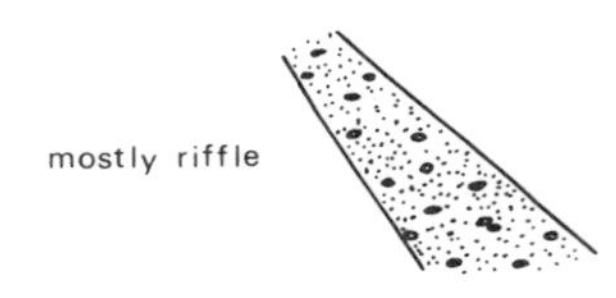

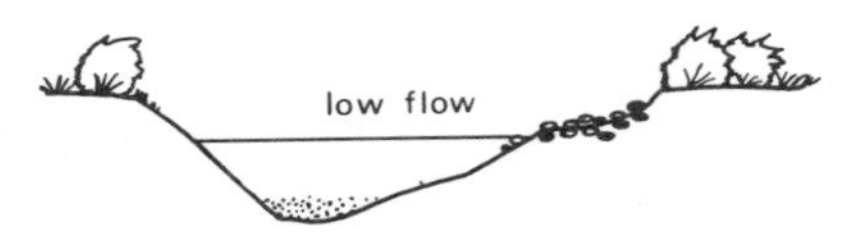

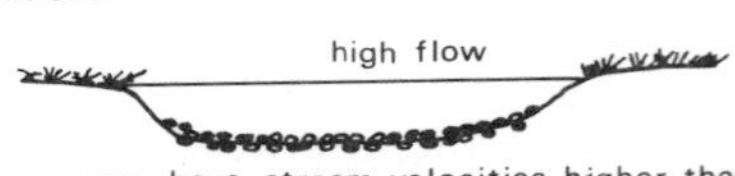

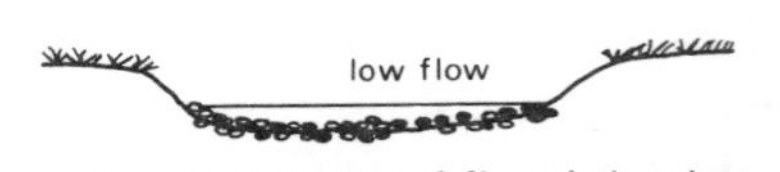

Figure 33. Comparison of the channel morphology and hydrology of a natural stream with a channelized watercourse. (Based on Corning, 1975. Reproduced by permission of Virginia Department of Game and Inland Fisheries.)

Time of recovery of the original sinuosity can vary from a few years, when the substrate is relatively fine-grained, to thousands of years when the substrate is hard bedrock. The ecosystem recovery process keeps pace with the physical recovery process. Predictions are therefore required of the time necessary for a stream to become reinhabited by its original species, assuming none were brought to ultimate extinction (Neuhold, 1981).

To understand the impact of river channelization on the flora and fauna of watercourses it is first necessary to outline the various habitat requirements and then to describe how these are destroyed or altered during construction. Research has identified a wide range of impacts of channelization on macroinvertebrates, fish, aquatic vegetation and riparian communities. The biological impacts of channelization have been reviewed by authors for various countries and regions, including the USA (Gaufin, 1959; Patrick, 1971; Barton *et al.*, 1972; Baker, 1973; Bradt and Wieland, 1975; Smith, 1975; Dale, 1975; Lee, 1976; Ellis and Whelan, 1978; Timbol and Maciolek, 1978; Maughan *et al.*, 1978; Schoof, 1980), the UK (Toms, 1975), Germany (Engelhardt, 1951), and the humid tropics (Santema, 1966).

The impacts on the invertebrates, fish and aquatic plants within a channel which are documented arise mainly from channel excavation and dredging, the lining of channels, clearing and snagging and weed cutting. The habitats of birds and mammals may also be destroyed by the removal during construction of bankside trees, bushes and plants. However, channelization works which are intended to facilitate drainage for agriculture may have far-reaching effects which extend into the floodplain. The draining of wetlands can have very serious consequences for mammals, amphibians, insects and birds. Few studies detail the effects of channelization on reptiles, although regulation of the Upper Rhine eliminated the turtle *Emys orbicularis* (Thienemann, 1950).

MACROINVERTEBRATE RESPONSE

Many factors which can be changed by channelization determine the occurrence and detailed distribution of benthic invertebrates. These include velocity of flow, temperature, including the effects of altitude and season, the substrate, including vegetation, and dissolved substances. Other important factors are liability to floods and drought, food supply, competition between species and shade (Hynes, 1970).

Habitat Requirements

Velocity controls the occurrence and abundance of species and therefore the whole structure of the animal community (Scott, 1958; Gore, 1978; Milner *et al.*, 1981; Jenkins *et al.*, 1984). Many macroinvertebrates rely on the current either for feeding or because their respiratory requirements demand it. However the majority of field studies correlate the species to a single measurement of velocity obtained at the time of sampling, and few take into account the variations of discharge over time which occur at a site. *Simulium* larvae move to sheltering positions with increasing discharge. Many researchers have found that particular species are confined to fairly definite ranges of velocity (Zahar, 1951; Phillipson, 1956). Where the velocities are persistently high at a site then almost all species may be absent; conversely slow-flowing or stagnant watercourses, which accumulate a large amount of silt, can only be tolerated by a limited number of species. The distribution of velocities at the bed is most critical to macroinvertebrates; certain species only colonize the protected downstream faces of cobbles and boulders. It is therefore impossible, other than in a very general way, to correlate a single measurement of velocity in a stream with the species present.

There are numerous ways in which the physical characteristics of the substrate influence invertebrate distribution (e.g. Percival and Whitehead, 1929; Sprules, 1947 and Jenkins *et al*., 1984). Several species are adapted to live in crevices beneath and between stones where turbulence is reduced. These species can only occur where the substrate is stony; for example they are particularly important in fast-flowing water for mayfly nymphs. Riffles are typically characterized by a greater density of invertebrates than in a pool, although this may also depend on the presence of plants. Mayflies, caddisflies, stoneflies and blackflies are more abundant on riffles. Other species, such as *Simulium* sp. and net spinning caddis larvae, are not found where the substrate is unstable or fine-grained. Species such as the burrowing mayfly nymph require fine particles, particularly coarse sand. Silt may be colonized by large numbers of burrowing worms.

In general the larger the stones, and therefore the more complex the substrate, then the more diverse the invertebrate fauna (Hynes, 1961, 1968; Townsend, 1980). Sand is a relatively poor habitat with few species and individuals. Fewer animals tend to live on loose stones than on embedded ones and it appears that this may be related to stability. On stony substrates the presence of silt reduces and changes the fauna. Other species live on wood. Silt-free riffles generally have richer faunas than silty pools in terms of the number of species and biomass. Detritus, partly caused by the accumulation of leaf litter from overhanging trees, is important for certain species (e.g. naidid worms) but reduces others. Vegetation also affects the fauna. Several studies have shown that there are more animals in moss, rooted plants and filamentous algae than there are on stones (Percival and Whitehead, 1929; Wright *et al*., 1983).

Oxygen and temperature are closely interrelated factors because the solubility of oxygen in water decreases steadily with rise in temperature. In unpolluted freshwater rivers oxygen rarely falls below critical levels. However, in slackwaters, such as backwaters or dense stands of emergent vegetation, oxygen may fall below critical levels during summer. Relationships are apparent between the respiratory rate and temperature for certain species. Some species of leaches require well-aerated water for survival because they have dependent respiration and can survive in stagnant places. A marked regularity in the downstream sequence of species of net-spinning caddis has been identified in British rivers (Townsend, 1980). This can be related to the general increase of temperature with distance downstream, each species having different respiratory requirements.

A few studies have reported a correlation between shade and the occurrence or abundance of particular species. Certain mayflies are more abundant in shady places than in exposed ones (Hughes, 1966). The differences observed may be due to the indirect effect of temperature or organic detritus from overhanging trees. The leaves and needles which fall into the stream also provide organic material, which contributes to aquatic insect production.

Impact of Engineering Works

The detrimental effects of various channel modification and maintenance practices on macroinvertebrates have been well documented (e.g. Haynes and Makarewi, 1982). The decline of shellfish (e.g. Ingle, 1952) and insect fauna such as dragonflies and stoneflies (e.g. Wene and Wickliffe, 1940) has been shown to be caused by modifications to river channels. Other animals such as chironomid midges and blackflies may be less sensitive to disruption and dominate immediately after dredging. Table XXVI lists a selection of examples.

Effects of Channel Excavation

Animals such as freshwater mussels may be physically removed during the process of excavation (Clark, 1944) or macroinvertebrates may fail to establish in the post-construction phase because of a changed substrate (Hansen and Muncy, 1971; Bou, 1977).

Silt deposition as a consequence of channelization may kill many benthic invertebrates (Apman and Otis, 1965). Silt screens out light and tends to hold extraneous substances, such as industrial wastes, on the stream bed. Sand or shifting silt on the bed may also eliminate shelter. Zimmer and Bachmann (1978) found that invertebrate drift was less in channelized reaches of tributaries of the upper Des Moines River in Iowa compared to natural reaches, and this was partly attributable to the absence of a coarse substrate. The larger number of drift organisms in a channelized section of the Little Sioux River, Iowa, was attributed to the lack of suitable attachment areas (Hansen, 1971a,b; Hansen and Muncy, 1971). Etnier (1972) found that channelization of the Middle Creek in Sevier County, Tennessee, resulted in decreased riffle habitat and gravel and boulder substrate. This was reflected in significant differences in macroinvertebrate density, species richness and composition, particularly a reduction in the abundance of Ephemeroptera, Trichoptera, and Plecoptera in sections which had been channelized for less than a year.

Standing crop, productivity, species diversity and numbers of macroinvertebrates were lower in channelized sections of the Luxapalila River even 52 years later (Arner *et al.*, 1975; 1976). This was attributed to differences in the substrate, pebbles being common in natural reaches and fine sand typifying channelized sections.

However, the effect of a changed substrate may be complex. In the Buena Vista Marsh, Wisconsin, channelization created an unstable shifting-sand substrate during high spring flows, which corresponded to low benthic biomass and density (Figure 34). By contrast, vegetation and silt detritus

Table XXVI. Examples of the impacts of channelization on macroinvertebrates (after Brookes, 1983)

Source	Location	Type	Date	Variables affected	Reduced parameters	Recovery
Clark, 1944	22 streams Auglaize County	Dredging	1940	Disrupted substrate; sediment deposited	Mussels eliminated by direct removal	Study 4 years after channelization
Morris *et al.*, 1968	Missouri River, Nebraska		1950s	Brush and piles and pools eliminated	Benthic area reduced by 67%; standing crop of drift reduced 88%	Study represents 15 years after channelization
Hansen and Muncy, 1971	Little Sioux River, Iowa			Lack of suitable substrate	Fewer macroinvertebrates; but more drift organisms	
Moyle, 1976	Rush Creek, California	Realigned	1969	Changed substrate; lack of pools and shade	Biomass of invertebrates per unit area reduced by 75%	Studied 4–5 years after channelization
Schmal and Sanders, 1978	Buena Vista Marsh, Portage Co, Wisconsin	Dredging	Vars.	Seasonal effect; substrate only unstable at high flow; at other times vegetation/silt	High invertebrate populations when substrate stable; vegetation and silt favours snails/midges. Elimination of stoneflies	
Crisp and Gledhill, 1970	Mill-stream Dorset, England	Dredging		Substrate disrupted	Population density of benthos reduced to 4000 individuals/m^2 but increased by 80–86% in 2 years	Before, during and after
Pearson and Jones, 1975	River Hull, Yorkshire	Resectioning		Substrate disrupted but largely unchanged	Benthos escaped the bucket and redistributed rapidly over the affected substrate	Before, during and after
Kern-Hansen, 1978	Gjern Stream Jutland, Denmark	Annual cutting of macrophytes		Habitat destroyed by removing macrophytes	99% increase of total drift density	Before, during and after

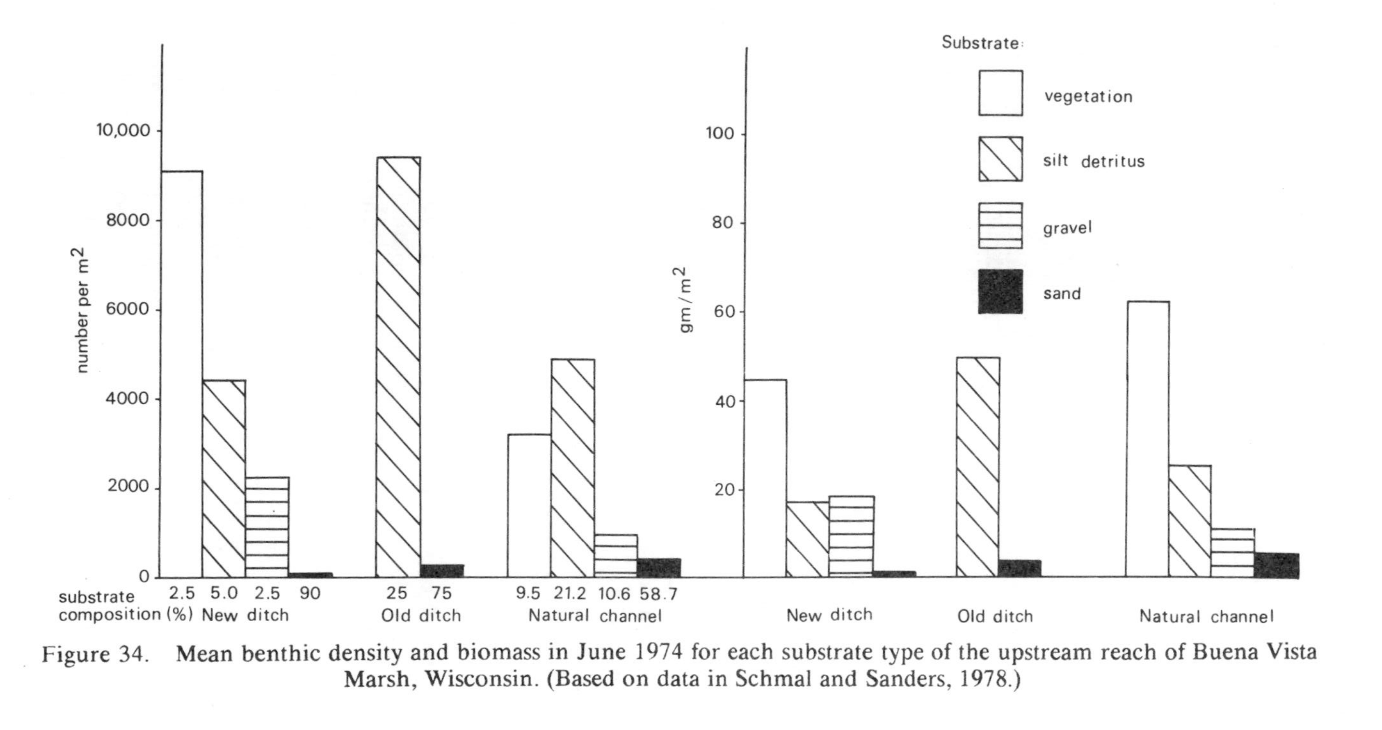

Figure 34. Mean benthic density and biomass in June 1974 for each substrate type of the upstream reach of Buena Vista Marsh, Wisconsin. (Based on data in Schmal and Sanders, 1978.)

substrates were predominant as spring flows subsided and invertebrate populations were high (Schmal and Sanders, 1978). These more stable conditions were favourable to Mollusca (snails) and Chironomidae (midges). More invertebrate taxa were collected from natural as opposed to channelized streams but differences were not statistically significant. However Plecoptera (stoneflies) were collected only from natural streams, and more genera of Ephemeroptera (mayflies) and Trichoptera (caddis-flies) were collected from natural streams.

Older channelized reaches in the Buena Vista Marsh, Wisconsin, were devoid of pools and point bars (Sanders, 1976; Schmal and Sanders, 1978). As a consequence habitat diversity was reduced, especially during high spring flows, and this may have increased the amount of drifting seston when the more productive substrates were unstable. Channelization appeared to affect invertebrate drift through its influence on drifting seston and benthos.

Recovery of the macroinvertebrate fauna following channelization has been shown to occur where there are no substantial changes in the substrate size and stability (e.g. Hortle and Lake, 1982). A study of the Bunyip River in Victoria, Australia, revealed no significant and consistent differences in the density, biomass and composition of the macroinvertebrates at channelized and unchannelized sites (Hortle and Lake, 1982). In contrast to other studies (e.g. Etnier, 1972; Arner *et al.*, 1976; Marzolf, 1978), two channelized sites of the Bunyip River had the most stable substrates composed of clay bedrock.

Dredging on a regularly maintained mill-stream in southern England had only a limited impact on the population density of benthos, with rapid recovery in approximately two years (Crisp and Gledhill, 1970), although Hill (1976) believed that such a rapid recovery may not occur where high sediment loads prevent the development of stable substrate conditions. A study by the Anglian Water Authority (UK) in 1982 showed a marked decline in invertebrate quality following dredging on the Great Ouse, which was attributed to silt covering the gravel bed. Only where the gravel bed subsequently became re-exposed did the invertebrate population recover, 1.5 years after dredging. Rapid recovery of a benthic community was also achieved on a chalk stream in Yorkshire following excavation of the gravel bed to a depth of 30–40 cm, and this was attributed to the behaviour of the animals which produced a rapid redistribution of the fauna over the available habitat (Pearson and Jones, 1975). The density of macroinvertebrates in the River Moy in Ireland was reduced by 90% following dredging operations and the rate of recovery was slow (McCarthy, 1975, 1977, 1980, 1981). Substantial densities of midges and mayflies were recorded after 2 years but caddis-flies were seriously affected. On the River Usk in Wales (UK) a variety of habitats including pools, riffles, tree roots and marginal vegetation were sampled at 10 sites before engineering works began (Brooker, 1985). A total of 80 taxa were recorded, compared to only 50 taxa collected at the same sites a year following channelization. The species absent after channelization were typical of low energy or vegetated reaches on the river.

For warm-water streams in Ohio and northeastern Indiana, Griswold *et al.* (1978) found that macroinvertebrate abundance, diversity and biomass was significantly lower in channelized areas of the Olentangy, Hocking Rivers and Rock Creek. Drift rates were higher in the unchannelized sections. Dominant macroinvertebrates in unchannelized areas were riffle species such as Hydropsychids, Heptageniids, Elmids and Psephenids, whilst in channelized areas the dominant species were burrowing forms adapted to living on soft substrates in slow-moving water (Oligochaetes and Chironomids). Channelization had involved widening of the bed, with a resultant reduced water velocity and therefore deposition of silt. By comparison no effects were noted for the Sandusky River, which may be explained by the presence of a bedrock bottom.

Channel modifications undertaken in 1972 following Hurricane Agnes on six coldwater streams in Pennsylvania (Fishing, Money, Beaver, Clover, Mill Creeks and Freeman Run) appeared to have no long-term effect on the number of taxa, number of organisms, density, standing crop and diversity of benthic communities (Duvel *et al.*, 1976) (Table XXVII).

Table XXVII. Summary of benthic macroinvertebrate data (September)

Parameter	Natural sites	Channelized sites
Mean number of taxa	8.97	8.95
Mean number of organisms	53.31	40.08
Mean diversity	2.18	2.31
Mean dry weight (mg)	96.38	80.45
Mean density (number m^{-2})	566	431
Mean standing crop (gm m^{-2})	1.04	0.87

From Duvel *et al.* (1976). Reproduced by permission of the American Water Resources Association

The lack of difference between natural and channelized sites was explained by the availability of a suitably diverse substrate composed of cobbles in the channelized reach and gravel, silt, bedrock and boulders in the natural reach. Although benthic habitat varied from channelized to natural sites, there was sufficient diversity of habitat in both instances. Similarly, Whitaker *et al.* (1979) found that macroinvertebrate communities of rural warm-water streams on the Delmarva peninsula in Delaware stabilized within a year after channel modification. There were no significant correlations between the time elapsed after channel modification (1 to 30+ years) and various diversity indices (Whitaker *et al.*, 1979). Channel modification did not cause enough stress to alter the diversity of macroinvertebrate communities.

Dragline operations in the River Hull (UK) had mainly short-term effects on benthic fauna. It was suggested that much of the substrate and its fauna escape the bucket and many animals probably fall out during the lifting operation (Pearson and Jones, 1975). The timing of dredging may influence the degree of impact, a spring to summer dredge having the least effect, since breeding of most species occurs shortly after redistribution.

The narrowing of rivers by dikes to form stable navigation channels reduces the water surface area and eliminates backwaters (Funk and Robinson, 1974). These alterations can decrease habitat diversity (Groen and Schmulbach, 1978). Channelization by the installation of pile dikes on the Missouri River in Nebraska reduced the width from 720 m to 240 m, and consequently the benthic habitat was decreased by approximately 67% (Morris *et al*., 1968). Much of this reduction in benthic area was accomplished by eliminating the relatively productive slack-water areas. However, during the study there was little difference between standing crops of benthos in the channelized and unchannelized rivers. However, the average standing crop of drift was reduced by 88% in the channelized reach.

Substrates created by construction, or subsequent adjustments, may be more suitable habitats for selected species. Channelization can provide a more suitable habitat for *Simulium* by providing a stable substrate with swift laminar flow (Müller, 1953).

Channelization of the Lower Mississippi caused current velocity to increase resulting in decreased silt substrates and decreased numbers of *Hexagenia*. Dikes create mosaic-like substrates, which vary with discharge. Beckett *et al*. (1983) found that although dikes are part of the channelization process they create habitat (mud substrate) suitable for *Hexagenia*. Sand and gravel areas are characterized by the clam (*Corbicula fluminea*) and Chironomids; mud is dominated by *Limnodrilus* sp. and C. *punctipennis*; and clay by burrowing mayflies and caddis-flies. Whilst the area of pools, chutes and sloughs within the dike fields all increased between 1962 and 1976, secondary channels, which are less desirable aquatic habitats, decreased (Nunnally and Beverly, 1983).

Clearing and Snagging

The removal of snags from stream channels allows deposits of leaves, twigs and fine-grained sediments to be washed downstream (Bilby and Likens, 1980). These deposits are an important habitat for many benthic species and in channels with sandy, shifting substrates form the only suitable habitat.

Effects of Weed Cutting

Recovery of macroinvertebrates from a capital works scheme will depend on the frequency with which a channel is subsequently maintained. The annual cutting of weeds has been shown to have a dramatic impact on the invertebrate fauna. The immediate effects of weed cutting on the River Hull in northern England on the macroinvertebrate fauna were the removal of large numbers of animals in the weed, an increase in the activity of many of the benthic species, and increased drifting of some plant-dwelling animals (Pearson and Jones, 1978). Increased activity was attributed to the disturbance of the general habitat and to the supplementation of populations on the bed of the channel by animals escaping from the cut weed.

For the Lower Schierenseebrook in West Germany, species diversity was higher during and after cutting in November 1974 and the number of individuals also increased (Statzner and Stechman, 1977). Values were always higher within the weed cut reach than further downstream. *Gammarus pulex* L., *Aphelocheirus aestivalis* (Fab) and *Hydropsyche* spp. were able to settle out of the drift. Certain species could not actively leave the drift and were therefore especially affected, notably *Dixa* spp., *Chaoborus flavians* (Meig), *Unionicola aculeata* (Mull), *Limnephilus rhombicus* L. and *Christatella mucedo* Cuv.

Weed cutting on the Gjern stream in Jutland (Denmark) increased the total drift density by 173 times, to a maximum of 24,722 invertebrates per 100 m^3 (Kern-Hansen, 1978). The dominant plants were *Ranunculus peltatus* and *Sparganium simplex*. In particular, species of invertebrates such as Hydroptilidae, Empidae and adults of Dytiscidae and *Elmis* were found only in drift samples from reaches affected by cutting. Even several days after cutting the drift density of many species (e.g. Oligochaeta, *Gammarus pulex, Baetis, Caenis, Heptagenia*, Corixidae, Sialis and larvae of Haliplidae and Dytiscidae) was still significantly higher than before cutting. These changes were attributed to the loss of macrophytes as habitat, and to very unstable conditions in the substrate following macrophyte cutting.

It has also been demonstrated that the macroinvertebrate fauna can rapidly recover after weed cutting, and that the community composition changes very little (Pearson and Jones, 1978). The substrate was not disturbed and recovery was therefore rapid. However, the timing of such disturbances can be very important for individual species. A June weed cut on the River Hull in Yorkshire (UK) would have affected the hatch of insects because some species such as Chironomidae and *Caenis horaria* attach their eggs to plants. By contrast a cut in July came when numbers were already declining (Pearson and Jones, 1978). It is probable that in streams with annual cutting susceptible species may already have disappeared.

FISH AND FISHERIES

In North America the number of papers and reports concerning the effects of channelization on fish and fisheries far exceeds those dealing with other biological consequences. Effects on fish are evident almost immediately after construction, whereas morphological adjustments generally take place in the longer term (Brice, 1981). Figure 35 depicts the number of papers and reports documenting the effects of channelization on fish and fisheries for the period 1940–85. Over 80% of this literature arises from the United

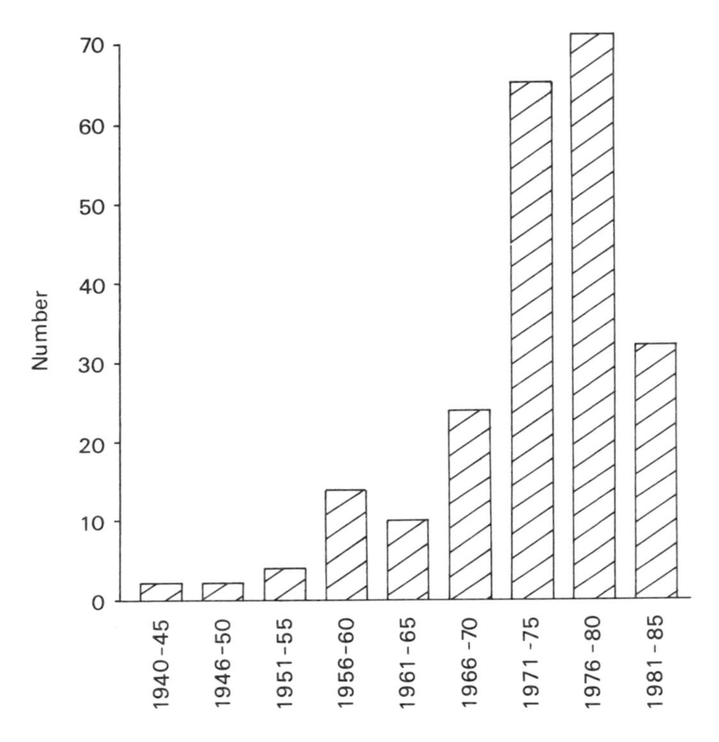

Figure 35. The growth in the number of publications (1940–1985) concerned with the effects of river channelization on fish and fisheries (including papers, departmental, agency and similar reports).

States of America, with only a handful of papers originating from other English-speaking countries. The peaks occurring between 1971 and 1980 reflect increased environmental awareness in the early 1970s, stimulated by government involvement. Research interest has continued to the present day and the majority of papers in the period 1981–85 are from countries outside the United States of America.

The effects of channelization are varied, depending upon the nature of modification, intensity and extent, and subsequent morphological adjustment. Table XXVIII outlines a number of the studies which have been undertaken. Emphasis has been placed on those types of channelization which involve the replacement of a meandering channel by a straight course or widening and deepening of an existing channel (e.g. Rees, 1959; Burkhard, 1967; Welker, 1967; Rich, 1969, Kendle, 1970; Witten and Bulkley, 1975). Only one study has been concerned solely with the effects of lining channels with concrete (Parrish *et al.*, 1978) and only a handful of papers document maintenance practices such as weed cutting and clearing and snagging.

Table XXVIII. Examples of the impacts of channelization on fish (after Brookes, 1983)

Source	Location	Type	Date	Variables affected	Reduced parameters	(%)	Recovery
Bayless and Smith, 1967	Lower Piedmont & coastal Plains, North Carolina		1930s	Loss of habitat	no. of game fish Weight Standing crop	90 85 80	Values given are for 40 years after construction
Belusz, 1970	Blackwater River, Missouri	Straightened			Biomass	77	Studied about 50 years later
Bulkley *et al.*, 1975	Various streams Iowa	Resectioned/ Realigned		Loss of habitat (pools/riffles)	No. species reduced		Recovery in 10–15 years if pools develop
Congdon, 1971	Chariton River Missouri	Resectioned	1940s	Loss of habitat/ cover for fish	No. species Total standing crop Standing crop	38 87 98	Values given are for 30 years after construction
Edwards *et al.*, 1984; Edwards, 1977	Olentangy River Ohio	Widened, deepened and straightened	1950	Shallow, silty, banks unstable	No. of species reduced	22	Studied 24–25 years after channelization
Elser, 1968	Little Prickly Pear Creek, Montana	Resectioned		Loss of pool/ riffle sequence; removal of cover	No. of trout No. of non-trout	78 100	Rapid recovery only with rock deflectors
Etnier, 1972	Middle Creek, Sevier County, Tennessee	Resectioned/ Realigned	1967	Loss of pool/riffle sequence; unstable substrate	Nos. of 3 species	36–95	
Graham, 1975	Ruby River, Montana	Resectioned		Loss of pool/riffle sequence; removal of cover	Biomass of trout	59	
		Riprap			Biomass of trout	50	

Table XXVIII (*continued*)

Source	Location	Type	Date	Variables affected	Reduced parameters	(%)	Recovery
Groen and Schmulbach, 1978	Missouri River	Various incl. pile dikes		Lack of niches & habitats (e.g. pools)	Annual catch Harvest rate Standing crop	47 48 63	
Hansen and Muncy, 1971 Hansen, 1971 a,b	Little Sioux River, Iowa	Realigned	1905–20 1962	Higher turbidity; loss of substrate/ higher temperatures	Fewer species esp. unfavourable to larger game fish		Movement of fish from nearby natural reaches prevented drastic differences of standing crop
Headrick, 1976	Buena Vista Marsh, Portage County, Wisconsin	Resectioned		Reduced cover/ increased silt/ unstable substrate/ increased temperature/ changed flows	Annual production Brook trout Angler success	 56 75	Recovery between 52 and 62 years
Huggins and Moss, 1974	Mud Creek, Douglas County, Kansas	Resectioned	1971	Destruction of pool/ riffle sequence/ bedrock substrate	No. of fish No. of species Biomass of fish Diversity	19 35 82 49	2 years after
Lund, 1976	St Regis River, Montana	Realigned and Resectioned		Loss of cover			Recovery after 1 year with mitigation
Moyle, 1976	Rush Creek, Modoc Co., California	Realigned	1969	Destruction of pool/ riffle sequence; reduced invertebrates	Trout biomass Total number Total biomass	86 36 49	After 5 years

Stroud, 1971	Flint Creek, Montana	Resectioned	1956	Destruction of habitat	Total biomass	93	After 11 years
Stroud, 1971	Idaho	Various			No. whitefish Biomass	98 93	
Tarplee *et al.*, 1971	North Carolina coastal Plain streams	Resectioned		Reduced stream cover; higher temperatures; lack of pools	No. of game fish Species diversity Carrying capacity Weight of game fish	75 28 75 75	After 15 years with no maintenance
Trautman, 1939	Lost & Gordon Creeks, Ohio	Resectioned/ Realigned	1887–1938	Lack of pools & riffles; uniform depth of water; no bottom cover; bedrock; turbidity increased	Loss of 7 species; lack of nos. of certain species		
Trautman & Gartman, 1974	Lost Creek, Ohio	Resectioned/ Realigned			Reduced nos. of species/loss of species		
Wharton, 1970	Tippah River, Mississippi	Resectioned		Destruction of habitat; bankside and instream cover for fish removed	Reduced standing crop	98	
White, 1975	48 streams	Various		Destruction of habitat	44 cases where the loss of nos. of trout and salmon exceeded 75%	75	

Gorman and Karr (1978) attempted to apply community ecology theory to stream fish communities. Fish tend to specialize on specific habitat types and the fish community characteristics of a segment of stream are determined by the complexity of habitats present in the area, especially the horizontal heterogeneity components. Periodic phenomena, such as low-flow and water quality characteristics, are also important in determining fish community structure. Within a habitat type there will be some specialization on food resources, if such sources are readily available.

The major habitat requirements of fish are summarized in Table XXIX. The principal abiotic factors are temperature, both directly and indirectly through the influence on oxygen consumption, rate of flow and discharge fluctuations, and the availability of suitable shelter (Hynes, 1970; Fraser, 1975; Milner, 1984). Each of these factors has been shown to have been altered by channelization. Diversity within stream channels is essential for the movement, breeding, feeding and shelter of fish.

Table XXIX. Major habitat requirements of fish

A. Movement and breeding

- migrations and other movements either up- or downstream require an absence of insurmountable obstacles and adequate water depth and velocity
- fish require a suitable substrate for spawning, requirements varying with individual species. Many species require a gravel or stony substrate (e.g. salmonids). Other species require rock surfaces or aquatic plants
- incubation of eggs requires a stable substrate, adequate water movement, oxygen supply and temperature

B. Feeding habits

- availability of suitable food (e.g. invertebrates which require suitable bed material or plant species)

C. Shelter and cover

- at high flows shelter is required from abnormally high water velocities; cover is needed for protection against predators. At all flows fish require suitable resting places. These conditions are provided by pools, overhanging banks, boulders, instream and bank vegetation, tree roots, debris

Salmonids

Salmonids (e.g. Atlantic salmon, brown trout, sea trout) require definite nest-sites in gravel substrates. By vigorous movements of the tail the female displaces stones after fertilization to bury newly laid eggs (Jones and King, 1950; McCart, 1969). These locations are known as 'redds'. Salmonids usually select places at the downstream end of pools where there is a downward movement of water into the gravel, or at the tail-end of riffles where there is an upward flow of water (Milner *et al.*, 1981). At such locations the buried eggs are constantly washed and supplied with oxygen. It should be noted that any silt (e.g. derived from a maintenance dredge upstream) could settle in the interstices between the gravel, thus preventing the movement of water and this invariably proves fatal. Spawning usually occurs in the period October to January, typically restricted to the upper reaches of river systems where suitable sites are found. The eggs hatch between March and April.

Trout require an adequate supply of food and feed on invertebrates which are themselves dependent on a suitable habitat, e.g. gravels or attached to aquatic plants or bankside vegetation. Morphological variability of channels is essential to trout, to provide areas of shelter from high velocities whilst the trout waits for prey. Higher velocity areas may also be important for carrying food from upstream. Overhanging vegetation can be important as a source of food; it can provide shade from excessive temperatures and provides cover for fish (Gibson and Power, 1975). Grayling also require gravel and stones but do not construct nests. The eggs tend to be buried by the general excitement of the fish, causing turbulence and disruption of the substrate.

Coarse Fish

Many coarse fish also breed on gravel and include barbel, chub and dace. Silt is not used by any of these species and most fish move upstream to gravel or spawn on aquatic or marginal vegetation. This includes many species found in the slower flowing waters of the UK and Europe, such as carp, bream and pike (Huet, 1962).

Weed beds also function as nurseries for fry and juveniles, providing protective cover. Morphological diversity is essential to provide suitable habitats for a range of plant species and slack-water areas suitable for the growth and development of young fish. Most fish have well-defined diets; they consume either plants, invertebrates, fish, detritus, or combinations. The roach and carp eat substantial amounts of aquatic and terrestrial plant material. Carp also eat seeds which have fallen into the water. Many cyprinids supplement an invertebrate diet with plant material. The majority of coarse fish eat macroinvertebrates with the exception of pike, which at a young age becomes purely fish-eating. Invertebrate feeders tend to eat larger prey as they grow.

Unpolluted running water is usually sufficiently oxygenated and is not a limiting factor for fish. Oxygen levels are related to the velocity, plant growth and decay. However, for species such as trout the rate of metabolism rises sharply with temperature. Since oxygen concentration falls with increasing temperature of the water, and because the oxygen consumption of fish is higher, excessive water temperature may be deleterious to this species by reducing its activity. Brook trout are also intolerant of temperatures above 25 °C. The tolerance of species to warm water varies: perch have an upper limit of tolerance of about 30 °C and European cyprinids perish at about 29–31 °C; the pike dies at about 29 °C. For trout to breed the temperature must fall below about 14 °C at some time of the year to allow them to breed. Temperature and oxygen are very much dependent on shade provided by bankside vegetation.

Clearly by providing a diversity of morphology in a channel, a range of flow and substrate conditions can be maintained which enable the successful breeding, feeding and movement of fish. Shade provided by bankside vegetation is very important for cold-water streams. The greater the variability then the greater the diversity of type and size of fish. Structurally diverse natural streams typically have a buffering capacity: meanders tend to moderate flood effects, pools serve as refuges for fish during dry periods and the shading effect of trees protects against excessive temperatures (Gorman and Karr, 1978). Natural streams in Indiana and Panama had increased stability of community composition and habitat structure (Gorman and Karr, 1978). By contrast channelized streams may offer little or no buffering capacity.

Approaches for Fish Management

Several approaches have been adopted by researchers to evaluate the effects of drainage works on fish and fisheries (Swales, 1982a). These include: (a) comparison of natural and channelized reaches within the same river (e.g. Golden and Twilley, 1976; Hansen and Muncy, 1971; Moyle, 1976); (b) comparison between a channelized river and a nearby natural river (e.g. Huggins and Moss, 1974; Tarplee *et al*., 1971); (c) the historical approach where documented records which have been made many years prior to channel alterations are consulted (e.g. Bayless and Smith, 1967; Trautman and Gartman, 1974); and (d) the approach whereby the channel is monitored 'before and after' alteration (e.g. Etnier, 1972; Whitney and Bailey, 1959; Swales, 1982a). Although a number of factors have been invoked as causes of differences in fish populations, few studies have attempted to quantify the relationship (cf. Hortle and Lake, 1982).

The majority of studies do not state with certainty whether the reductions in abundance are temporary, being mainly due to fish moving away as a behavioural response to the disturbance created by channel excavations, or whether they are more long-lasting as a result of major changes in the physical features of the habitat. Luey and Adelman (1980) reported that natural reaches downstream can act as refuges for fish displaced by channelization. Hansen (1971b) suggested that the impact may be lessened when favourable habitats are available immediately upstream and possible movements downstream into the channelized area. Short-reach channelization (less than 0.5 km) may have a relatively small impact, which is difficult to quantify. However, in the longer term it may be that substantial lengths of channelization may alter river zonation (Hawkes, 1975). As a result of straightening streams in Sweden to facilitate the transport of logs by water, grayling extended its zone to occupy reaches previously occupied by trout (Hynes, 1960).

The initial effects of channelization may be a reduction in the numbers and biomass of resident fish populations whilst the longer-term effect depends on whether populations can recover by adapting to new conditions. Losses have mainly been expressed in terms of numbers of fish, species diversity and standing crop, although a few studies have quantified the impact in terms of angling catch, harvest rate or annual loss to fishermen.

Initial Impacts

Channelization has been shown to greatly reduce the standing crop and diversity of fish populations of streams in several regions of the United States of America. A study of 23 channelized and 36 natural streams in North Carolina revealed that channelization reduced the number of game fishes (over 15 cm in length) by 90% and reduced the weight by 80% (Bayless and Smith, 1967). Only limited recovery was observed 40 years after channnelization. Also in North Carolina Tarplee et al. (1971) reported that the standing crop of channelized streams for all fish species was 32% and for game species only 23% of that found in natural streams (Figure 36). The average weight of game fish per hectare (ladyfish (*Elops saurus*), American shad (*Alosa sapidissima*), redfin pickerel (*Esox americanus americanus*), chain pickerel (*Esox niger*), white perch (*Morone americana*), largemouth bass (*Micropterus salmoides*), warmouth (*Lepomis gulosus*), Green sunfish (*Lepomis cyanellus*), spotted sunfish (*Enneacanthus gloriosus*), pumkinseed (*Lepomis gibbosus*), redear sunfish (*Lepomis microlophus*), redbreast sunfish (*Lepomis auritus*), Bluegill (*Lepomis macrochirus*), mud sunfish (*Acantharchus pomotis*), yellow perch (*Perca flavescens*) and spot (*Leiostomus xanthurus*)) was over 400% greater in natural streams than in channelized streams. Game fish were more adversely affected than non-game fish and the number of harvestable game fish was reduced by more than 75% by channelization. The average size of fish in channelized streams was found to be smaller than the average size of fish in natural streams. The overall quality of streams, as based on species diversity, was reduced by 27.5% following channelization.

Channelization of the Chariton River in Missouri by straightening 320 km of river to form a new route 165 km long, reduced the number of fish species from 21 to 13, which corresponds to an 83% reduction of total standing crop from 340 to 59 kg ha^{-1} (Congdon, 1971). The standing crop of catchable size fish was reduced 209 to 30 kg ha^{-1}. Figure 37 shows that all species of fish were affected, including carp (*Cyprinus carpio*), river carpsucker (*Carpiodes carpio*), channel catfish (*Ictalrus punctatus*), small mouth and bigmouth buffalo (*Ictiobus bubalus* and *Ictiobus cyprinellus*) freshwater drum (*Aplodinotus grunniens*), northern redhorse (*Moxostoma macrolepidotum*) and flathead catfish (*Pylodictis olivaris*). In the unchannelized sections carp was the dominant species by weight (50% of total), whilst channel catfish was the dominant species (59%) in the channelized sections. There were

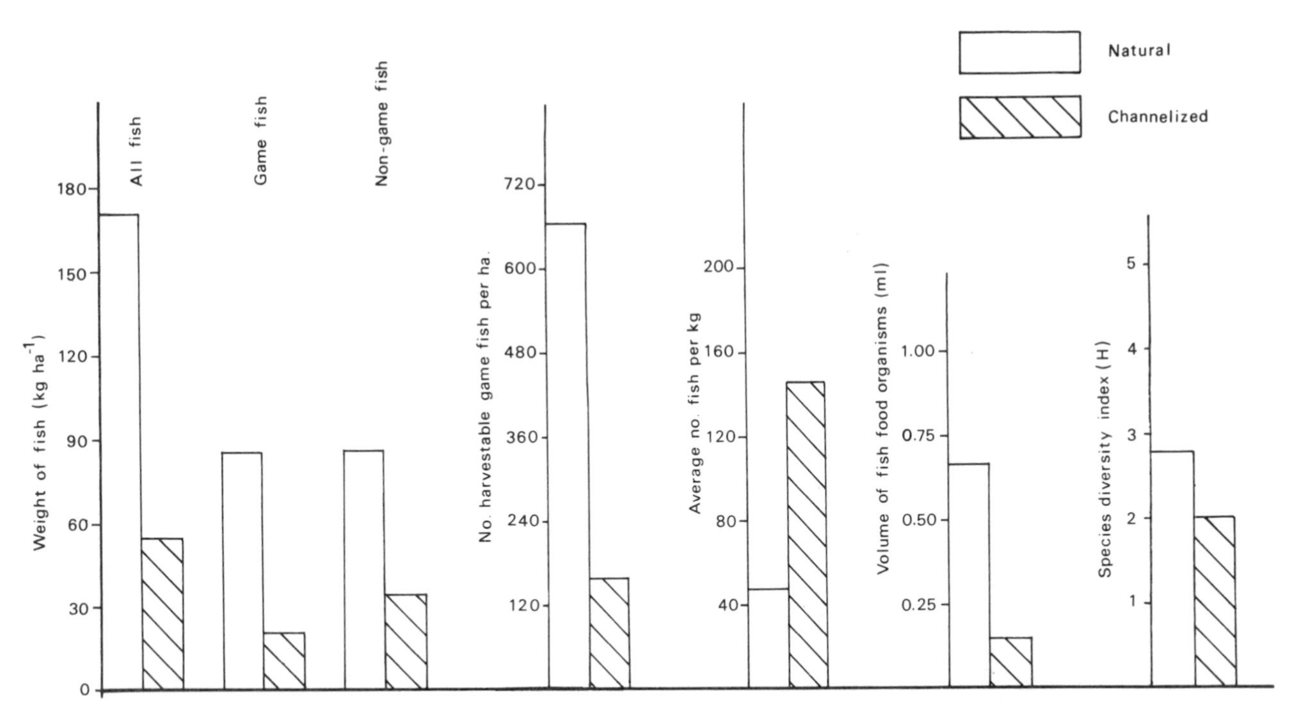

Figure 36. The effects of channelization on fish populations in North Carolina's coastal plain streams. (Adapted from Tarplee *et al*., 1971. Reproduced by permission of North Carolina Wildlife Resources Commission.)

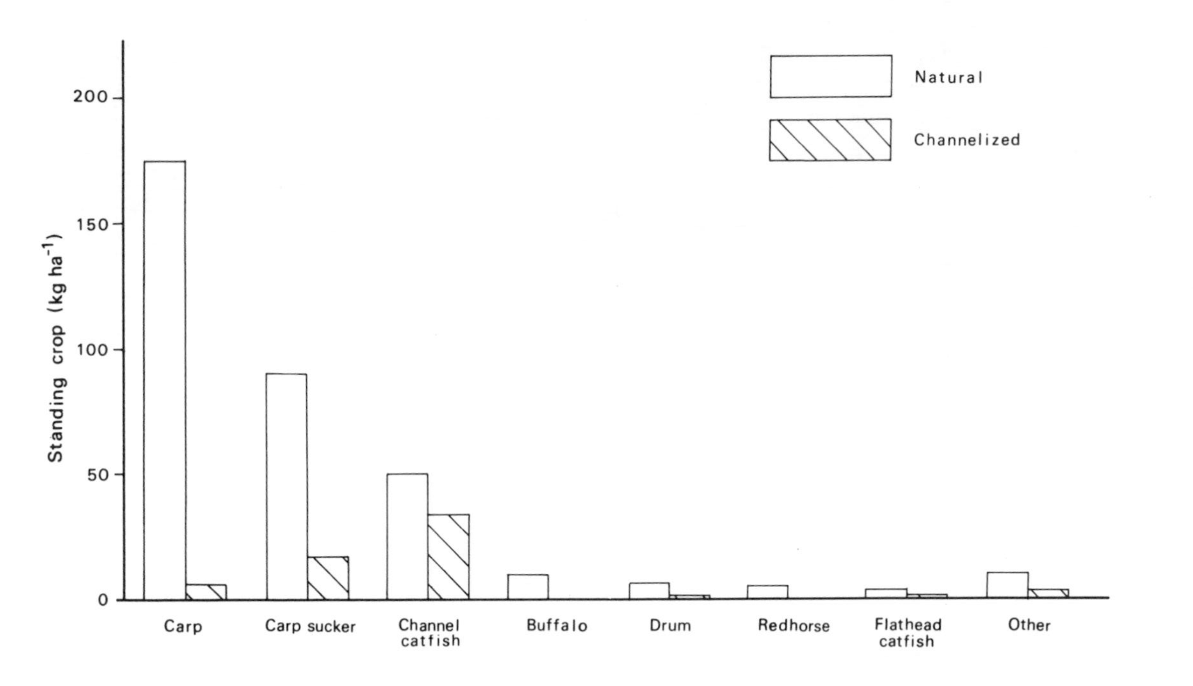

Figure 37. Estimated standing crop of the fish population in channelized and non-channelized sections of the Chariton River, Missouri. (Based on Congdon, 1971. Reproduced by permission of the American Fisheries Society.)

only four species of catchable size fish in the channelized section compared to six species in the unaltered control reach. In the unchannelized section, carp comprised 81% and channel catfish 15% of the catchable-size fish by weight. The remaining four species (4%) were freshwater drum, white crappies (*Pomoxis annularis* Rafinesque), flathead catfish and largemouth bass (*Micropterus salmoides* Lacepede). In the channelized section channel catfish comprised 82% of the catchable-size population by weight and carp 15%. The remaining 3% of the population was made up of freshwater drum and white crappies.

On the Little Bighorn River in Montana a total of 1987 channel alterations had been made to 1230 km of stream (Peters and Alvord, 1964). Unmodified channel lengths produced 5.5 times more trout and 10.5 times as many whitefish as the channelized reaches. Also working in Montana, Whitney and Bailey (1959) found a 94% decrease in the number and weight of large-size game fishes one year after channelization. The number of small-size game fishes was reduced by 85% and the weight by 76%.

Stroud (1971) found 7 times as many trout and 60 times as many whitefish in natural streams as in channelized streams, this being equivalent to a weight differential of 14 : 1. In 29 streams in Idaho game fish production was reduced by 87% (Gebhards, 1973). There was 1.5 to 112 times more weight of fish in natural sections than in altered parts of the same stream. In 3 of the 29 streams the altered sections contained no game fish at all.

A trout stream, Flint Creek, in Montana, which had been dredged, cleared and straightened, had only 0.7 kg of fish compared with 9 kg in unaltered reaches (Stroud, 1971). In the Blackwater River in Missouri only 147 kg ha^{-1} of fish were found at a channelized site, 503 kg ha^{-1} at a slighly channelized site and 633 kg per ha^{-1} in a natural reach. Up to 90% reductions in fish population densities occurred in the channelized Middle Creek, Tennessee (Etnier, 1972).

Different types of channel modification have varied effects on fish populations. A natural reach of the Ruby River in Montana had a trout content of 200 kg km^{-1} of channel (Graham, 1975). Following bank stabilisation by riprapping, fish production was reduced to 100 kg km^{-1}, and where the stream bottom had been bulldozed and reshaped the fish population further dropped to a level of only 80 kg km^{-1}.

Losses have also been expressed in terms of the value of sport fisheries. An estimated 750 fishermen days per km per year were lost on the California bank of a channelized length of the Colorado River (Beland, 1953). The value of the fishery was reduced by 85%. Six channelized streams in Pensylvannia no longer supported legal-sized trout and an annual average of 310 fishermen days per km were lost (Duvel *et al.*, 1976). Assuming that fishermen did not move to non-channelized areas then this represents a loss of $5.2 million (at 1976 prices). In the Buena Vista Marsh in Wisconsin angler success was reduced from 1.02 trout per hour in natural reaches to only 0.26 trout per hour in dredged sections (Headrick, 1976). The channelized reach of the Middle Missouri River supported an annual average catch of 0.38 fish per hour and an annual harvest rate of 0.26 fish per hour compared with values of 0.72 and 0.50 respectively for the unchannelized reach (Groen and Schmulbach, 1978).

Channelization may have a very severe effect on larger fish, which in turn affects the rate of production. In 29 channelized streams in Idaho the number of catchable-sized trout was reduced by 6.5 times and the number of catchable whitefish was approximately about 10% of the number in natural streams (Irizarry, 1969). The average fish weight was 7.5 times greater and the numbers 5.7 times greater in natural than altered streams. In the Blackwater River, Missouri, there were only 13 kg of fish per ha over 30 cm in a severely channelized reach, compared to 334 kg per ha in slightly channelized reaches and 452 kg per ha in unaltered sections (Belusz, 1970). Fish became smaller and the weight of fish was less in the channelized reaches. Hansen and Muncy (1971) reported that channelization of the Little Sioux River in Iowa reduced the larger sizes of catfish that could be supported whilst the number of smaller catfish was the same as unchannelized reaches.

Over 40% of 151 km of channel which had been modified in Hawaii were lined with concrete and these reaches were particularly damaging to the fish populations (Parrish *et al.*, 1978). Introduced species, notably poeciliid fishes, were most abundant in heavily channelized sections, whilst the more valuable native species (*Awaous genivittatus, Awaous stamineus, Eleotris sandunicensis, Sicydius stimpson, Lentipes concolor*) were almost entirely absent.

European Studies

There are few published reports concerned with the impacts of channelization on fish in the UK (Brooker, 1985). A large proportion of the resident fish community of the River Soar (UK) moved away during dredging, and one month later a 90% reduction in the population density of the total community and an 85% reduction of standing crop were recorded (Swales, 1980). Figure 38 depicts the mean population densities and standing crop estimates recorded in the pre- and post-drainage periods. Following the drainage scheme, a reduction in population density was recorded for all species and ranged between 30 and 100%, with population standing crop estimates of between 15 and 100%. The mean density of the total community decreased by 70% and the mean standing crop decreased by 76%. Of the cyprinids, dace (*Leuciscus leuciscus*), chub (*Leuciscus cephalus*) and roach (*Rutilus rutilus*) were the predominant large-sized species present during the pre-drainage period with occasional brown trout (*Salmo trutta*). The mean reduction in fish population density caused by channelization was 73.5% for dace, 87.5% for chub, 30.5% for roach and 100% for brown trout. Three months later fish populations had improved slightly but were still substantially below pre-drainage levels. Swales (1982a) concluded that there was a

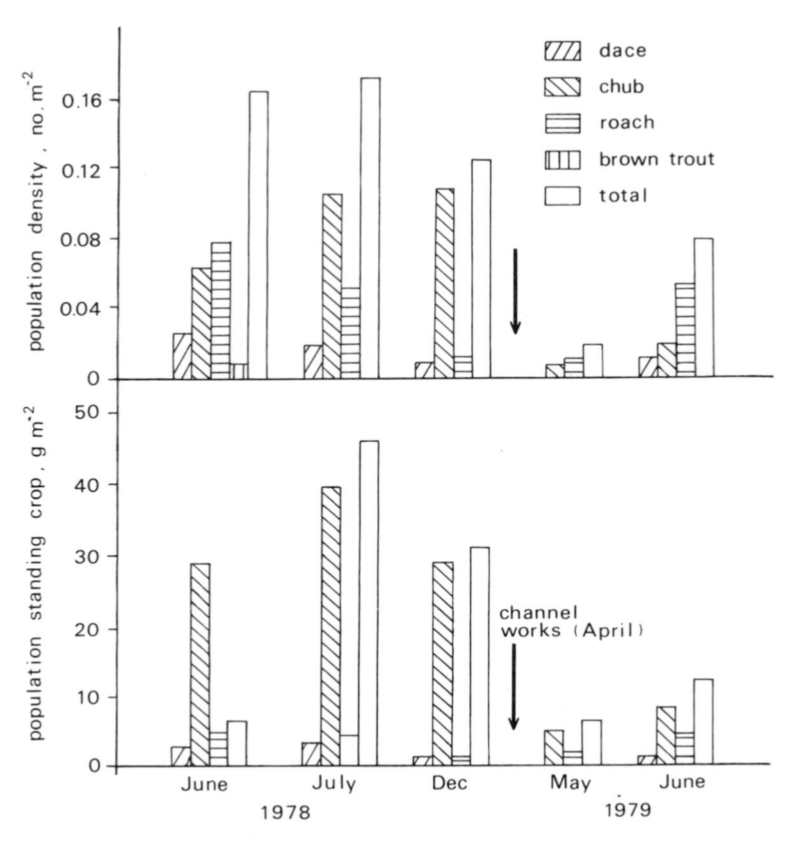

Figure 38. Population density and standing crop estimates of fish before and after channel works on the River Soar, England. (From Swales, 1982a. Reproduced by permission of Blackwell Scientific Publications Ltd.)

need for many more studies, covering a range of physical environments in upland and lowland areas.

After dredging and deepening of two sites on the River Camowen in County Tyrone, Northern Ireland there was a greater number of salmonids of one year and older than before (Kennedy *et al.*, 1983). By contrast in a reach which was shallower after the channel works the numbers of one year old and older fish had been greatly reduced. This area was dominated by fry.

Studies of channelization involving realignment and resectioning in France (Clavel *et al.*, 1978; Bouchard *et al.*, 1979; Cuinat, 1983), Central Germany (Tesch and Albrecht, 1961) and Scandinavia (Müller, 1962) have all reported adverse impacts on fish populations.

Studies in Australia

There are relatively few published studies of the effects of channelization on fisheries from the Southern Hemisphere, although a number of writers have expressed concern. In the Bunyip River in southern Victoria, Australia, the total number of fish caught and the total biomass of fish and species richness were reduced by channelization (Hortle and Lake, 1983). The numbers of common jollytail (*Galaxias maculatus*) were much greater at the channelized sites, whilst eels and trout were more abundant and attained a larger size at the unchannelized sites.

Habitat Changes Caused by Channelization

The initial cause of fish populations moving out of a channelized reach may be a behavioural response (Swales, 1982a). However, whether this is temporary or not will depend on the subsequent habitat changes caused by the works. Since channelization may affect several of the habitat features listed in Table XXIX (p.126) it is difficult to isolate individual causes for changes in fish populations. A number of studies have identified primary causes, whilst others have listed several factors which act in combination. Although various causes have been invoked, few studies have attempted to quantify the relationship between changes in the physical environment and reductions of fish density. However, studies suggest that changes in the physical rather than the chemical form of the habitat are mainly responsible for changes in fish populations. pH values, dissolved minerals and gases may undergo little change (Gebhards, 1973). Changes of sediment loads, temperature and physical habitat elements may be important (Duvel *et al*., 1976). The majority of studies indicate that reductions of habitat diversity created by channelization are responsible for changes in fish populations (Table XXVIII pp. 123–125). Straightening of stream channels may substantially reduce the available habitat (e.g. McClellan, 1974).

Sediment released during dredging may have a detrimental effect on salmon fisheries in Scotland, UK (Graesser, 1979). Although fish can tolerate substantial amounts of suspended sediment, silt deposits can blanket portions of the stream bottom, eliminating potential spawning grounds and reducing the available food by killing bottom-dwelling organisms (Apman and Otis, 1965). Channel works on the Rivers Loire, Allier and Dore in the Massif Central, France, had a major impact on fisheries (Bouchard *et al*., 1979). The river bed became uniform and unstable, together with an increase in suspended sediment concentrations and siltation. The mortality of trout eggs and fry increased, together with a reduction in the hatching of trout eggs and a decrease in the numbers and biomass of fish populations. Benthic invertebrate populations had declined and this could be correlated with the decline of fish which usually feed on these organisms. Increased suspended sediment concentrations (600 mg per litre) arising from dredging on the River Moy in Ireland did not affect runs of adult fish but the number of fry in the affected area decreased by 30% (Toner *et al.*, 1965). Increased turbidity during construction may not only affect fish spawning but may also interfere with angling (Gebhards, 1973). Sedimentation was not an important factor affecting fish populations in six streams in Pennsylvania (Duvel *et al.*, 1976).

A major reason for sparse adult trout populations in channelized reaches appears to be the lack of cover provided by undercut banks, overhanging shoreline vegetation, deep pools and other obstructions such as logs and boulders (Duvel *et al.*, 1976; Chapman and Knudsen, 1980). All trout, in particular brown trout, exhibit some degree of negative phetotropism and they are therefore found under features which provide concealment. Channelization often destroys these entities.

The relationship between salmonid populations and protective cover has been investigated by several authors (Greenly, 1936; Tarzwell, 1937, 1938; Shetter *et al.*, 1946; Bryant 1983). These studies have demonstrated that the addition of artificial cover can cause an increase in number and size of trout in a given section of stream. Addition of brush cover to four sections of a stream in Gallatin County, Montana, caused a 258% increase in the weight of fish, whilst removal of natural brush from two sections led to a 40.5% decrease in the fish population (Boussu, 1954). The natural vegetation consisted of a heavy cover of sedge (*Carex* sp.) and willow (*Salix* sp.). At a further two sections an undercut bank was removed, causing a 33% decrease in the numbers and size of fish.

The removal of overhanging vegetation as a consequence of channelization has been cited as a significant factor affecting the distribution of fish in several studies (e.g. Swales, 1982a; Hortle and Lake, 1983; Scarnecchia, 1988). A further effect of removing the bankside and instream vegetation cover may be to create excessive illumination and water temperatures (Gray and Edington, 1969; Corning, 1975). Temperature plays a significant role in the occurrence and distribution of fish species, particularly cold-water fish such as trout. Rises in water temperature may not only affect the fish directly but may also eliminate invertebrates on which fish feed (Gebhards, 1973). Where a substantial length of channelization exists then a rise in water temperature may occur. Daily fluctuations of water temperature in July were greater and maximum daily water temperatures averaged 1.3 °C higher. The mean daily water temperature was 0.3 °C higher in the channelized section than in the unchannelized section of the Little Sioux River in Iowa (Hansen, 1971a). Duvel *et al.* (1976) showed how channelization of the Fishing Creek in Pennsylvania in 1975 raised the water temperature by 1.7 °C above that of a well- shaded natural site upstream. Such temperature increases are likely to occur on clear sunny days during mid-summer when the absence of natural shade is most significant. Similarly water flowing through a number of channelized reaches can become progressively warmer with distance downstream (Duvel *et al.*, 1976). In a channelized section of the Buena Vista Marsh in

Wisconsin, mid-summer water temperatures reached upper lethal levels for brook trout (*Salvelinus fontinalis*) and were partly responsible for reducing the annual production by 56% (Headrick, 1976). Increases of channel width may create shallow water depths, further increasing temperature (Hansen and Muncy, 1971).

Removal of near-stream vegetation may result in reduced invertebrate and fish populations as a result of the loss of allochthonous (terrestrial) energy inputs (e.g. Gelroth and Morzolif, 1978). Areas without deciduous vegetation often have a poor diversity and low numbers of invertebrates (Johnson, 1970). Organic matter can be directly ingested by invertebrates, which shred the matter and digest the micro-organisms growing on it. In turn, fish rely on invertebrates as a source of food.

A commonly stated reason for change is the loss of a natural pool–riffle sequence which provides a variety of flow conditions suitable as cover for fish and for the organisms on which fish feed. Shelter areas are required at high flows to protect fish from abnormally high water velocities and these conditions may be absent where a meandering stream, with an abundance of long pools, separated by short riffles is converted into a straight stream composed mainly of riffle. Alteration of the width and depth variables in a channel may create shallow and unnatural flows, which result in an unsuitable habitat for fish and may present topographical difficulties for fish migration (Keller, 1976). Natural sorting of bedload materials on riffles and point bars is important in providing an environment in which bottom-dwelling organisms can thrive, thereby providing a source of food for the fish.

Larimore and Smith (1963) reported that dredging of streams in Champaign County, Illinois, eliminated many of the pools essential for fish production and brought about an increase in silt accompanied by a decrease in aquatic vegetation. These factors reduced the fish from 90 species in 1929 to 74 species in 1959. A four- to six-fold reduction in total fish biomass in the channelized Mud Creek, Douglas County, Kansas, was attributed to the almost complete elimination of pool habitat area together with a decrease in the quantity and quality of riffle crests (Huggins and Moss, 1974). Natural bedload materials were replaced by bedrock.

Numbers of large trout may increase in those situations where the depth and cover are increased by channelization (Saunders and Smith, 1962; White, 1972; Kennedy *et al*., 1983). Although channel modifications in general may have a deleterious impact on trout populations, there may be no long-term adverse effects on forage fish species. For warm-water streams in Ohio and northeastern Indiana (Olentangy, Hocking Rivers and Rock Creek), Griswold *et al*. (1978) found that bottom- and detritus-feeders such as catostomids and cyprinids dominated channelized areas where widening had caused a reduction in velocity and subsequent deposition of sediment.

For Gordon Creek in northwestern Ohio, the populations of species requiring sand and gravel substrates, rooted aquatic vegetation and/or well-defined pools and riffles and a variety of flow conditions were either reduced or eliminated by the destruction of their habitats through channelization (Trautman and Gartman, 1974). These included the central mudminnow (*Umbra limi*), grass pickerel (*Esox americanus vermiculatus*), goldenshiner (*Notemigonus crysoleucas*), hornyhead chub (*Nocomis biguttatus*), mimic shiner (*Notropis volucellus*), tadpole madtom (*Noturus gyrinus*) and pirate perch (*Aphredoderus sayanus*). By contrast, species which were tolerant of a more or less uniform current flow, such as that produced by channelization, were generally distributed and more numerous (Trautman and Gartman, 1974). These included the creek chub (*Semotilus atromaculatus*), common shiner (*Notropis cornulus*) and spotfin shiner (*Notropis spilopterus*).

An earlier study of the Lost and Gordon Creeks revealed that there had been a drastic change in the numbers of various fish species between 1887 and 1938 in channelized reaches (Trautman, 1939). Species which were most affected by a loss of habitat through dredging included the sucker (*Catostomus commersonii*), suckers of the genus *Moxostoma*, bullheads of the genus *Ameiurus* and the majority of sunfish. Forage fish displaying the greatest decrease in numbers were those requiring clear water, constant flow, well-defined pools or aquatic vegetation (e.g. *Nocomis biguttatus* and *Rhinichthys atratulus*). Forage fish displaying the greatest numerical increase were all tolerant of dredged conditions, namely increased turbidity and accumulation of clean-swept sand, rapid fluctuations in height of flow and water velocity, lack of rooted aquatic vegetation and other cover. Species included *Ericymba buccata*, *Phenacobius mirabilis* and *Pimephales promelas*.

A study of six trout streams in Pennsylvania revealed that populations of non-game species (excluding suckers) in channelized stream reaches were generally equal to or greater than those in natural reaches (Duvel *et al.*, 1976). Larger populations of non-game fish may be attributed to the lack of predatory pressure and the presence of a suitable stream bottom consisting of rocks with a relatively uniform size (Figure 39).

Channelization of the Buena Vista Marsh in Portage County, Wisconsin, improved the habitat for the white sucker (*Catostomus commersoni*); (Headrick, 1976). Dredging created a deep, silt-bottomed pool, with slow-flowing water. A newly channelized section of the Luxapalila River in Mississippi was dominated by number and weight by migratory fish such as the spotted sucker (*Minytrema melanops*) and smallmouth buffalo (*Ictiobus bubalus*) (Arner *et al.*, 1975). By contrast the unchannelized reach possessed a variety of sport, commercial and/or migratory species.

In channelized reaches of the Kaneohe and Manoe streams in Hawaii exotic species predominated and native species were considerably reduced compared to a natural watercourse (the Waiahole stream) (Norton *et al.*, 1978). The altered bottom of the channelized reach served as a nursery for exotic poeciliids *Poecilia mexicane* and *P. recticulata*. Concrete-lined channels in Hawaii were characterized by an average pH value (yearly mean mid-afternoon) of 9.9 compared to 7.2 in natural streams (Parrish *et al.*, 1978). Conductivity and dissolved oxygen levels increased and lined channels

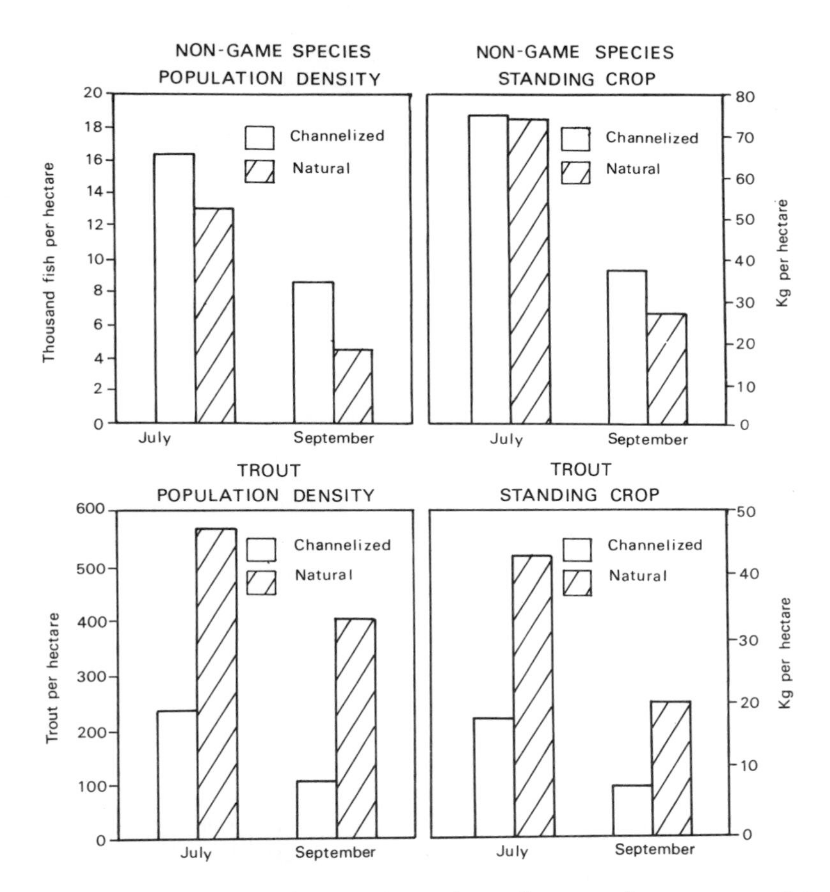

Figure 39. Average population density and standing crop for non-game fish and trout in channelized and natural stream reaches in Pennsylvania, during July and September 1974. (From Duvel *et al.*, 1976. Reproduced by permission of the American Water Resources Association.)

had daily temperature fluctuations of 17.8 to 36.2 °C compared with a range of 19.5 to 26.8 °C in natural streams. Exotic species were able to tolerate these conditions, whilst native species were not. However, species diversity was lower in reaches with artificially lined beds.

Straightening of a stream eliminates meanders and reduces the area of habitat. Although some attempts are made to keep oxbows open, they often fill up with silt after dredging, because of the absence of flow, and may become covered with aquatic plants (Little, 1973). Beland (1953) reported a 100% loss of fisheries in the backwaters of a section of the Colorado River. Elimination of these feeding and spawning grounds inevitably affects the diversity of species.

Water velocities may increase due to changes in slope and bed topography (Gebhards, 1973) and there may be rapid fluctuations of discharge (Trautman, 1939). Channelization of the Lower Colorado River in 1951 converted a meandering stream with pools and riffles into a straight, dredged channel with highly turbid water (Beland, 1953). Fishery losses were attributed to destruction of the vegetation cover, elimination of pools and riffles, increased turbidity, accelerated bank erosion and reduced spawning areas. Decreased standing crops in the channelized Chariton River, Missouri, were attributed partly to the loss of space for the fish population to occupy, caused by a 55.2% reduction of channel length by realignment (Congdon, 1971). Other factors were the lack of a pool–riffle sequence in the new straight channel, creation of a shifting sandy bed during floods, and greater fluctuations of water temperatures.

Channelization of the Little Prickly Pear Creek in Montana resulted in a 40% reduction of meanders (Elser, 1968). In the newly straightened channel there was no pool–riffle sequence and 87% of the water was shallow and fast-flowing as opposed to 49% in unaltered reaches. As a consequence of the lack of slow-flowing water, trout were 78% less abundant in the altered reach. This represents a loss of 4700 trout with a weight of 1000 kg. There were no non-trout species in the altered channel.

Recovery

The time taken for recovery after channelization is an important aspect which has rarely been quantified (Table XXX). Recovery may depend on a number of factors, including the intensity and extent of the channelization scheme. Recovery depends on an improvement of habitat, whether this be naturally or artificially induced. Maintenance of channelized reaches tends to offset recovery (Little, 1973; Swicegood and Kriz, 1973).

Table XXX. Examples of the recovery of fish and fish populations (after Gebhards, 1973)

River	Time elapsed since channelization	Observations
Yankee Fork, Idaho	30 years	97% less productive than natural areas of the same stream
North Carolina	40 years	fish populations 80% below normal levels
Blackwater River, Missouri	50 years	fish production down 77%
South Fork, Coeur d'Alene River	77 years	fish production down 99%
Portneuf River, Idaho	86 years	fish production down 83%

Discrepancies in the rates of recovery of fish populations are evident from the various studies. The type and extent of channelization appears to be an important factor, recovery being slowest where a channel is realigned over a considerable distance and replaced by a straight channel. Projects which involve dredging, but retain the original sinuous course of a channel, may recover more rapidly if pools and/or natural substrate materials reform (Crocker, 1960; Stuart, 1960; Kennedy *et al.*, 1983).

Recovery of salmonids was fairly rapid following drainage works in 1962 on the Bunree, a tributary of the River Moy in Ireland (McCarthy, 1981). Immediately after dredging there was a 30% drop in the number of young salmonids and a 93–100% mortality of salmon in the spawning season when the dredgers were still working. However, rapid recovery was achieved because lowering of the bed during dredging removed peat and exposed glacial gravels which increased the spawning potential. Furthermore the programme of dredging was only short-term, lasting for one year.

Kennedy *et al.* (1983) found that salmonid density and biomass initially declined as a result of dredging on the River Camowen, Northern Ireland, in 1968 (Figure 40). However, fry densities increased from 1975 onwards although recovery occurred progressively downstream, with fry densities taking up to 6 years to improve at the site furthest downstream. These changes could be related to progressive stabilization of the substrate which was initially unstable in the post-drainage period, making it unsuitable for spawning. Yearling and older fish recovered to pre-drainage levels more rapidly than fry, probably as a result of the deepening effects of dredging.

Slow recovery of salmonid populations was recorded on the Trimblestown River, a tributary of the River Boyne in Ireland, after channelization in 1972 (McCarthy, 1981). Prior to drainage the river was an important salmonid nursery with a large population of brown trout and juvenile salmon, together with resident fish, namely eels, sticklebacks and stoneloach. The ratio of salmonids to other fish was 14 : 1 before drainage. Immediately after drainage stoneloach and minnows were the dominant species and the ratio of coarse fish to brown trout was 20 : 1. Six months later, small coarse fish outnumbered salmonids by 5 : 1 and impeded re-establishment of wild salmon stocks. Studies in all major tributaries of the River Boyne which had been modified showed little sign of recovery even 10 years later (McCarthy, 1981).

Some streams in Iowa continued to degrade after channelization, whilst others had fish populations similar in abundance and composition to those in natural stream sections (Bulkley *et al.*, 1976). These differences were attributed to variations in the quality of habitat affected by channelization.

Short lengths of channelization may recover more rapidly (Bulkley *et al.*, 1976). For North Carolina's coastal plain streams it has been suggested that recovery may occur in 15 years provided that no further alterations are made to the stream bed, bank, forest canopy or aquatic vegetation (Tarplee *et al.*, 1971). In North Carolina, species diversity increased with corresponding

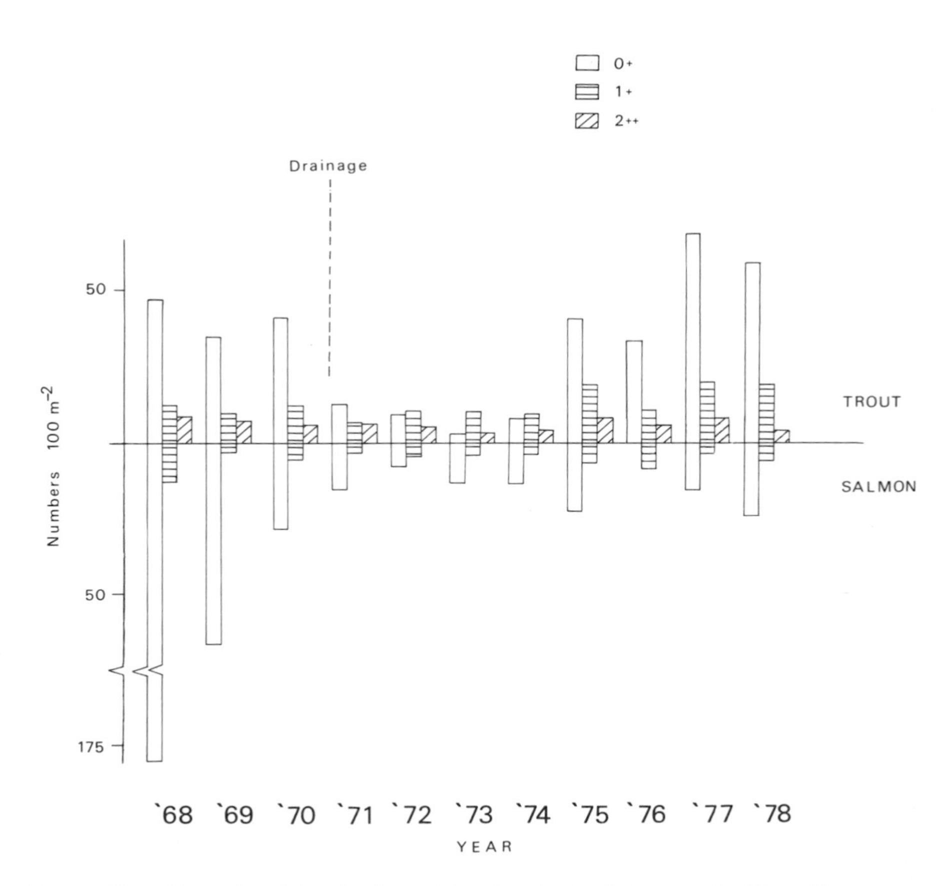

Figure 40. Mean densities of salmonids at the drained sites on the River Camowen, Northern Ireland, during July of each year 1968 to 1978. (From Kennedy *et al*., 1983. Reproduced by permission of Blackwell Scientific Publications Ltd.)

increases in cover and time since channelization (Figure 41). Channelized rivers in Iowa of between 10 and 15 years old had as many fish as natural reaches of the same rivers where brush piles and trees had been allowed to accumulate in the channel. After 30 years of 'healing' on a resectioned reach of the Chariton River in Missouri the channel had reverted from a uniform width and depth back to natural conditions with meanders and pools, whilst logs increased instream cover (Congdon, 1971). Work on the Lower Piedmont and coastal plain streams of North Carolina revealed that the numbers of game fish and biomass were considerably reduced even 40 years after channelization (Bayless and Smith, 1967). Likewise Arner *et al*. (1975) found that there was little recovery of fish productivity 43 years after channelization.

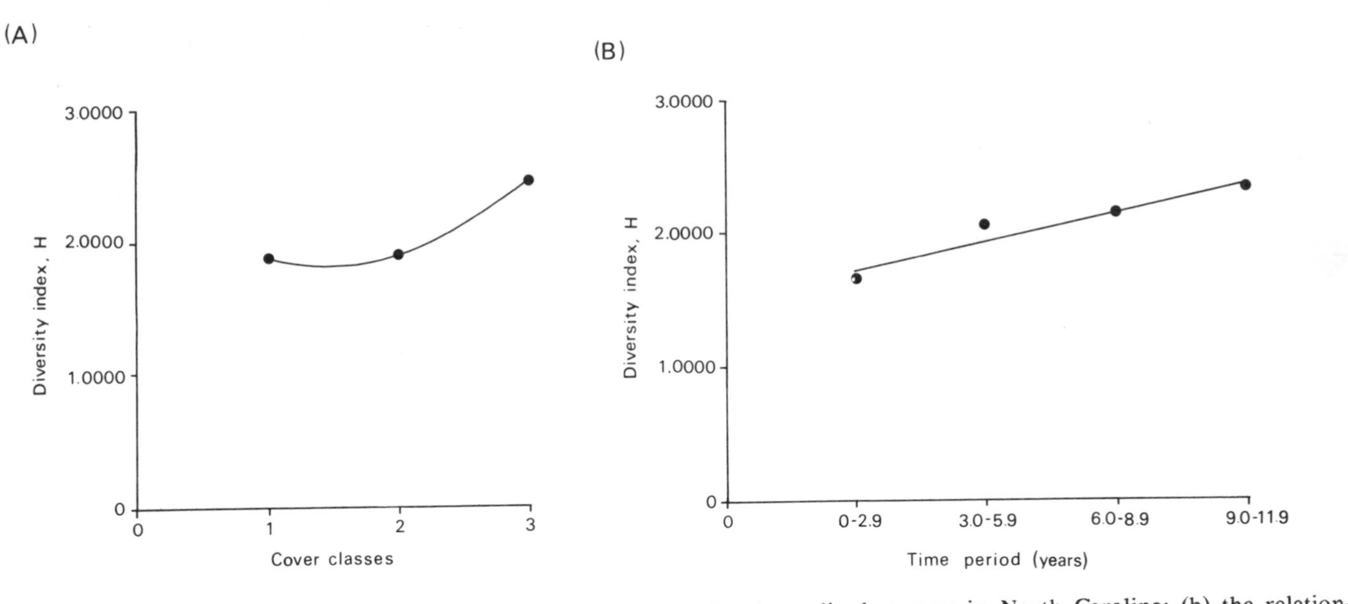

Figure 41. (a) the relationship between diversity index and cover in channelized streams in North Carolina; (b) the relationship between diversity index and time since channelization. (From Tarplee *et al.*, 1971. Reproduced by permission of the North Carolina Wildlife Resources Commission.)

Recovery of the River Camowen in Northern Ireland was partly attributed to the dislodging into the river of boulder riprap from the outside of bends, which diversified the instream morphology (Kennedy *et al*., 1983). Rock riprap placed to protect the bank of a realigned reach of the Wildcat Creek in Oregon had moved or had been undercut, and provided some resting pools for cutthroat and steelhead trout. Since recovery depends on re-establishment of the physical conditions that promote fish production it is possible that under extreme conditions of realignment adjustments may take up to 1000 years (Ryckborst, 1980).

After disturbances such as channelization it has been suggested that seasonal peaks in species diversity of fish may attain levels typical of unmodified streams, but that stability of the fish community is lower because seasonal changes in stream quality are high in disturbed reaches (Gorman and Karr, 1978). Re-stocking of channelized reaches has been undertaken on other rivers, although there is a lack of data describing the results (e.g. McCarthy, 1981).

Channel Maintenance

Fish populations may recover in the absence of maintenance. The effect of annual maintenance on the fish population of the Middle Creek, a tributary of the Little Pigeon River (East Fork) in Sevierville, Sevier County, Tennessee, has been determined by mark and recapture (Etnier, 1972). In 1967, 8 km of channel was resectioned, regraded and straightened for flood control purposes and this length was regularly maintained. Channelization eliminated the pool–riffle sequence and annual rechanneling prevented recovery.

The post-rechanneling absence of pool habitats and decreased habitat variability resulted in downstream winter migration of darters until a pool habitat was located below the study area. However for the darters *Etheostoma simoterum* and *E. stigmaeum jessiae* it is unlikely that the low populations observed after rechanneling (Table XXXI) were the result of migration to pools because specimens were found on riffles in neighbouring streams. It is apparent that these had relatively little or no reproductive success in the channelized reach. By contrast *Hypentelium* populations remained stable

Table XXXI. Mid-January estimates for three fish species for the Middle Creek, Sevier County, Tennessee

	1966	1967*	1968	1969	1970	1971
Etheostoma simoterum	222	496	23	97	79	13
Etheostoma stigmaeum jessiae	228	228	41	7	—	2
Hypentelium nigricans	20	129	82	59	88	34

* 1967—year of channelization.
From Etnier, 1972. Reproduced with permission of the American Fisheries Society.

and species may have migrated to the area. Other species such as *Etheostoma kennicotti, Cottus carolinae* and *Etheostoma blennioides* have either been extremely rare or absent following rechanneling. There was also a marked decrease in the diversity of invertebrate fauna and this may have contributed to the fish fauna changes.

There have been relatively few studies of the impact of maintenance activities such as weed cutting and clearing and snagging on fish populations. Qualitative evidence has suggested that weed cutting can have severe effects on river ecology and cause a decline of fisheries over several years. Boussu (1954) showed aquatic vegetation to be valuable as cover in a Montana trout stream, both whilst rooted and when free-floating after becoming detached. The impact of weed cutting on trout fry has been demonstrated for a small stream in Denmark (Mortensen, 1977). A study of a 400 m length of the River Perry in Shropshire (UK), attempted to quantify the impact of weed cutting (Swales, 1982b). However, the results were inconclusive, primarily because it was difficult to distinguish between natural fluctuations in population levels and those which may arise from the effects of weed cutting.

Weed cutting may affect fisheries by disrupting fish feeding, reproduction and normal behaviour patterns. Since aquatic macrophytes and algae form a major source of invertebrate fish food organisms, removal of these may be significant. Considerable numbers of invertebrates may be removed during weed cutting (Statzner and Stechman, 1977; Pearson and Jones, 1978; Kern-Hansen, 1978). Weed cutting may destroy potential spawning areas for phyophilous species such as roach, pike and perch and may remove incubating eggs and eliminate slack-water areas required for the development of fry. The reduction in water level through the removal of weeds caused incubating eggs, attached to moss growing on vertical pilings in a chalk stream, to be exposed (Mills, 1981). Low levels of roach populations in the River Perry, England, may have been a consequence of annual weed cutting (Swales, 1980). During the weed cutting operation there may be considerable movements of fish out of an area, leading to a reduction of fish stocks, although this may not be permanent because of rapid colonization.

Vegetation growing within the channel is just one of the major habitat requirements which can be eliminated by channel excavation, and in turn affects those fish and macroinvertebrate species which are dependent upon it. The purpose of the following discussion is to review in detail both the short- and long-term changes in the standing crop and species composition of aquatic plants resulting from different types of channelization.

PLANTS

Habitat Requirements

Plants in the riverine environment include submerged species which grow within the channel, emergent species growing at the margins of the channel and on the banks and vegetation adjacent to the banktop, including trees and shrubs (Peltier and Welch, 1969; Haslam, 1978). These plants play vital roles in either providing shelter and protection for macroinvertebrates and fish, by shading watercourses from excessive water temperatures, providing an input of organic leaf litter, or as oxygen producers. Plants also diversify local conditions by altering the detailed flow patterns within a channel and by inducing silt deposition.

Plant patterns vary across the channel according to differences which are physical (flow, substrate), chemical (nutrients in the silt) or biotic (e.g. shading by larger plants or grazing by animals). The morphology of natural channel cross-sections varies in shape from symmetrical to asymmetrical. Potentially river plants can live on all angles of substrate and at all water depths commonly found in Britain. Plants cannot grow in deep water, either because the substrate is unstable or because the turbidity of the water prevents light from reaching the bed. Heavy shading from overhanging trees and shrubs prevents the growth of aquatic plants. Submerged species may grow within the channel and tall monocotyledons and fringing herbs at the channel edges. Floating or emerged species can grow in mid-channel. Above the channel's edge is the bankside which is dry and characterized by land plants such as herbs, grasses and shrubs. Banks vary in steepness as well as height and this can influence the vegetation; for example, narrow channels with steep banks 1–2 m high may be completely shaded by tall herbs. Finally, trees may grow on the banktop. Individual plant species can be correlated with physical variables, especially flow type and substrate, width and depth, bankside shading, and with nutrient status.

The Impact of Dredging on Aquatic Plants

Dredging may be undertaken as part of a capital works scheme to form a new channel or to deepen an existing one, but may involve the removal of accumulated sediment as part of a maintenance scheme. Dredging often reduces the morphologic variability of a channel bed and banks and this can reduce plant diversity. If the hard bed is unbroken by dredging then plants recover quickly; if the bed is destroyed but silt accumulates then plants may invade (Haslam, 1978). However, if the substrate is unstable following works, and there is no associated sediment deposition, then the vegetation returns very slowly. Disrupting the bed removes underground plant parts, thereby delaying recovery which may not occur for 5–10 years. Generally species return to a site within 2 years of dredging, and the amount of vegetation is back to normal after 3 years (Haslam, 1978). However, if dredging removes silt then the species composition may change, e.g. *Ranunculus* will invade a gravel bed exposed by dredging.

Tall emerged monocotyledons recover relatively slowly after dredging because most have complex root systems which need to establish before they can dominate. Other plants, which initially regrow, are later shaded-out by tall species. For example, regularly dredged brooks, which are normally characterized by *Sparganium erectum*, are often dominated by short species (e.g. *Sagittaria sagittifolia*) for approximately 2 years after dredging. The rate of recovery following excavation appears to depend on the proportion of plant propagules (fruits, stems, rhizomes, etc.) which have been removed. Following shallow dredging recovery can be within 1 year; in more extreme cases recovery takes several years.

Brookes (1987d) assessed the short-term recovery of vegetation at sites in south-central England in terms of standing crop and species composition. The rate and nature of recovery depended on a number of factors. Recovery of vegetation was slow in the first growing season following dredging of chalk streams. Figure 42 shows the recovery of the Wallop Brook in Hampshire over two growing seasons following dredging in December 1981. Regrowth occurred from plant rhizomes remaining in the channel bed after shallow excavation and was relatively rapid, but in most cases regeneration appeared to depend on the invasion of plant particles and propagules moving

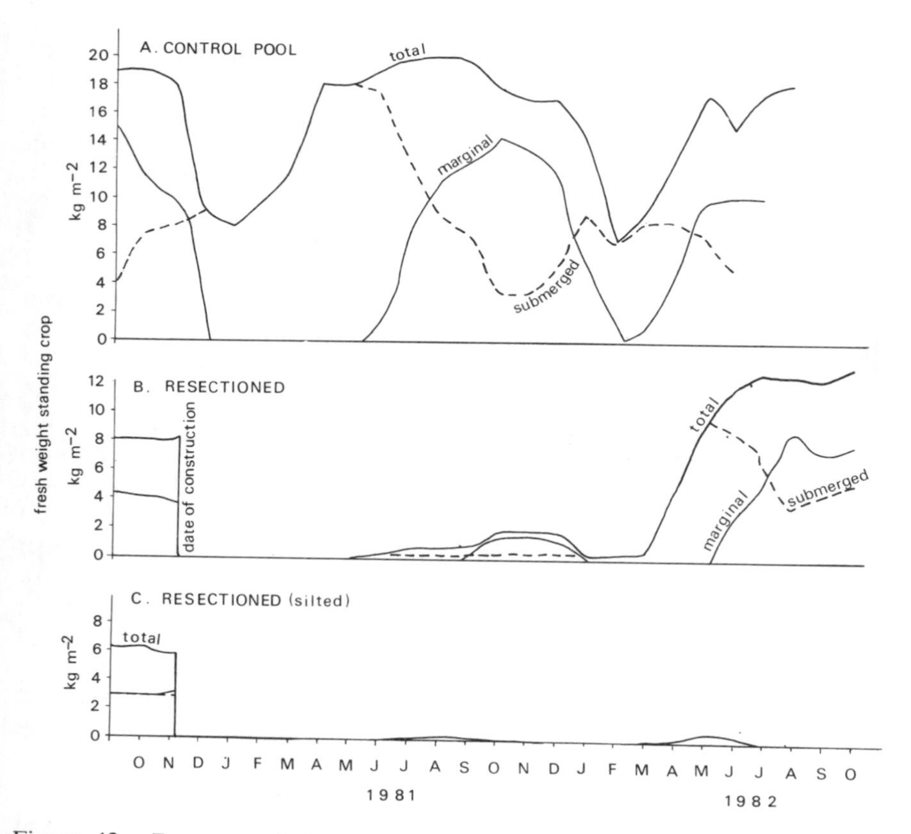

Figure 42. Recovery of the standing crop of aquatic vegetation in resectioned reaches of the Wallop Brook, Hampshire, England, during 1981 and 1982. (From Brookes, 1987d. Reproduced by permission of Academic Press (London) Inc.)

from upstream. However, by the second growing season the standing crop of the excavated channel generally attained the value recorded in the control reach. Valley gravels extended for several metres below the beds of the chalk streams studied and therefore the substrate available for colonization by plants was not substantially changed by excavation. The species composition in chalk streams which had been excavated was considerably different by the second season of growth. Fringing herbs such as *Apium nodiflorum, Oenanthe fluviatilis* and *Berula erecta* grew prolifically during the second season, whilst *Nasturtium officinale* dominated the control reaches.

Relatively little recovery of vegetation occurred in two years after excavation of a clay-bed stream, where the original substrate had been destroyed and replaced by bedrock. The gravel bed remained unstable after completion of works on a New Forest stream and recolonization by plants was limited (Brookes, 1987d). Over-deepened reaches within channelization works may function as permanent silt traps, thereby covering gravels and limiting the growth of submerged species. Silt may favour the luxuriant growth of emergent plants and may contain plant propagules and parts necessary for regeneration. Very deep pools had been created within the channelized reaches of three streams and these precluded the growth of vegetation by excluding light from the bed. However, enlarged channels with a slightly greater depth than previously can support a larger amount of vegetation provided that sufficient light can penetrate to the channel bed.

Weed Cutting

A severe hand cut on the Candover Brook in November 1972, a chalk stream in Hampshire (UK), was shown to change the percentage coverage of each plant species compared with the previous year and the growth pattern of the most dominant species was changed (Soulsby, 1974). The cut reduced the plant coverage from 48% to 8% and the mean dry weight standing biomass from 168 gm^{-2} to 32 gm^{-2}. *Callitriche* was the most dominant plant prior to the cut (Figure 43) and consequently its coverage was reduced from 32.8% to 4.3%. Although *Callitriche* became the dominant plant again in 1973 it only covered 50% of the area prior to the cut. In the growing season of 1973 *Callitriche* could only establish itself and the plant did not overgrow other plant beds, particularly *Ranunculus*, as it would do normally following overwintering. This was partly because substantial amounts of silt had been washed out after the cut. This enabled *Ranunculus* to grow in the year following the cut. The coverage of *Berula* was reduced from 4.9% to 2.5% after the cut but quickly recovered by December, probably growing from roots in the bed of the channel.

It would appear that since individual plants have distinctive growth patterns, then there is a particular time of the year when a cut would be most effective in controlling one plant and not another. It has been shown that a spring cut of *Ranunculus* during its growing season stimulates further

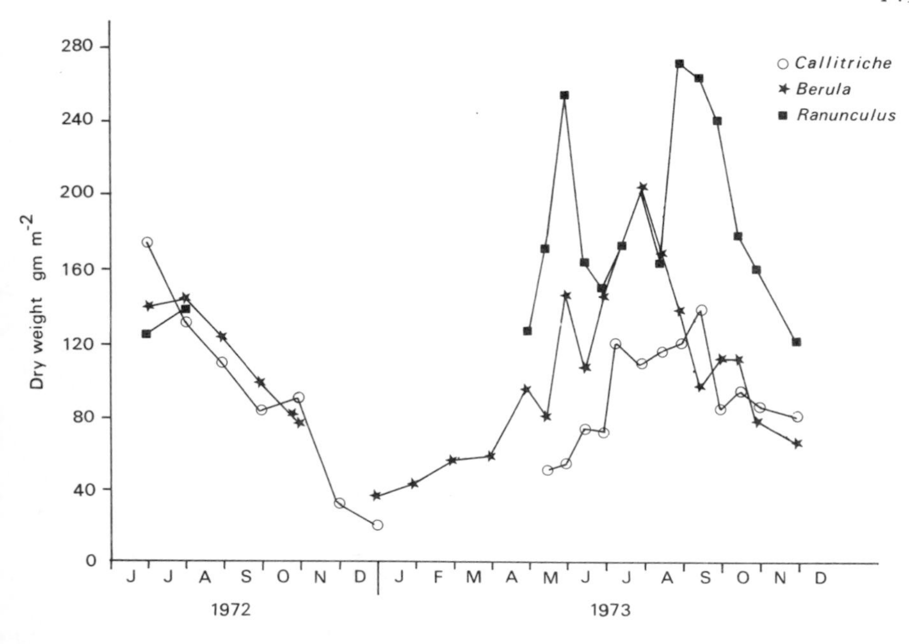

Figure 43. Mean standing biomass of aquatic plant species during 1972 and 1973 for the Candover Brook, Hampshire, England. (Based on Soulsby, 1974. Reproduced by permission of Blackwell Scientific Publications Ltd.)

growth rather than controlling it (Westlake, 1968a; Westlake and Dawson, 1982; Dawson, 1986).

Long-term Effects

The effect of increased dredging activity and herbicide sprays (introduced in 1960) has been dramatic in drainage channels in the Monmouthshire Levels in South Wales (Wade, 1977, 1978; Wade and Edwards, 1980). By reconstructing the flora of the channels for the past 150 years (1840–1976) using published records, it was found that 55 of 100 species known to have occurred in the drainage channels have remained either widespread or restricted in their distribution. Six species had disappeared, 11 species had recently appeared and others had changed their distribution. However, the overall long-term changes in the aquatic flora had been small and could reasonably be attributed to intensification in the management of the drainage system, which included dredging every 5.5 years and annual spraying with herbicides. The maintenance programme suppressed the development of emergent species but it is likely that the disappearance of species was related to salinity fluctuations, shading and species introduced by man. Dredging may be important in initiating hydroseral succession in channels. Increased maintenance may have encouraged *Ranunculus* sp. and *Ceratophyllum demersum* in the Monmouthshire Levels.

Table XXXII. Impact of changed substrate on aquatic macrophytes. (From Brookes, 1987d)

River	Plant community*	Date of works	Survey date	Nature of modification studied	Standing crop			
					Control reach		Excavated reach	
					kg	kg m^{-2}	kg	kg m^{-2}
Cherwell, Oxon	Lowland clay	1967	07/1981	Bedrock replaced gravels	4.80	1.10	0.00	0.00
Caldew, Cumbria	Resistant rocks	1959	06/1982	Boulders and cobbles removed	0.84	0.20	0.00	0.00
Clow Beck, Durham	Hard limestone	1963	04/1981	Cobbles removed	0.60	0.50	0.00	0.00
Monks Brook, Hants	Soft sandstone	1966	06/1982	Concrete replaced gravels	0.50	0.16	0.00	0.00
Erewash, Notts	Coal measures	1958	07/1982	Concrete replaced gravels	1.40	0.28	0.00	0.00
Meon, Hants	Chalk	1957	06/1981	Concrete replaced gravels	6.50	5.10	10.00	4.00
Win, Dorset	Chalk	1970	07/1982	Concrete replaced gravels	7.00	2.12	0.00	0.00

*After Haslam and Wolseley (1981).

Reproduced by permission of Academic Press Inc. (London) Ltd.

Table XXXIII. Impact of removal of bankside vegetation on aquatic macrophytes. (From Brookes. 1987d)

River	Plant community*	Date of works	Survey date	Nature of modification studied	Standing crop					
					Control reach			Excavated reach		
					ϕ	kg	kg m^{-2}	ϕ	kg	kg m^{-2}
Lynher, Cornwall	Resistant rock	1962	06/1982	Trees removed	−4.0	0.38	0.09	−4.2	13.50	2.70
Washford, Somerset	Hard sandstone	1974	05/1982	Trees removed	−3.4	1.60	0.39	−3.5	7.60	2.20
Severn, Powys	Resistant rock	1974	06/1982	Trees removed	−3.4	10.30	0.73	−3.3	55.50	3.96

*After Haslam and Wolseley (1981).

ϕ is mean grain size of sediment.

Reproduced by permission of Academic Press Inc. (London) Ltd.

Vegetation surveys were undertaken at 22 sites throughout England and Wales where channel works had been completed up to 24 years previously (Brookes, 1987d). The destruction of the original substrate, significant changes of width or depth, and the removal of bankside vegetation had longer-term impacts on the vegetation. Where the substrate was substantially unchanged by excavation then the plant species were identical to those found in the upstream control reach. However, deepening of alluvial rivers had often replaced the original mixed sediment substrate with expanses of bedrock and the regrowth of vegetation was found to be inhibited, as exemplified by the River Cherwell in Oxfordshire (Table XXXII). Boulders and gravels had been removed from upland rivers (e.g. River Caldew, Clow Beck) thus destroying the habitat for bryophytes. Exceptionally, the original substrate had been replaced by concreted sections, which were generally not colonized by plants.

The new width and depth dimensions of an excavated channel are also important, and a widened section may support a greater total standing crop because of the increased area of bed suitable for colonization, whilst a sufficiently deepened section may preclude the development of plants by reducing or eliminating the light available for plant growth at the bed (Brookes, 1987d). In over-deepened pools the accumulation of silt deposits may permanently obscure the substrate, thereby encouraging the growth of silt-loving plant varieties. Where the widened part of a channel has a similar substrate to that existing in the remainder of the channel, then these areas become colonized by similar plants. The total standing crop in these reaches is greater because of the increased habitat area. However, a channel which has been over-widened has a reduced water depth at normal flows, and consequently the column of water in which plants can grow is reduced.

The removal of bankside trees and other vegetation is a feature of many channelization works because of the requirement for access to machinery. Unshaded channelized reaches of three sites with different plant communities had standing crops of between 4.4 and 29 times greater than those measured in naturally shaded control reaches (Table XXXIII); (Brookes, 1987d). In the absence of trees more light could penetrate to the channel bed and therefore encourage plant growth.

RIPARIAN AND WETLAND ECOSYSTEMS

Effects of Channelization on Bankside Ecology

The construction of channelization schemes typically results in the destruction of vegetation, often for a distance of between 10 and 15 m either side of the channel, but sometimes up to 100 m (Hill, 1976). If land use changes follow channelization then widespread destruction of natural plant communities may occur. For example, extensive clearance of bottomland hardwoods has occurred in the Mississippi delta (Little, 1973), in Tennessee (Barstow,

1970, 1971) and Georgia (Wharton, 1970). Bankside vegetation which is not cleared may be significantly affected by altered surface and soil water levels arising from channelization. Where surface waters are drained then aquatic plants will be eliminated and replaced by terrestrial species.

Impacts on Birds

There have been comparatively few studies concerned with the impacts of channelization on the avifauna and mammals which inhabit river banks. Carothers and Johnson (1975) indicated that the severity of vegetational manipulation in the riparian habitat determines the degree of impact on species composition and total avian productivity in the southwestern United States. Work on the Gila and Verde Rivers Basins in southern and central Arizona and the Lower Colorado River showed the importance of riparian habitat for bird populations. Figure 44 compares the structure of bird populations

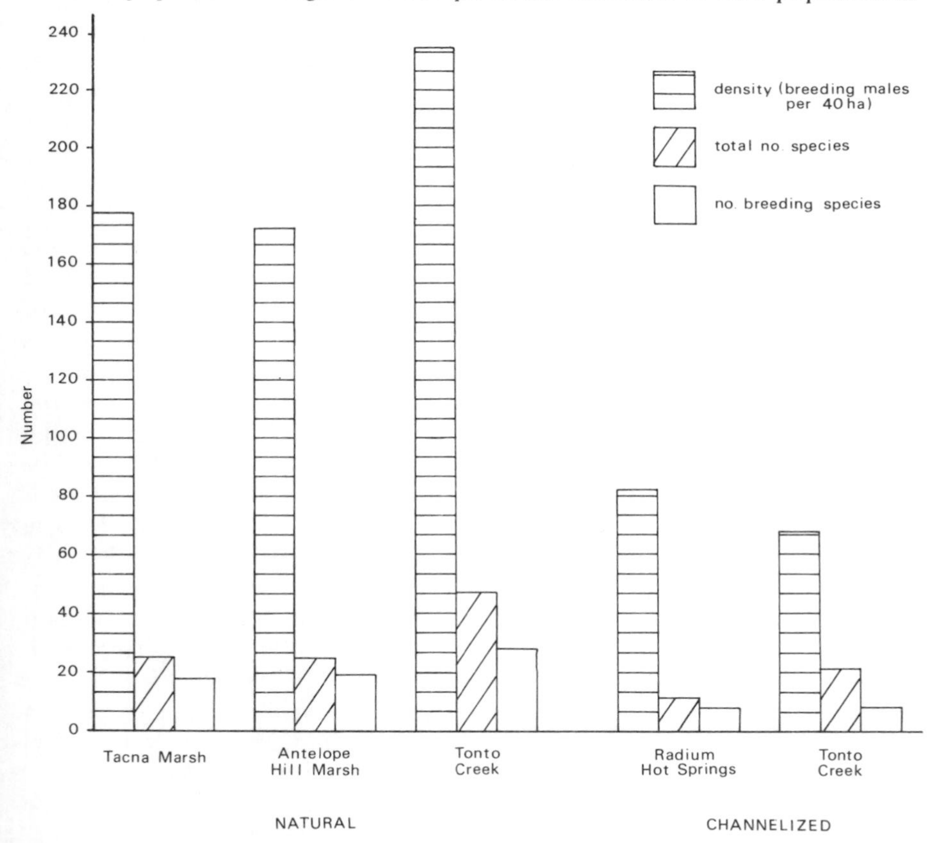

Figure 44. Comparision of breeding bird density, number of breeding species and total number of species observed in channelized and natural areas of the Gila River and Tonto Creek (after Carothers and Johnson, 1975.)

during the breeding season on natural and channelized areas along the Gila River and Tonto Creek, in Gila County. The most sparse bird populations were found in those areas which have been subject to vegetation removal, especially trees, shrubs and herbs, through channelization. Forty-nine per cent of species contributing to the population along a channelized length of the Gila River were determined to be exclusively dependent upon riparian vegetation, and an additional 2% showed a preference for riparian vegetation. On the channelized area of Tonto Creek 70% of the birds were found to be exclusively dependent upon the riparian vegetation and a further 12% showed a preference for this vegetation type. Natural riparian habitats were characterised either by standing or slow-flowing water surrounded by marsh or emergent vegetation or by flowing water with small amounts of marginal vegetation and low-density deciduous riparian trees.

Several rare or protected species of birds were found to be dependent on the riparian habitats of the southwestern United States, including the southern bald eagle (*Haliaeetus leucocephalus*), the yuma clapper rapil (*Rallus longirostris yumanensis*), northern black-bellied tree duck (*Dendrocygna autimnalis fulgens*), the zone-tailed hawk (*Butco alticaudatus*), northern gray hawk (*Buteo gallus a. anthracinus*), and American osprey (*Pandian haliaetus carolinensis*). In addition some of the largest population densities of non-colonial nesting birds in the US occurred in mature riparian (streambank) cottonwood forests in central Arizona.

Songbirds in the White River watershed of Vermont (drainage area: 185 sq km) were found to be adversely affected by channelization for flood control in summer 1973, although not involving continuous removal of vegetation, straightening, regrading and dumping of spoil on the banks (Dodge *et al*., 1977; Possardt and Dodge, 1978). Overall there were significantly fewer birds in the channelized reaches (39% less). This was true for autumn (fall) of 1974 (50% less) and spring 1975 (50% less) although no statistically significant differences were observed for the early or late summer of 1985. Fewer species of birds were found in channelized areas in autumn/fall 1974 (23 vs 38) and early summer 1975 (25 vs 37). The greater diversity of birds in non-channelized areas during spring 1975 was highly significant. Only 11 species of breeding birds were found in channelized areas compared to 17 in undisturbed reaches. The most abundant species collected in non-channelized areas were the yellow throat (*Geothlypis trichas*), song sparrow (*Melospiza melodia*), black-capped chickadee (*Parus atricaplilus*), chestnut-sided warbler (*Dendroica pensylvanica*), American redstart (*Setophaga ruticilla*), belted kingfisher (*Megaceryle alcyon*), ruby-throated hummingbird (*Archilochus colubris*) and catbird (*Dumetella carolinensis*). The most abundant species found in channelized areas were the song sparrow, eastern phoebe (*Sayornis phoebe*), ruby-throated hummingbird, yellow throat, belted kingfisher, barn swallow (*Hirundo rustica*) and tree swallow (*Iridoprocne bicolor*). Figure 45 groups the birds according to diet and/or methods of procuring food. These differences in the numbers and diversity of birds reflect the loss of streambank vegetation from channelization work. The degree to which various species of birds are affected depends on their diet or method of procuring food. Swallows, which are aerial insect catchers, and spotted pipers benefited by the increased aerial space and gravel areas created by channelization. Since aquatic insect populations recovered rapidly following channelization the food source for these birds (aerial flycatchers) increased. Hummingbirds and fringillids were unaffected because flowering and seed-producing animals on which they feed recovered rapidly after channelization.

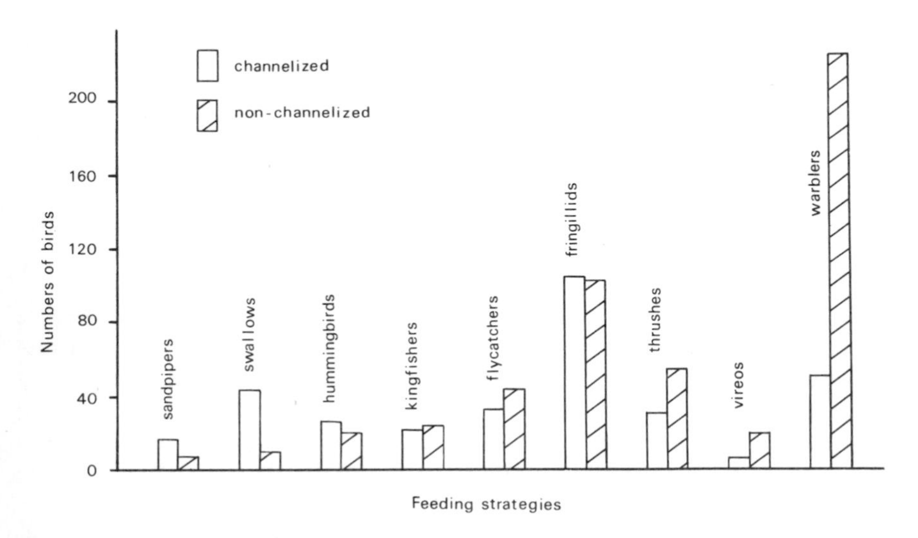

Figure 45. Use of channelized and non-channelized sections of the White River watershed, Vermont, USA, by birds of various taxonomic groups utilizing nine different feeding strategies. (From Possardt and Dodge, 1978. Reproduced by permission of The Wildlife Society.)

However the smaller number of thrushes in channelized reaches was attributed to the absence of the lower herbaceous layer which provide berries and habitat for insects (Possardt, 1975; Possardt and Dodge, 1978). Vireos and warblers feed principally on insects from the stems and leaves of the shrub and tree canopy, and were the most adversely affected by channelization. Clearly the most dramatic impacts occur during periods of nesting and fledging and dispersal of offspring when the parent birds need to obtain a large volume of food with minimum energy expenditure. The recovery of bird numbers and species diversity two years after channelization corresponded with regrowth of the vegetation, although trees and shrubs were expected to take several years to re-establish.

A study was undertaken of the impacts on birds of a flood alleviation scheme on the River Stort in Hertfordshire, England, which involved increasing the capacity to a 1 : 70 year flood of 23.5 cumecs by resectioning and regrading in summer 1979 to March 1980 (Kite, 1979; Weeks, 1982). The immediate impact of the works was to destroy marginal plant communities and these were extremely slow to recolonize within the first year. This disruption resulted in the number of species and individuals becoming reduced, in particular coot, dabchick and mute swan became only 'occasional visitors'. The decline of the moorhen population could be directly related to the loss of bankside vegetation, especially tree, shrub and sedge cover.

Williamson (1971) reported the effects of dredging and bank clearance on seven waterside birds on the River Stour in Dorset (UK). The numbers of kingfishers and reed and sedge warblers were still depleted three years after completion of the work. On the River Wye near Hereford (UK) major differences in bird populations were observed between a managed and an adjacent unmanaged reach (Williams, 1980). Twenty-five breeding species and 146 territories per 10 km were found in the unmanaged reach, whilst only 11 species and 43 territories per 10 km were found in the heavily managed channel.

Mammals

The status of the otter (*Lutra lutra*) in England and Wales is of considerable concern because recent surveys have shown that the species is rare or extinct over much of the country (Macdonald, 1983). It appears that the continuing loss of this species is related to increasing pressures on suitable habitat caused by channelization, especially bankside tree clearance. Macdonald and Mason (1982) indicated that the total number of sites within a river reach showing evidence of otters was related principally to the occurrence of sycamore and ash trees, these being species with wide, eroded root systems suitable as otter holts. Macdonald *et al*. (1978) found that reaches of the River Teme with otters had more bankside vegetation, a higher density of trees and suitable holts provided by ash and sycamore trees. In the UK the impact of channelization on other mammals has received relatively little attention. Major works on the River Roding in England had little effect on water vole burrows (Weeks, 1982).

For the White River watershed in Vermont, Possardt and Dodge (1978) showed that the number of small mammals was lower in channelized areas in the fall of 1974 (26 compared to 48 in unmodified reaches) and in early summer 1975 (64 compared with 100). Species diversity was also less in channelized areas in the fall of 1974 (5 vs 7 species). Figure 46 compares the use of channelized and non-channelized sections of the White River watershed by mammals, grouped according to diet or method of procuring food.

The white-footed mouse (*Peromyscus leucopus*) population recovered rapidly, probably because this species has a more general diet and habitat requirements and was able to exploit channelized areas. By comparsion

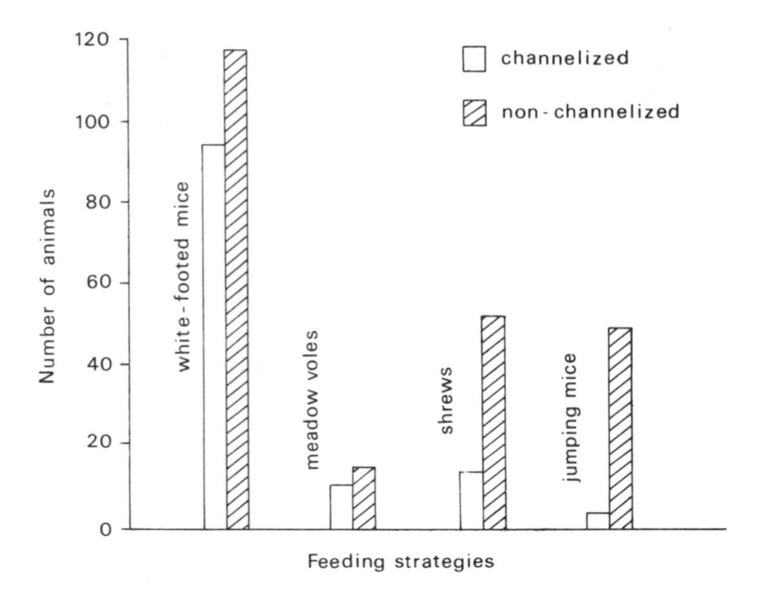

Figure 46. Use of channelized and non-channelized sections of the White River watershed, Vermont, USA, by small mammals grouped according to diet and/or food procurement habits (From Possardt and Dodge, 1978. Reproduced by permission of The Wildlife Society.)

jumping mice and shrews (*Sorex cinereus; Blarina brevicauda*) continued to be non-existent or greatly reduced 2 years after channelization, because they are more specific in their habitat requirements. The meadow jumping mouse (*Zapus hudsonicus*) is restricted to meadows and the woodland jumping mouse (*Napaeozapus insignis*) to woodland associated with streams, and their diet is principally grass seed. These requirements were not provided by the channelized reaches.

The impacts of channel modification in the Luxapalila River near the Alabama—Mississippi border on fur-bearers have been examined by Arner *et al.*, 1975. Both old channelized segments (about 50 years old) and segments undergoing channelization at the time of the study were examined. Mink

Table XXXIV. Number of fur-bearers in channelized and natural reaches of the Luxapalila River (data based on night sampling undertaken by Arner *et al.*, 1975)

	Newly channelized	Old channelization	Unaffected reaches
Raccoon	2	9	7
Mink	0	1	0
Beaver	2	12	29
Opossum	0	3	1
Muskrat	1	0	12

(*Mustela vison*), muskrat (*Ondatra zibethicus*) and beaver (*Castor canadensis*) could be considered as species dependent upon a riparian habitat. Preliminary studies found considerably larger numbers of beaver and muskrat in the unaffected reaches compared to the old and newly channelized segments (Table XXXIV).

Case Study: Investigations of the Rush and Wildhorse Creeks in the Southern Grasslands of Oklahoma

This is one of the studies funded by the US Fish and Wildlife Service under the National Stream Alteration Study Team (Barclay, 1980). The principal objective was to obtain quantitative information that could be used to predict the consequences of stream alteration and associated land use changes on the vegetation, amphibians, reptiles, birds and mammals. Channelization, principally by straightening, of the Rush and Wildhorse Creeks in the 1920s and 1930s greatly facilitated the clearing, drainage and subsequent farming of the bottomland areas. Much aquatic and riparian habitat has been lost. Channelization resulted in a reduction of 31% of channel lengths and a 500–900% increase in channel capacity due to subsequent erosion. Between 1871 and 1969 there was an associated 93% and 84% reduction in bottomland forest for Rush and Wildhorse Creeks respectively. This process was aided by channelization. There had also been complete drainage of the once extensive wetlands. By comparing channelized versus unchannelized forested sites, Barclay (1980) demonstrated marked differences (Table XXXV).

Vegetation, large mammals and birds on sites channelized prior to 1933 showed no evidence of recovery to existing levels on unchannelized sites with comparable land use. In some instances values for old channelized sites were equal to or below those of recently channelized sites.

Floodplain Wetlands

Attention has so far concentrated on the direct impacts of channel works on the biology of stream channels and the bankside habitat. Much of the earlier concern of the US Fish and Wildlife Service was for the consequences of channelization on fish and wildlife communities. However, it was soon recognized that most stream alterations could not be considered separately from changes in the floodplain vegetation and animal communities. A growing body of literature on riparian ecosystems suggests a strong dependency between stream and floodplain processes (e.g. Johnson and McCormick, 1978; Warner, 1979; Brinson *et al*., 1981; Cowardin, 1982). Wetlands are lands transitional between terrestrial and aquatic ecosystems (Hook, 1987).

It has been estimated that 50 million hectares, or 6% of the land area in the United States (excluding Alaska), is subject to the 1:100 year flood and therefore potentially supports riparian ecosystems (Brinson *et al*., 1981). However, it was estimated that only 9 million ha exists in a natural

Table XXXV. Effects of channelization of the Rush and Wildhorse Creeks, Oklahoma (after Barclay, 1980)

	Causal factors
Amphibians and reptiles	
69 species (snakes, turtles, lizards, frogs and toads); species richness, diversity and numbers reduced	Meanders, pool and overhangs lost. Forest converted to cropland. Reduced flooding; lowering of groundwater levels. Fewer niches due to reduced food and cover
Birds	
19 species; richness, density and diversity of species reduced. Species composition changed	Removal of forest resulted in loss of breeding habitat of wood pewees, warblers, ducks and vireos. Due to decreased vegetation complexity especially during the critical winter and migration periods. Birds favouring open ground or savannah found after modification (e.g. killdeer); absence of tree-cavity nesters
Small mammals	
11 species(e.g.white-footed mice and fulvous harvest mice) Reduced numbers and diversity	Shrews and moles depend on litter/humus layer; land clearing caused loss of microhabitats and increased vulnerability to predation (reduced food and cover). Species richness and numbers of individuals are related to habitat complexity
Larger mammals	
16 species (e.g. badgers and beaver)	No overall effect on species richness. More common where recently channelized because habitat conditions are more favourable
e.g. armadillo, raccoon, beaver and coyote	More common on unchannelized sites because of reduced habitat quality on channelized sites

From Barclay (1980). Reproduced with permission.

or semi-natural forested condition. It is estimated that 70% of the original floodplain forest has been converted to urban and agricultural use. Up to 95% loss of natural vegetation has been observed for the Lower Mississippi, Sacramento and Missouri Rivers. Riparian lands are one of the most severely affected ecosystems in the United States (Holder, 1970a,b). Wetland drainage has also been extensive throughout Europe (Van Der Voo, 1962; Baldock *et al*., 1984) and the USSR (e.g. Averyanov *et al*., 1971).

Floodplain and swamp channelization usually accompanies mainstream channelization and tends to drain lakes, sloughs and swamps and to lower the water-table (Choate, 1971, 1972; Darnell *et al*., 1976). This reduces or eliminates the annual flood and the sediment load normally deposited in such environments. Spoil banks and levees prevent peak flows from flooding the adjacent land and therefore large quantities of nutrients and valuable freshwater is lost. Both the abundance of water and nutrient supply are particularly responsible for maintaining the productivity and vitality of riparian ecosystems. Aquifer and groundwater recharge is also reduced. Wetland becomes dryland. Channelization is frequently accompanied by extensive timber cutting and the dry lowlands are invaded by agriculture or other land-related activity. Soil erosion is now a problem. If the area is in the low coastal plains, salt-water penetration may occur (Darnell *et al*., 1976). Typically these effects are inextricably bound with other human changes such as the installation of land drainage, changes of land use and construction activities, which usually follow a channelization scheme (e.g. Green, 1973, 1979; Darnell *et al*., 1976). However the following discussion focuses on those floodplain changes arising as a direct consequence of channel works.

The effect of artificial drainage is to increase the efficiency with which available moisture is collected and carried to watercourses. This removal of surface and gravitational water will have an impact on the level of the water-table. Areas of groundwater discharge to the surface are often associated with permanent wetlands. In these situations the level of excess water on or in the soil corresponds to the regional water-table. Channel deepening or the construction of new drainage works will permanently lower the water-table. Such works have caused the regional water-table over considerable areas of southern Michigan and the Mississippi delta to be lowered by 1–2 m (Zumberge, 1957; Little, 1973). The impact of drainage works on groundwater levels also depends on the hydraulic conductivity of the soil. A limited impact can be expected in impermeable clay soils; whilst highly permeable sandy soils transmit water readily and ditches can be expected to collect water from considerable areas (Found *et al*., 1975, 1976: Hill, 1976). The impact on organic soils is related to the size of the pore spaces. Small pore spaces inhibit water movement and Burke (1969) found that 1 m deep ditches through blanket peat bogs in Ireland had little impact on water tables 1–8 m from the ditch. By contrast the water-table in a sedge bog in Newfoundland drained for agricultural purposes was significantly affected 11 m from the ditches (Rayment and Cooper, 1968). Working in Minnesota,

Boelter (1972) found that the water table declined 50 m from a ditch cut into fibric peat, which has large pore spaces, compared to only 5 m in a moderately decomposed hemic peat.

A considerable amount of research has been undertaken into the impacts of arterial drainage on the unique flora and fauna of wetlands. Wetlands provide cover, food resources and breeding sites for many types of wildlife (Whitney *et al*., 1972). In channelized inland marshes and fens, marshland vegetation is succeeded by dryland vegetation (McCarthy and Glass, 1982). Wetland plant communities have been observed to disappear in many areas of the world as a result of land drainage (e.g. Aus, 1969; Sukopp, 1973). Joyal (1970) reported how a ditch cut in 1938 altered water levels in a bog near Ottawa, Canada, allowing the invasion of plants such as aspen and poplar. However research has shown that drainage of wetland forests may improve tree growth. In Scandinavia, drainage of bog areas has markedly improved the growth of coniferous forest (Seppala, 1969). By contrast the drainage of hardwood forests may be detrimental, causing the death of older trees and a general reduction in growth rate where the water table is permanently lowered (Klawitter, 1965, 1970; Makin *et al*., 1980).

The ecosystems of natural wetlands range from relatively simple to complex systems. Certain plant species may grow in water for most of the year, whilst others require short periods of inundation. Riparian ecosystems provide essential habitat requirements for a large diversity of vertebrate species. Although catastrophic flooding may temporarily reduce the abundance of terrestrial vertebrates these species are adapted to rapid recolonization after flooding (Brinson *et al*., 1981). Riparian ecosystems are unique habitats characterized by flowing water, high plant productivity, and nutrient rich conditions. Following drainage these species are replaced by species which thrive under less moist conditions. As a consequence bird nesting habitats and food sources are destroyed. Most research has been concerned with waterfowl because of their significance for recreation (Hill, 1976). Other animals such as beaver, mink, muskrat and salamanders, which are dependent on wetland vegetation for cover, are unable to compete with other species which colonize following drainage. Drainage of wetlands greatly reduces species diversity. Little (1973) described the drainage of wetland areas in the Kissimmee River Project in Florida which almost totally eliminated migrating waterfowl. The habitat of marsh birds such as herons (*Ardea* spp.), bitterns (*Botaurus* spp.), coots (*Fulica* spp.) and rails (Rallus spp.) may be destroyed. Fur-bearers such as muskrat (*Ondatra zibethicus*), beaver (*Castor canadensis*), otter (*Lutra* spp.) and mink (*Mustela vison*) are also reduced or eliminated by the disappearance of a suitable habitat.

For the UK, The Royal Society for the Protection of Birds (1983) concluded from a variety of surveys that land drainage improvements are catastrophic for breeding and wintering wetland birds, arising from changes in the water status of the catchment area but also as a result of subsequent changes in land use practices (see also Williams *et al*., 1985).

Buckner (1955a,b) showed how the drainage of a bog in Canada resulted in increases of the saddle-backed shrew (*Sorex arcticus*) and the red-backed vole (*Clethriconomys gapperi*). By contrast the masked shrew (*Sorex cinereus*), the dominant insectivorous mammal in the wet bogs, declined. It was suggested that increases in the sawfly (*Pristiphora erichsonnii*) population occurred because both the saddle-backed shrew and red-backed vole were less discriminating in their choice of food, whilst the masked shrew selectively predated the sawfly. The destruction of cover may also affect species which are not directly dependent upon the aquatic environment. Game such as deer, pheasants (*Phasianus colchicus*), grouse (*Bonasa umbellus*) and partridge (*Perdix perdix*) have been adversely affected (Hill, 1976).

Removal of areas of surface water by draining a wetland can substantially reduce or eliminate frog (*Rana* spp.) populations and other amphibians, together with insects (Darnell *et al*., 1976). This reduces the food available to many other types of wildlife and the effects are therefore often complex. Beland (1953) found that the draining of backwaters in the Colorado River between Needles and Topack, California, as a consequence of straightening, caused a 100% loss of fisheries in those areas. Floodplain ponds which connect to the river during floods may also be eliminated by drainage. These ponds may be important nursery and breeding areas for fish and invertebrates.

Small marshes with a few hectares of shallow open water, surrounded by a fringe of various types of emergent aquatic plants, are excellent habitats for ducks. The prairie pothole region of North America is the prime example of this wetland type, where small lakes are found in the numerous depressions of glacial moraine areas (Munroe, 1967). This wildlife habitat is threatened by land drainage (Aus, 1969; Burwell and Sugden, 1964; Kiel *et al*., 1972).

The seasonally flooded bottomland hardwood swamps of the Mississippi and the swamps of the Atlantic and Gulf coastal plains are also important for waterfowl (Hill, 1976). These are major production areas for wood ducks (*Aix sponsa*) and the principal wintering areas for mallards (*Anas platyrhynchos*) in North America (Sincock *et al*., 1964). Channelization and associated clearance of the bottomland hardwoods therefore poses a major threat to waterfowl. East (1973) anticipated that channelization of the Cache River in Arkansas would eliminate the wintering habitat for 500,000 mallard.

Large marshes have also been the subject of major reclamation for agriculture throughout Europe and North America. For example it has been estimated that drainage of the marshes bordering the Caspian and Black Seas over a period of 30 years reduced the wintering area for millions of ducks, geese and coots to 10% of its former size (Osakov *et al*., 1971).

The objective of this chapter has been to review the biological impacts which have arisen from various types of channelization. The majority of studies have documented the detrimental effects which have arisen, whilst others have endeavoured to assess the extent of biological recovery in the years following the initial impact. These range from impacts on the macroinvertebrates, fish and aquatic vegetation within the channel, to the mammals and birds which are reliant on bankside vegetation, but the effects can also extend to wetlands in the adjacent floodplain. However, the repercussions of channelization can also extend to the natural reaches downstream and these consequences are the subject of the next chapter.

CHAPTER VI

Downstream Consequences

Connectivity in the fluvial system means that repercussions of any man-induced change at any given location can be transmitted over a wide area, especially in the downstream direction. The integrated nature of the downstream effects in relation to adjustments occurring within realigned reaches and upstream has been demonstrated in Chapter IV (e.g. Parker and Andres, 1976). The issue of downstream effects was one of seven topics described in a national report on channelization in the USA submitted to the Council on Environmental Quality (Little, 1973). Straightening of stream channels in Iowa has achieved drainage and flood control of land along the modified reaches but has often resulted in damage to property downstream by increasing the discharge (Campbell *et al*., 1972). Frequently landowners downstream have to repair or improve their part of the channel and this is undoubtedly one of the main reasons for the piecemeal development of channelization works. Where a catchment-wide drainage scheme is implemented it is typical for engineers to work upstream from a major tributary junction, thereby avoiding the risk of flooding properties downstream.

Litigation can occur between drainage districts, with downstream districts claiming partial reimbursement of construction, repair and maintenance costs from the districts usptream. During 1955 the US Army Corps of Engineers undertook a programme to straighten certain river segments and to remove all vegetation along river banks and on islands over a 15 km reach of the Susquehanna River to prevent further flood losses in the City of Binghamton, New York (Figure 47; Coates, 1976a). These works had the express purpose of increasing the river velocity, thereby shortening the residence time of flood waters on adjacent land. During the spring of 1956 three floods eroded river banks and partially destroyed adjacent properties. In a case against the State of New York, Paul Demoski claimed that the loss of land and destruction of a retaining wall on his property resulted directly from the construction activities (New York State Supreme Court, 1957). The government claimed that the erosion was normal and to be expected. However, evidence was produced to show that the retaining wall, built 11

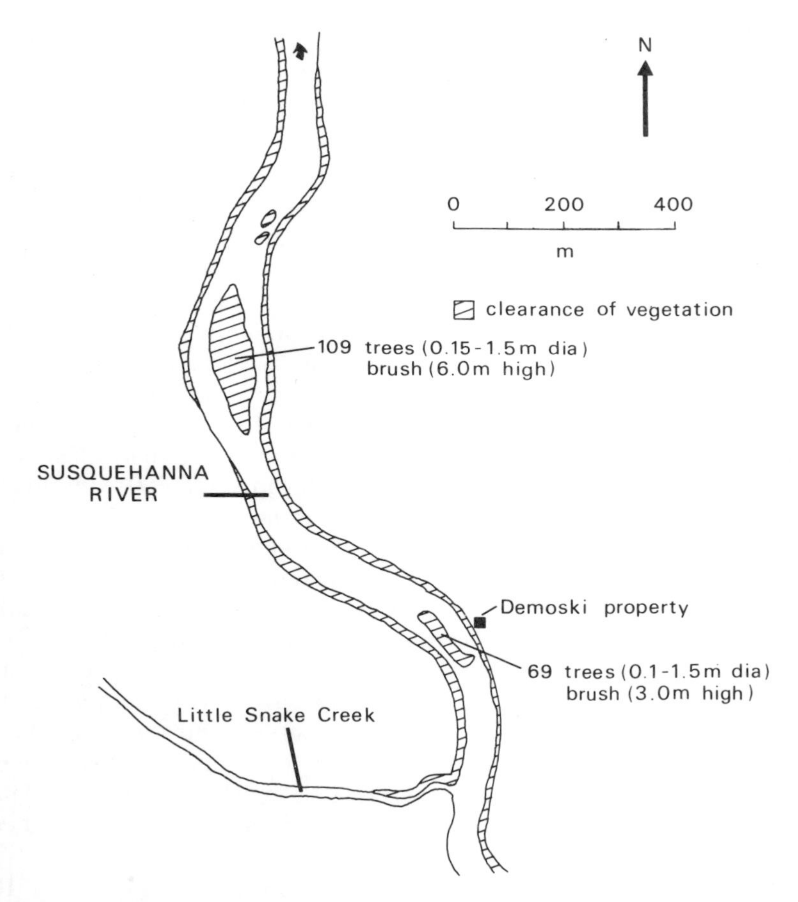

Figure 47. Plan of the Susquehanna River near the Demoski property. (After Coates, 1976a. Reproduced with permission.)

years previously, had suffered 23 floods, several of which were higher than the three which caused the damage. The decision of the court was that the erosion could be directly attributable to the works.

In terms of scientific research the downstream consequences of river channelization have received less attention than impacts occurring within modified reaches. However the literature can be conveniently grouped into hydrological, morphological, water quality and biological impacts.

> The removal of restrictions (channelization) makes for lower stages and faster getaway of the water upstream. This also results in a faster delivery of water to downstream points. The reduction in channel storage by shortening the channel and reducing upstream stages can serve only to increase peak rates of flow below the project reach.
>
> (Linsley *et al.*, 1949)

The evolution of flood control engineering on the Mississippi River has been the subject of a book by W. E. Elam, 1946, entitled *Speeding Floods to the Sea*. There have been few quantitative studies detailing the effects of channelization works on the timing and magnitude of downstream flood flows (e.g. Hung and Gaynor, 1977). Numerous observations suggest that downstream flooding may be a consequence of widening, deepening or straightening of channels (Table XXXVI). However, at 31 of 42 sites studied in the United States for the Council on Environmental Quality flooding was declared to be insignificant (Little, 1973). The lack of impact was attributed to very slight changes of gradient and velocity or because the channelized reach terminated at a junction with a mainstream or canal. Downstream flooding was most pronounced at sites of major realignment (Little, 1973).

Channel improvement works may only have a local effect on water velocities and levels upstream (Heneage, 1951). The prevention of flooding does eliminate a certain amount of local storage such that water which would previously have spread onto an adjoining floodplain is now contained in the channel and is passed downstream. Peak and near-peak flows are therefore increased. It should be noted that the impact would be considerably reduced further downstream if the peaks from the mainstream and a tributary did not coincide. Channel improvement works which accelerate the flow in either the mainstream or a tributary may therefore be beneficial if flood peaks are desynchronized. Numerous improvements to streams in a watershed may substantially affect flooding downstream.

Hillman (1936) calculated that the effect of protecting an area of 1012 hectares of the River Trent through Nottingham (UK), which normally flooded to a depth of 0.76 m, was to increase the flood depth downstream by 10–13 cms. Physical model tests showed that the effect was negligible a few kilometres downstream.

The effects of straightening on flood flows have been examined for the Boyer River in western Iowa, where the total channel length was reduced from approximately 400 km to 160 km between 1900 and 1950 (Komura, 1970; Campbell *et al.*, 1972). Using a unit hydrograph approach and flood routing procedures it was demonstrated at a total of 36 cross-sections that straightening increased the peak discharge in the range 90–190%, depending on the roughness value of the floodplain (Campbell *et al.*, 1972). The time base of the discharge hydrograph was significantly shortened (Figure 48) and the time of travel of the flood wave down the river was greatly

Table XXXVI. Downstream hydrological consequences

Type of channelization	Area of study	Evidence	Remarks	Source
Channel improvement	Britain	Inferred from experience	Flooding arising from single works and from general basin improvements	Heneage, 1951
Embankment	River Trent (UK)	Analysis of hydrograph	Elimination of floodplain, increasing flood levels immediately downstream	Hillman, 1936
Realignment	Sangamon River, Illinois; Fox, Wyconda and Salt Rivers, Missouri	Observation	Downstream flooding	Lane, 1947
Realignment	Blackwater River, Missouri	Eye-witness accounts	Downstream flooding	Emerson, 1971
Unspecified	Bayou, Cocodrie and N. Bayou, Louisiana	Observation	US Army projects required below Soil Conservation Service schemes	Callison, 1971

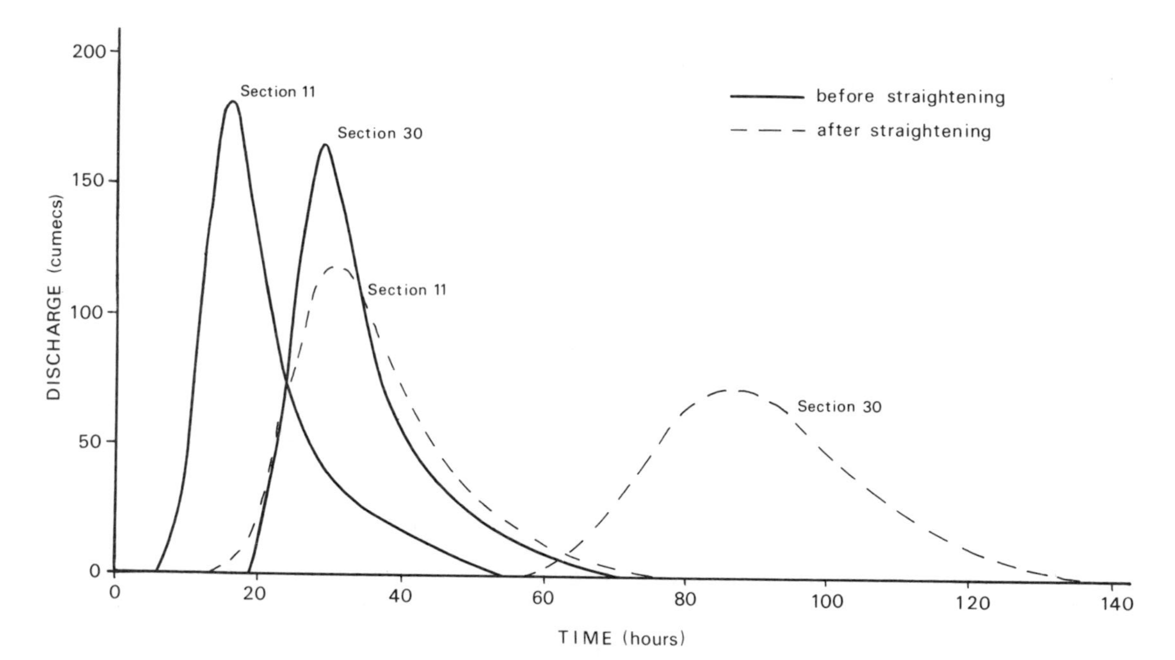

Figure 48. Effect of channel straightening on the movement of a single flood wave consisting of 2.5 cm (1 inch) of surface runoff. (Based on Campbell *et al.*, 1972. Reproduced by permission of American Society of Agricultural Engineers.)

reduced. Flooding, which persisted from 30 to 40 hours before the channel was straightened, had been eliminated.

The effects of extensive arterial drainage works on flood magnitude have been assessed for 12 catchments in Ireland (Bree and Cunnane, 1980). The works, which were completed between 1955 and 1972, involved increases of cross-sectional area by up to 100% over channel lengths ranging between 146 and 1200 km. Drainage works increased flood discharges, the value of the three-year flood being increased by between 3 and 118% for all 12 catchments. For seven of these catchments the three-year flood increased between 57 and 65%. These changes were statistically significant and were attributed to the elimination of floodplain storage and the general improvement in the hydraulic efficiency of the channels. Variations in the increase of flood discharges between catchments could not be adequately accounted for.

Permanent wetlands such as swamps and bogs are important regulators of streamflow, since organic soils can store large amounts of water during wet periods and sustain baseflow during dry periods (Hill, 1976). Sammel *et al.* (1966) calculated that during the spring more than 17 billion litres of water was retained by the floodplain swamps of the Ipswich River in the northeast United States, thereby reducing downstream floodpeaks. The Massachusetts Water Resources Commission (1971) estimated that a reduction of only 10% of the wetland flood storage of the Neponset River basin would cause a 46 cm increase of flood stage. Channel clearance and peat drainage had a marked but temporary effect on the water balance of the Glenullin Basin, in County Londonderry (Wilcock, 1977). Both the total streamflow and magnitude of low flows increased in the short term as a result of the withdrawal of water from storage.

In a report intended to update the UK Flood Studies Report of 1975, Bailey and Bree (1981) indicated that channel improvements increased the three-year flood value by 60%. The unit hydrograph time to peak was reduced. The effects of channel works on downstream flood levels can be modelled by sophisticated mathematical techniques such as ONDA, developed in Britain by consultant engineers, Sir William Halcrow & Partners. However research is currently being undertaken at the Institute of Hydrology at Wallingford to investigate the hydrological impacts of a range of channel works selected from throughout the UK.

Downstream Channel Morphology

Whilst scour has been recognized by design engineers as a potential threat to stability at the tail-end of concrete walls, which needs to be considered at the design stage, relatively little attention has been given to the geomorphological effects of river channelization which may occur over considerable distances downstream. Parker and Andres (1976) described how steepening of a water surface by a cutoff in erodible materials increases sediment transport capacity and results in scour of a bed. Excess sediment load is then deposited downstream from the cutoff because transport capacity in this reach is unchanged from its original capacity. Trapezoidal or rectangular concrete-lined channels cause erosion of the channel for short distances downstream, forming bulbous caving re-entrants (Dunne and Leopold, 1978). Morisawa (1976) examined the downstream consequences of channelization along several small tributaries of the Susquehanna River near Binghamton, New York. Significant bed erosion occurred between 1974 and 1975 directly below a length of channel on Willow Run which had been riprapped. This was attributed to increased velocity and energy, and the downstream channel had widened and downcut by varying amounts (Morisawa and Vemuri, 1975). Erosion of the downstream channel also caused large deposits of sediment. Brice (1981) described effects occurring in reaches of natural channel adjacent to relocated segments of 87 sites at which prior stability was assessed. Effects were identified at 21 sites, the most common adjustment being an increased rate of bank erosion at bends (19 sites), which was not discerned at a distance greater than 20–30 channel widths from the ends of the relocated segment. Degradation was observed at 5 sites and aggradation at 1 site. Below straightened reaches of the Sangamon River in Illinois and the Wyconda, Fox and Salt Rivers in Missouri, the stream beds were raised by the deposition of sediments, which reduced channel capacities. Together with the increase in peak discharges due to the reduced storage, this effect caused considerable flood damages (Lane, 1947).

Brookes (1987a) described the nature of channel adjustments downstream from a total of 46 channelization works located in low and high energy environments in England and Wales. Channel changes were identified principally by the method of field survey and by reconstructing the original positions of the eroded beds and banks. Use was also made of maps, aerial photographs and engineering drawings of different dates and the technique of space-for-time substitution was applied.

Field surveys undertaken between 1980 and 1983 revealed that the majority of high stream-power sites had undergone some form of erosive adjustment downstream from channelization works (Brookes, 1987a). By contrast the majority of low stream-power sites in lowland areas had not changed. Table XXXVII lists channel changes evident below a variety of different types, sizes and dates of channelization. Types were included which changed the capacity of an existing channel, namely resectioning by widening and/or deepening, embanking and comprehensive works which combine more than one type.

Increase of channel capacity ranged from a minimum of 18% for the River Severn in Powys to a maximum of 153% for the Clow Beck in Durham. A maximum increase in width of 90% (29 m) was calculated for the River Ure in Yorkshire, whilst the largest increase of depth was 67% (0.50 m) for the River Conder in Lancashire. Figure 49 shows typical recession of both banks, together with collapse of fences, downstream from resectioning works on the River Lune in Westmorland. The dominant channel characteristic to change at each site is listed first in Table XXXVII. Width changed

Table XXXVII. Summary of channel adjustments downstream from channelization works at high stream-power sites in England and Wales

River	Location	Grid reference	Drainage area (km^2)	Date	Type of works	Length of works (m)	Length of effects (m)	Maximum capacity increase (m^2)	Maximum capacity increase (%)	Type of change
Aln	Abberwick	NU123142	107	1947	Resectioned	7760	575	17.9	85	w^+ d^+ P
Caldew	Mosedale	NY361343	41	1959	Resectioned	3475	1952	5.4	46	w^+
Clow Beck	Hutton Magna	NZ127127	27	1963	Resectioned	3500	656	4.2	153	d^+ w^+
Coly	Colyton	SY249943	82	1980	Resectioned	1183	0	0	0	—
Conder	Galgate	SD481552	29	1965	Comprehensive	520	265	6.3	85	d^+ w^+
Exe	Exford	SS855383	23	1958	Comprehensive	380	221	3.9	39	w^+
Ithon	Penybont	SO117632	145	1971	Resectioned	1200	685	15.7	49	w^+ d^+
Kent	Kendal	SD517906	188	1979	Comprehensive	5494	0	0	0	—
Lune	Wath	NY684052	16	1970	Resectioned	2432	728	6.3	125	w^+ d^+
Lynher	Bathpool	SX284747	46	1964	Resectioned	325	280	9.8	82	w^+
Pickering	Pickering	SE792824	72	1962	Embanked	2786	120	3.5	73	d^+ w^+
Ribble	Studfold	SD812700	82	1967	Resectioned	1619	0	0	0	—
Severn	Newtown	SO118922	460	1974	Comprehensive	1840	466	17.0	18	w^+
Ure	Kilgram	SE198555	514	1954	Embanked	6643	1645	172.0	150	d^+ w^+
Usk	Brecon	SO044281	434	1969	Comprehensive	585	1350	47.0	53	w^+
Wansbeck	Morpeth	NZ204863	303	1974	Comprehensive	1912	0	0	0	—
Waren Burn	Lucker	NU154303	26	1949	Resectioned	800	350	12.6	110	w^+
Washford	Roadwater	ST039394	35	1974	Resectioned	2094	168	3.4	64	w^+

w^+ = width increase.
d^+ = depth increase.
P = planform adjustment.
From Brookes (1987a). Reproduced by permission of John Wiley & Sons Ltd.

preferentially to depth at a majority of sites, and out of a total of 14 sites where the method of channelization involved an increase of channel capacity, 7 sites underwent change of width (w^+) only, and at a further three sites width changed preferentially to depth (d^+). At only four sites were depth increases more evident than width increases. Preferential changes of channel width have been identified in other studies of human impact such as urbanization (e.g. Wolman, 1967; Mosley, 1975). A controlling factor appears to have been perimeter sediment, and bed armouring or underlying rock could have restricted bed degradation. Below channelization works on the Waren Burn and River Severn bedrock could have prevented incision, whilst large boulders naturally occurring on the beds of the Rivers Caldew, Lynher and Lune could have restricted downcutting, as evidenced by the similarity between actual and reconstructed values of depth (Table XXXVII). Armouring had occurred below only two schemes on the Pickering and Clow Becks, where the original mixed-sediment beds had been destroyed and only cobbles remained. The Rivers Wansbeck, Kent and Ribble all had downstream reaches incised into bedrock and therefore did not undergo change.

Figure 49. Erosion of both channel banks below resectioning works on the River Lune, England. Looking in the direction of flow. (From Brookes, 1987a. Reproduced by permission of John Wiley & Sons, Ltd.)

Figure 50 plots bankfull discharge against water slope for natural reaches in the vicinity of channelization works, with lines of equal specific stream power superimposed. Eroded sites had specific powers within the range 25 to 500 W m^{-2}, whilst the remaining sites without change had specific powers between 1 and 35 W m^{-2}. Channel enlargement was explained in terms of increased flood flows causing higher stream velocities, which in turn caused erosion, thereby increasing channel width and/or depth. Examination of flow records for 35 stations revealed flood events which would formerly have spread overbank but were subsequently confined by the channelization work. At sites with downstream change the energy of increased flows is likely to have been sufficient to exceed the threshold required for erosion of the perimeter sediments. The absence of change at a majority of sites in low-energy lowland areas could be a reflection of both the incompetence of increased flows to erode and resistance provided by perimeter sediments. None of the eroding sites appeared to have attained a new equilibrium.

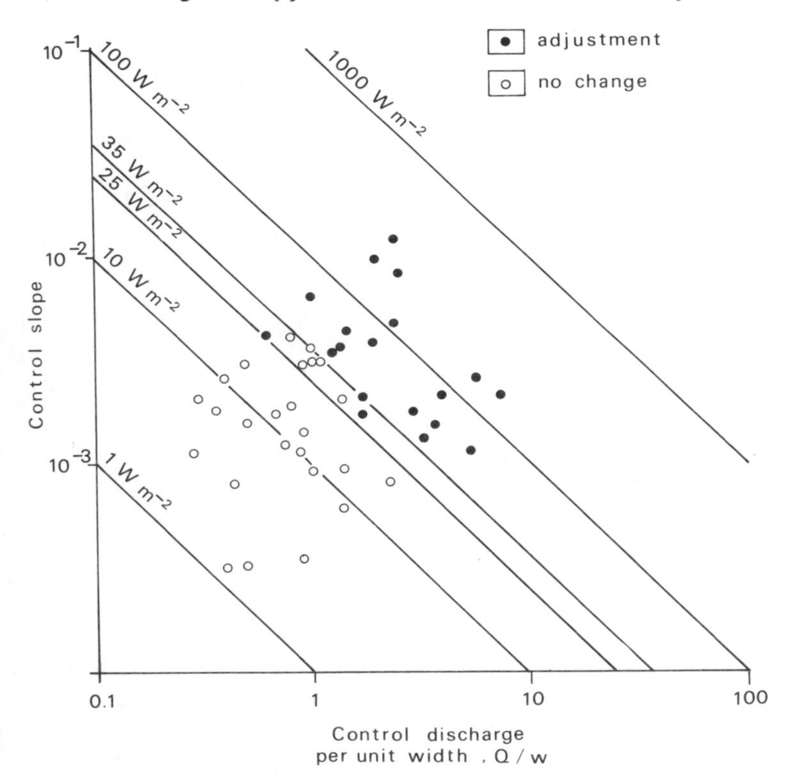

Figure 50. Relationship between bankfull discharge per unit width and water slope for natural control reaches of eroded and non-eroded sites of channelization in England and Wales. Lines of equal specific stream power are superimposed. (From Brookes, 1987a. Reproduced by permission of John Wiley & Sons, Ltd.)

The extent of downstream effects ranged from 120 m on the Pickering Beck to 1950 m on the River Caldew in Cumbria (Table XXXVII). Channel enlargement varied spatially below each channelization work, although there was no well-defined trend in the decline of magnitude of effects with distance downstream. Downstream variability in the amount of erosion was not attributable to local differences in the composition of bank sediments or to longitudinal variations caused by the pool–riffle sequence. Tree roots were a major factor which had locally restricted the recession of banks by binding the sediment. Downcutting, evidenced by destruction of the original substrate and associated bank collapse, was determined to be a significant form of adjustment on the Rivers Ure, Conder, Pickering and Clow Becks, and of minor importance on the Rivers Lune, Ithon and Aln (Table XXXVII). At each of these sites downcutting has been restricted to a short length of channel immediately below the channelization work, where thresholds for the entrainment of bed sediments may have been exceeded.

By utilising the technique of stepwise regression, a series of equations were produced in an attempt to predict the magnitude of downstream effects in the UK (Brookes, 1985b). In one equation 88% of the variation in the dependent variable, the volume of material eroded from the downstream reach, could be accounted for by five easily measured independent variables:

$$\log_e Y_2 = 12.5 - 0.63 \log_e X_1 + 0.33 \log_e X_2 - 2.47 \log_e X_3 + 0.38 \log_e X_4 - 0.008X_5$$

where Y_2 = volume eroded (m^3)
X_1 = excess specific stream power ($W\ m^{-2}$)
X_2 = length of channel works (m)
X_3 = average shear strength of downstream bank sediments ($kN\ m^{-2}$)
X_4 = drainage area at downstream limit (km^2)
X_5 = age of scheme for which effects are being estimated (months).

These variables can be derived from simple field measurement, from standard water authority plans and from Ordnance Survey maps. The excess specific stream power is defined as the difference between the bankfull stream power for the channel existing prior to channelization and the maximum power for the designed channel. This crude variable incorporates all the factors which are altered by channelization, including discharge, roughness and channel slope. The equation has been used to predict the volume of erosion at additional sites with a reasonable degree of accuracy.

A diversion channel, which is used to divert some or all of the flood flow away from the main channel, may have an impact below the diversion where it re-enters the mainstream (Lindner, 1952). If the bedload is unchanged, but the discharge in the mainstream is reduced by the diversion then deposition may occur to reduce the cross-section. Bars may be built and pools filled and deposition will progress downstream. If the bedload is reduced by the diversion without a proportionate reduction in discharge then the bed and banks may erode to enlarge the cross-section. This erosion may progress downstream until the channel attains a new equilibrium (Lindner, 1952).

There remains a requirement for further research throughout the world to identify the range of morphological effects occurring below a variety of different types of channelization works in various environments.

WATER QUALITY IMPACTS

Environmental impact analyses of channelization projects frequently predict temporary increases in sediment loads but do not state the magnitude or significance of those changes. There are relatively few studies which quantify either the temporary sedimentation effects or changes in other aspects of water quality. A review of published studies was made by Shields and Sanders (1986) (Table XXXVIII). It is suggested the effects on particular water quality variables are site specific, reflecting watershed land use, the severity of modification and the length of the recovery period.

Sedimentation

The modification of suspended sediment concentrations as a result of land use changes which occur over a whole basin such as the removal of vegetation for cultivation, forest clearance through burning, afforestation, and the construction of urban areas, has been one of the most studied human impacts (Chapter I, p.3). However, the effects of channelization over a specific length of a river on downstream sediment loads has received less attention. There have been many qualitative observations, perhaps one of the earliest being that described by Lauder (1830) when erosion of a newly made artificial cut on the River Dorback in Scotland turned the waters downstream to a 'liquid yellow mud'. The report prepared by Little (1973) revealed that erosion both during and after channelization caused large amounts of suspended sediment, which subsequently deposited downstream to the detriment of aquatic life. Hill (1976) stated that increased sediment loads downstream from channelization works are probably at a peak during dredging and in the immediate post-construction phase when erosion of unvegetated banks is at a maximum. Parrish *et al*. (1978) found that channelization of streams in Hawaii increased the turbidity. In the Kaneohe Stream on Oahu the suspended sediment load was up to 265 times greater than the natural stream and 5.3 times greater than concentrations found during a flood. Dredging of the River Bunree in Ireland increased silt concentrations from 8 to 79 ppm but decreased as the machine moved upstream. The impact of forest drainage operations upon upland sediment yields has recently been quantified for catchments in the UK (Robinson and Blyth, 1982; Burt *et al*., 1983, 1984).

Below a reach of the Blackwater River in Johnson County in Missouri, the river was reduced in length from 53.6 km to 29 km in 1910, and this caused extensive flooding downstream (Emerson, 1971) according to the local residents. Overbank deposits have caused burial of fenceposts and it estimated that about 2 m of deposition occurred in 50–60 years. Downstream flooding is aggravated by the reduction in channel capacity due to sedimentation.

Table XXXVIII. Impacts of channelization on water quality

Setting	Variable	Effect of channelization	Remarks	Source
Four channelized and four unchannelized North Carolina coastal plain streams in small agricultural watersheds with similar land use and no point sources. Samples 2 years, 26 times/year.	Mean TKN, NH_3-N, TSS, temperature, pH. Mean velocity, DO, Total P, NO_3-N TOC, COD	Nil. (*F* test, $P > 0.10$) Increase. (*F* test, $P > 0.10$) Decrease. (*F* test, $P > 0.10$)	Major effect was increase in NO_3-N (380%), presumably due to loss of swampy riparian areas that attentuate NO_3-N by denitrification and uptake	Humenik *et al*., 1980
Two natural and two modified (various types of development over the last 50 years) streams in SW Minnesota in small agricultural watersheds sampled monthly for 2 years.	Discharge, turbidity, TSS, temperature, DO, alkalinity, Hardness, pH, CO_2	Nil. (*F* test, $P > 0.10$)	Neither magnitude or timing of discharge affected. Turbidity correlated with discharge ($r > 0.70$) for each stream	Marsh and Waters, 1980
Two stations upstream of and two stations within channelized reach of Big Muddy Creek, Kentucky. Last channel maintenance 50 years ago. Single sampling in August, 1974.	NH_3, Total P DO	upstream—0.12, 0.16 mg/litre within—0.16, 0.24 mg/litre upstream—5.7, 6.4 within—6.2, 3.7	Channelized reach received lumber mill sawdust pile drainage	Golden and Twilley, 1976

Four small agricultural watershed streams in central Illinois, two with riparian vegetation, two without. Base flow sampled monthly at 7–12 sites/watershed for 17 months.	TSS	Increased magnitude and variability	Increased TSS due to increased instream organic production	Schlosser and Karr, 1981
	Turbidity	Changed timing of peak		
	Total P	Changed timing of peak		
Black River, North Carolina coastal plain. Monthly samples from three sites for five years. Period covered pre-, during, and post-construction.	TSS, conductance, temperature, velocity, DO, pH, NO_3, NH_3, Org N, Total P, TOC	Increase. (Analysis of spatial covariance)	Increase in DO related to higher velocity and removal of organic substrate. Mean temperature increased 1 °C	Simmons and Watkins, 1982
	NO_2	Nil		
Seven small North Carolina coastal plain streams—four modified, three natural. Two–three stations per stream sampled at 2–6 week intervals for 2 years.	Mean turbidity, color, conductance, temperature, velocity, DO, pH, NO_3, TOC, total coliforms, Pb, Mg, Zn.	Increase. (Wilcoxon matched pairs signed-rank test)	Summer maximum temperatures were about 5 °C higher in unshaded channels	Schlosser and Karr, 1981
	Mean temperature	Nil		
	Fe, Mn	Decrease		

Key: DO, dissolved oxygen; COD, chemical oxygen demand; TSS, total suspended sediment; TOC, total organic carbon; TKN, total Kjeldahl nitrogen; NH_3, ammonia; NO_3, nitrate; NO_2, nitrite; CO_2, carbon dioxide; P, phosphorus; Fe, iron; Pb, lead; Zn, zinc; Mg, magnesium; Mn, manganese; N, nitrogen.

From Shields and Sanders (1986). Reproduced by permission of the American Society of Civil Engineers.

A preliminary study of sites undergoing channelization in south-central England reported suspended sediment concentrations, principally fine silt, downstream up to a maximum of 340 times in excess of those measured simultaneously in the natural channel above each of the works (Brookes, 1983). The concentrations were approximately equal to those found in storm events.

The River Wylye near Salisbury in Wiltshire (UK) (drainage area: 288 km^2) was realigned for a short distance of 95 m to facilitate road improvements over a period of 15 working days in January and February 1982. The spoil excavated from the new channel was used to backfill the old river course. Figure 51 depicts suspended sediment concentrations for the points monitored above and below the reach undergoing channelization, concentrations in the downstream reach exceeding normal levels by as much as 40 times (560 mg l^{-1}). The amount of sediment released was at a maximum during working hours but remained high both overnight and at weekends. Only those activities causing physical disruption to either the bed or banks contributed substantial amounts of sediment. Concentrations were as high during the first three days of construction (January 25–27) initially as a pilot channel was cut along the new course and then as this channel was progressively enlarged. Spoil from these activities was used to build a small embankment on the right-bank. Higher concentrations were recorded on January 28 when spoil excavated from the new course was used to backfill the old channel and was eroded by flowing water. It was not until all flow was eventually diverted to the new course by plugging the old channel that sediment concentrations declined (January 29 and February 1). Between February 3 and 11 the new channel was enlarged and suspended sediment values remained relatively low but higher than those recorded upstream. The new cut had a cross-sectional area 160% larger than the old channel, creating a very deep pool with low flow velocities unable to erode the new channel and remove sediment in suspension to the downstream reach.

Suspended sediment concentrations were found to decline with distance downstream during construction, which could be attributed to localized deposition of sediment and to the dilution effect of tributaries carrying relatively little suspended sediment entering the mainstream. Deposition downstream was at a maximum immediately below the construction site, generally declining with distance downstream, and sediment was thickest in pools. However, deposits were observed to be a short-term phenomenon, with a rapid supply depletion after construction had ceased. The total sediment load released to the downstream reach of the River Wylye as a result of construction was 514 tonnes (312 m^3).

Temperature

The reduction of shade immediately following construction, as a result of removal of trees/undercut banks, rocks and debris, usually results in an in-

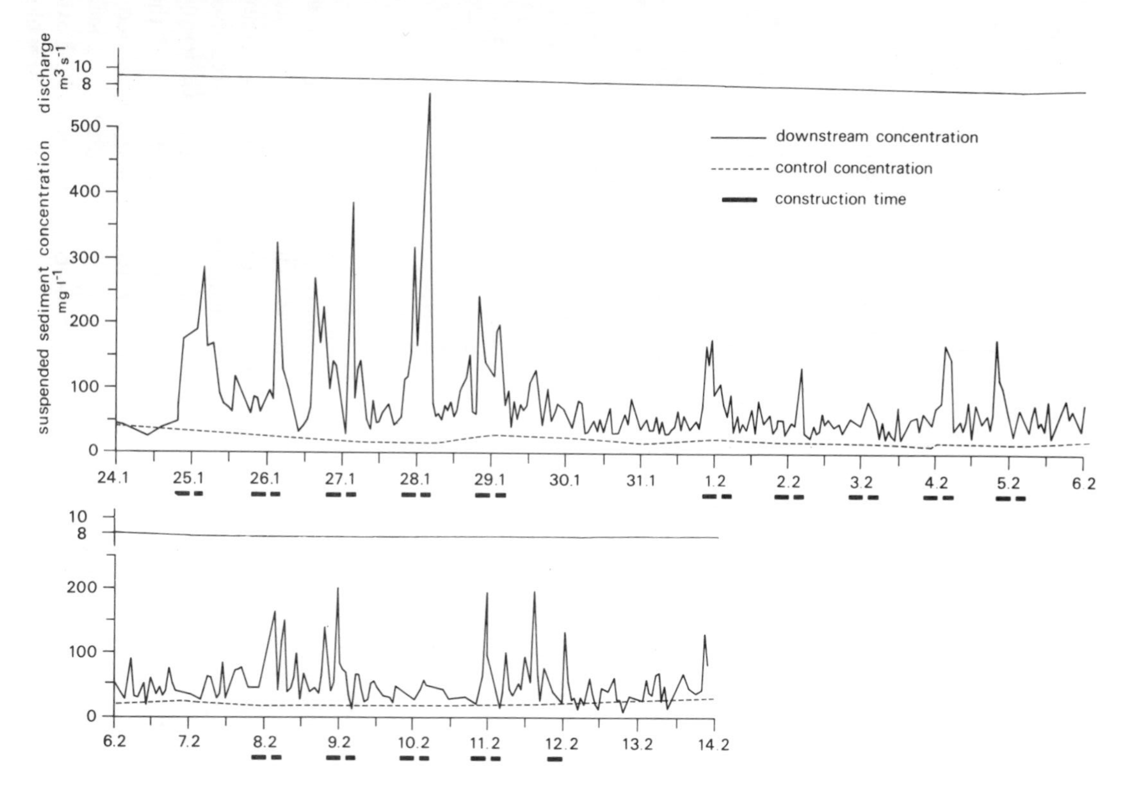

Figure 51. Suspended sediment concentrations measured during construction above and below a short reach of realignment on the River Wylye in Wiltshire, England, 24 January to 14 February 1982. (After Brookes, 1983.)

crease of mean and daily fluctuation of temperature (e.g. Gebhards, 1973; Corbett *et al*., 1978; Simpson, 1981). Duvel *et al*. (1976) indicated that where substantial lengths of modified channel exist a rise in water temperature will occur. Over a 1.6 km channelized length of Fishing Creek, Clinton County, Pennsylvania, the temperature rose by 1.7 °C in July 1975. Schmal and Sanders (1978) found that water temperatures in the Buena Vista Marsh were generally higher in the new and old ditches than in natural control streams which were shaded by vegetation (Table XXXIX). Water temperatures of up to 25 °C were recorded in the newly channelized downstream cold-water ditch, which exceeded the level tolerated by cold-water fish. This increase resulted from the lack of foliage and lower water velocities. For the Yellow Creek in northeast Mississippi, which was relocated over a distance of 9.6 km, the average daily maximum stream temperature was 4 °C greater after construction.

Table XXXIX. Monthly mean water temperatures (°C) for upstream and downstream sites of the Buena Vista Marsh, Portage County, Wisconsin between September 1974 and May 1975

	Upstream			Downstream		
	New ditch	Old ditch	Natural	New ditch	Old ditch	Natural
Mean of daily maxima	8.8	8.4	7.6	7.1	7.2	6.3
Mean of daily minima	4.1	3.5	3.7	4.4	4.1	3.8
Mean of daily temperature	6.1	5.9	3.8	5.6	5.7	4.9
Mean of daily fluctuation	4.7	3.9	3.8	2.6	3.0	2.5

Reproduced with permission from Schmal and Sanders (1978).

Water Chemistry

The oxygen balance and chemical characteristics of artificial drainage channels subject to maintenance have been studied (Marshall, 1981, 1984). However there have been few comprehensive studies concerned solely with channelization. Higher dissolved oxygen (DO) levels can result from the removal of organic substrate, increased velocities and more turbulent flows arising from channelization. Dredging may cut through the oxidized layer of the substrate and expose a deep unoxidized layer (Darnell *et al*., 1976). Sediments removed in suspension from this layer are in a chemically reduced state. They have very high chemical and biological oxygen demands. Toxic materials may be removed such as hydrogen sulphide, methane, organics and heavy metals.

Schmal and Sanders (1978) found that channelization appeared to increase alkalinity and calcium hardness in new and old ditches of the Buena Vista Marsh. In the upstream area, alkalinity, calcium hardness, and specific conductance were significantly higher in the ditches than in the natural stream and the same parameters were generally higher, although not significantly, in the ditches than in the natural stream in the downstream area. Approximately 90% of the stream flow in the Buena Vista Marsh originates as groundwater and therefore surface water hardness reflects these groundwater conditions. Arner *et al.* (1976) reported increases in total hardness in a newly channelized segment of the Luxapalila River in Mississippi and Alabama. By contrast Gebhards (1973) found that pH and dissolved minerals and gases underwent little change. Arner *et al.* (1976) suggested that with the exception of turbidity, water quality may not be adversely affected by channelization. In channelized reaches of the Buena Vista Marsh, water quality parameters such as NO_3, PO_4 and pH were similar to those found in the natural streams (Schmal and Sanders, 1978).

For the Yellow Creek in northeast Mississippi, Shields and Sanders (1986) observed that mean values of specific conductance, turbidity, colour, COD and total alkalinity, hardness, ammonia, phosphorus, sulphate, iron, lead and manganese were 50–100% greater during construction of a cut than before. Estimated average daily loadings of total metal, nutrients and dissolved solids were greater during construction. The changes of water quality at Yellow Creek reflected not only the increased sediment input but also the changed nature of the sediments. The changes in mean dissolved calcium, sulphate, chloride and total iron and manganese levels were probably a reflection of the character of the soils exposed by excavation of the new cut. The slightly higher BOD and COD levels probably reflected the small amount of organic matter present in the exposed soils. The shallow diversion channel may have also caused re-aeration.

Concentrations of faecal coliforms may increase as a result of channelization. Grimes (1975) reported that the disturbance and relocation of bottom sediments by dredging on the Mississippi River navigation channel caused a significant release of sediment-bound faecal coliforms.

Within channelized reaches water quality tends to fluctuate more through time (Kuenzler *et al.*, 1977; Benson and Weithman, 1980; Schlosser and Karr, 1981; Shields and Sanders, 1986). This has been attributed to the reduced travel time as a consequence of modification. Channel maintenance and further work following initial construction may cause further variations in water quality.

Downstream Ecological Changes

Downstream reaches may act as a temporary refuge for fish disrupted by excavation (Luey and Adelman, 1980). Downstream movement of fish from the channelized section of the Little Sioux River in Iowa possibly prevented drastic differences in standing crops of fish between channelized and unchannelized areas (Hansen, 1971a). The effects of inorganic sediments on fish and invertebrate fauna in streams are well documented (Ellis, 1936; Cordone and Kelley, 1961; Chutter, 1969). The majority of these studies deal with increased sediment concentrations arising from activities other than channelization, such as mining (Sumner and Smith, 1939; Cooper, 1956; Hamilton, 1961; Chapman, 1962; Gammon, 1970; Nuttall, 1972; Lewis, 1973a,b) deforestation (Burns, 1972) and road construction (Barton, 1977). Gravels will rapidly become silted in flows with low concentrations of suspended solids (Carling, 1984). Sedimentation simplifies the substrate and reduces species diversity. Productivity in aquatic ecosystems is reduced across all trophic levels (e.g. King and Ball, 1964). The deposition of sand downstream may fill pools and degrade riffles (Jackson and Beschta, 1984), thereby eliminating shelter. The direct effect of increased turbidity on fish has been difficult to quantify, although adult fishes can withstand normal high concentrations without harm. Only prolonged sediment concentrations in excess of 20,000 parts per million cause mortality in adult salmon and trout, primarily by clogging the opercular cavity and gill filaments (Wallen, 1951) and by preventing normal water circulation and aeration of the blood. Other species can tolerate ten times this amount (Apman and Otis, 1965). Whilst such concentrations are rare, the major impact appears to be on the reproductive cycle of fish (Cordone and Kelley, 1961). Sediment can adversely affect eggs and alevins, particularly where deposition of silt is excessive. Deposits may blanket portions of the stream bed thereby eliminating potential spawning grounds and reducing the available food by killing bottom-dwelling organisms. Ellis (1936) found that populations of mussels in the Mississippi River had been much reduced by being smothered by silt. In particular fine organic fractions (<1 mm) accumulate in the interstices of gravel beds (Iwamoto *et al.*, 1978; Beschta and Jackson, 1979). Biological recovery from siltation may occur (Hynes, 1973).

In Brant Creek Drain downstream siltation due to erosion both during and after construction obliterated pools and riffles and affected the fish populations, whilst clearing and snagging of the Wild Rice River in Minnesota gave rise to a large amount of sediment which was detrimental to aquatic life (Little, 1973). Deposits in a downstream reach on the North Fork of Broad River provided an unstable substrate which was unsuited as a habitat for several species (Soil Conservation Service, 1970). The dredging of alluvial materials from the upper Loire, Allier, Cher and Dore Rivers in the Massif Central of France was shown to be detrimental to the benthic invertebrate populations, salmonid eggs and fry. Mortality increased, whilst the numbers and biomass of fish decreased (Clavel *et al.*, 1978; Bouchard *et al.*, 1979). Bou (1977) described similar consequences of dredging the River Tarn, a tributary of the Garonne in France, particularly due to sedimentation of the substrate.

The effects of silting on vegetation are complex (Haslam, 1978). Suspended material may abrade the shoots of plants (Lewis, 1973a) and may also smother the leaves, thereby preventing photosynthesis. Silt may be deposited above gravel, thereby causing plants to root in unstable conditions, or may lead to the burial of aquatic plant communities (Edwards, 1969). When the organic content is high silt may create anaerobic conditions around the roots, although it will usually improve the nutrient status. Anaerobic mud is unfavourable to species such as *Glyceria maxima* (Westlake, 1968c), whereas *Schoenoplectus lacustris* thrives. Rates of deposition in the Norfolk Broads (UK) have increased by as much as 50 times since 1800, owing to the movement of sediment downriver, and this has been an important process in the elimination extensive stands of aquatic macrophytes (Mason and Bryant, 1975; Osborne and Moss, 1977; Moss, 1980).

A study of the impact of suspended sediment on vegetation has been undertaken on four small rivers in south-central England (Brookes, 1986). The Wallop Brook in Hampshire, Ober Water in the New Forest, the River Wylye in Wiltshire and Cale in Somerset were channelized between 1980 and 1982 over lengths ranging from 95 to 1400 m. A comparison was made of the downstream reaches affected by sedimentation with upstream control reaches for periods of up to 24 months.

Individual species of plant responded to sedimentation by either adjusting their rooting levels or alternatively they became smothered (Haslam, 1978). In the Wallop Brook plants were able to adjust to a rate of sedimentation of 130 cm in only 14 days. Plants which could not vary their rooting levels became buried due to the depths of sediment. Rooted plants which were dominant in the Wallop Brook prior to construction, and which were affected by sedimentation, subsequently recovered following wash-out of the sediment. However, other submerged species which could not vary their rooting levels were totally eliminated, the loss was prolonged and site diversity was decreased over a considerable length of time. Emergent plants such as *Sparganium erectum* may already be established in silt prior to construction and may therefore be unaffected by a thin layer of sediment. Plants already growing in the downstream reaches were not abraded by suspended sediment arising during construction. During high flows on the River Wylye, sediment remained in suspension and had no impact on the vegetation.

The effects of sedimentation on standing crops of two dominant species, *Nasturtium officinale* and *Ranunculus* sp., were investigated for the Wallop Brook, a small Hampshire chalk stream in England (Figure 52). The seasonal growth of vegetation in both the downstream reach and a control section upstream are compared with the mean monthly discharge (Brookes, 1986). In the natural reach upstream *Ranunculus* sp. grew from small rhizomes buried in the stream bed in autumn until a maximum was attained in late spring (May) with flowering. This species declined during the summer as *Nasturtium* spread and became dominant, coinciding with a reduced baseflow in summer. By contrast this pattern was different for the downstream reach affected by sedimentation. The cycle was normal until dredging in December 1980, with characteristic recovery of *Ranunculus* sp. as *Nasturtium* washed out by the seasonally increasing flows (cf. Dawson *et al*., 1978). Rapid deposition of sediment in the downstream pool caused a marked decrease of both emergent and submerged plants, such that in January

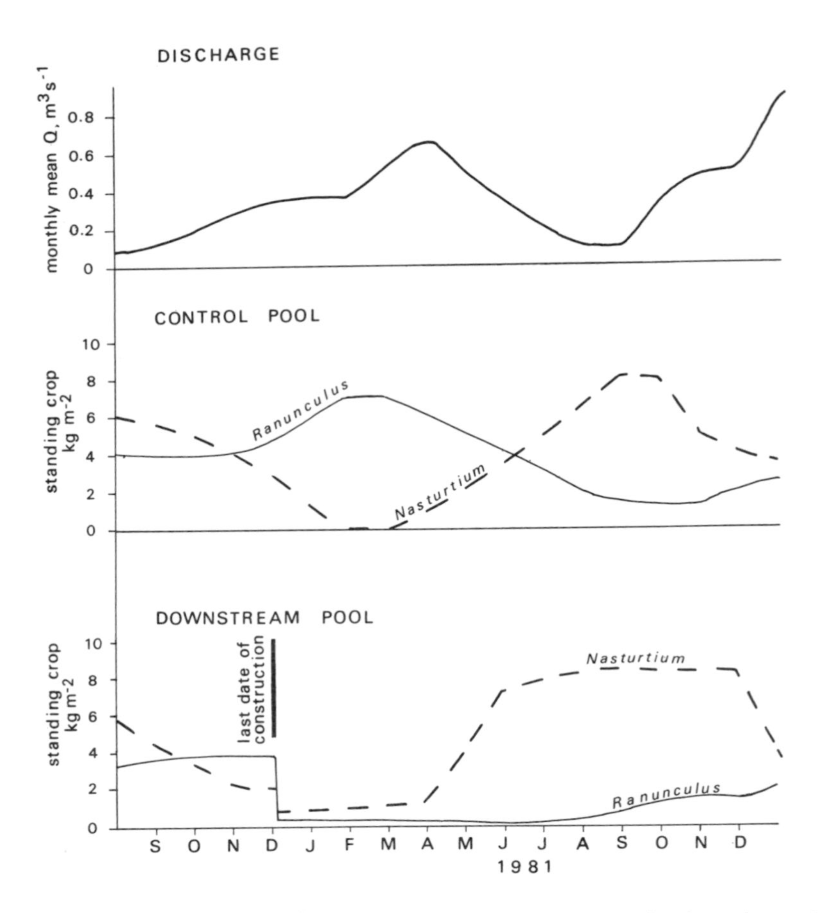

Figure 52. The standing crop of submerged and emergent aquatic plants in pools in the Wallop Brook, Hampshire, England, from August 1980 to January 1982. (From Brookes, 1986. Reproduced by permission of Elsevier Applied Science Publishers.)

the cover of *Ranunculus* was only 5% of that observed in the control reach. This plant could not vary its rooting level in response to sedimentation and was therefore smothered. The standing crop of *Nasturtium* was reduced by only 60% following construction, since the majority of these plants were able to adjust their rooting level. In the downstream pool there was no characteristic rise to a *Ranunculus* maximum in early winter, and only *Nasturtium* survived, subsequently attaining a seasonal maximum after most construction sediment had washed out by May 1981 with rising baseflow discharge. *Nasturtium* grew normally on silt which had been transported from the natural catchment above the works and deposited during the summer of 1981. *Ranunculus* recovered slightly during the summer and then increased normally in the following January after *Nasturtium* had washed out.

Significant deposits of sediment were found only in pools in the first 500 m below the works, and therefore vegetation growing on riffles or further downstream was unaffected. Along Ober Water in the New Forest during the first complete growing season following channelization in 1982, the downstream channel became choked with *Elodea canadensis* over a length of 175 m (Figure 53). This may have resulted from the growth of plant propagules which moved downstream after becoming dislodged from the bed during excavation.

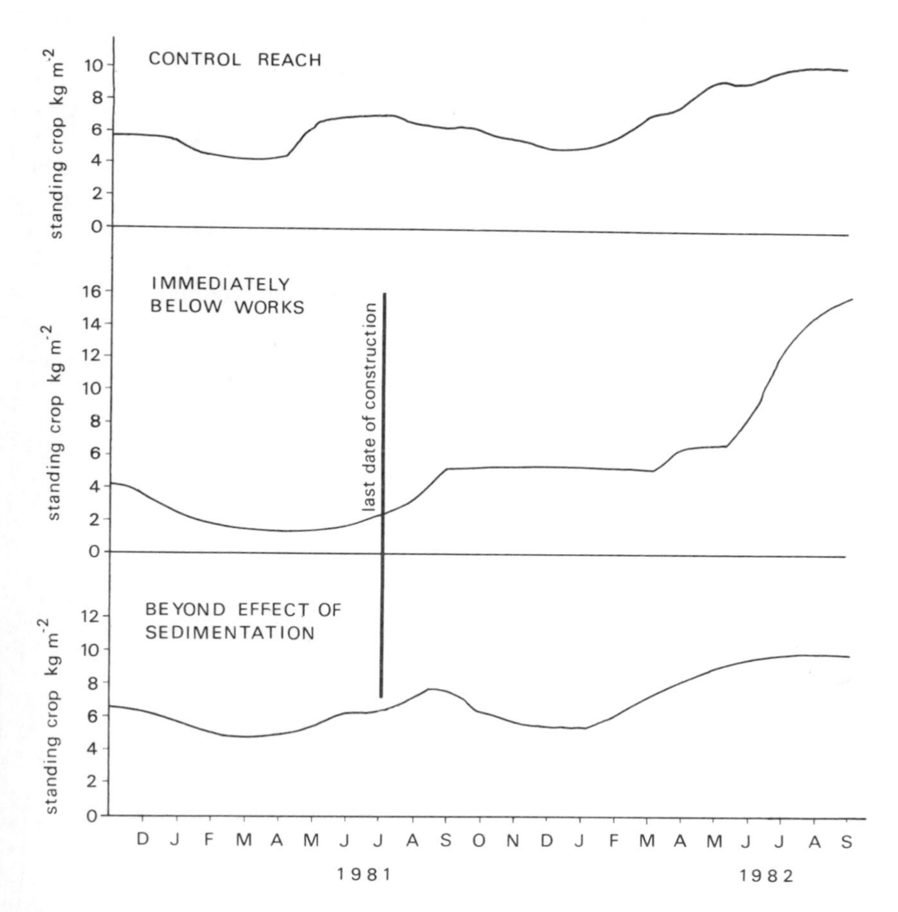

Figure 53. Standing crop of aquatic macrophytes at Ober Water, New Forest, England, from November 1981 to September 1982. (From Brookes, 1986. Reproduced by permission of Elsevier Applied Science Publishers.)

Recommendations for Reducing Sediment Loads

To reduce downstream sedimentation it has been suggested that bed and bank protection is essential in realignments to reduce stream erosion and high sediment loads (Apman and Otis, 1965). Temporary silt traps have been recommended to reduce the undesirably large increases in sediment loads resulting from forest drainage operations (Robinson and Blyth, 1982). Sand bedload sediments in Poplar Creek in Michigan have been removed by instream sediment basins and therefore kept fisheries spawning areas relatively free from sediment (Hansen, 1973; Hansen *et al*., 1983). However sediment basins are not suited to all streams. On steeper gradient streams the removal of the bedload could upset the equilibrium and create excessive degradation further downstream, particularly if there are no controls on the bed (Heede, 1980). They are potentially suited to low gradient streams similar to those found in the midwest of the USA.

A series of recommendations for minimizing the impact of sedimentation on the downstream environment were formulated by Brookes (1986). Sediment is released wherever the bed and banks of a channel are disturbed by machinery in flowing water and can be minimized by: (1) limiting disruption to the bed and banks (ensuring that tracked vehicles are kept out of the channel) and by minimizing the excavation and the length of works; (2) rapidly plugging the old channel at a site of realignment to reduce the duration of contact between water and easily eroded loose sediment used as backfill; and (3) carefully selecting the time of year when the work is carried out. High winter flows may cause more sediment to be released but this may remain in suspension rather than being deposited downstream. The impact on aquatic macrophytes may be least apparent during the winter months when standing crop is at a minimum in most rivers.

Recommendations

CHAPTER VII

Revised Construction Procedures

INTRODUCTION

Following the realization that conventional types of channelization may have undesirable consequences a major development has been a number of studies which have explored alternative methods of restraining rivers, and these have developed around the emphasis upon the need to work with a river rather than against it and to minimize the aesthetic degradation of a river channel and its environs. These studies exemplify the 'design with nature' school of thought proposed by McHarg (1969) and are a means of achieving a 'reverence for rivers' (Leopold, 1977). At the planning stage of channel works it is increasingly possible for the environmental scientist to predict the potential physical and biological impacts. Adopting a basin-based approach to river channel management will identify the causes of channel changes which can affect the capability of a channel to accommodate discharges. Alternative techniques include relatively simple modifications to the design and construction procedures of conventional engineering practices, new techniques which ameliorate certain adverse effects of channelization, or structures or measures undertaken at some time after construction for the same purpose. Several alternatives have been developed by geomorphologists who recognize the river as an open system in which there is a balance between channel form and process and this constitutes part of the concept of 'geomorphic engineering' proposed by Coates (1976b). A greater number of techniques have been developed by biologists, particularly in relation to fisheries, and many of these have become part of the discipline known either as 'bioengineering' or 'biotechnical engineering'. On the basis of both geomorphic and ecological findings many of the methods have been implemented in specific areas of the world by river managers, and in particular structural measures such as low-flow weirs have been incorporated at the design stage of channelization works with the aim of altering channel morphology (Swales and O'Hara, 1980).

Revised construction procedures and the need for a basin-based approach to river channel management are outlined in this chapter. Mitigation, enhancement and restoration techniques are encompassed in Chapter VIII. All of these procedures rely on an understanding of the fluvial processes and forms of natural channels.

Characteristics of Natural Channels

Geomorphological classifications show there to be considerable variation in the type of natural alluvial river channel (Chapter IV p. 82–84). River channel patterns range from almost straight to tortuous meandering, whilst other channels contain occasional islands, or may be classified as braided (Kellerhalls *et al.*, 1976; Palmer, 1976) and this variability may prevent the universal application of an individual alternative design. There are several fundamental components of natural river channels that may be preserved or recreated to reduce environmental degradation in channel works, including pools, riffles and point bars (Keller, 1976; Binder, 1979; Burke and Robinson, 1979; Walker, 1979; Binder *et al.*, 1983). Many of the concepts concerning the behaviour of streams depicted in Figure 54 have been insufficiently used by engineers.

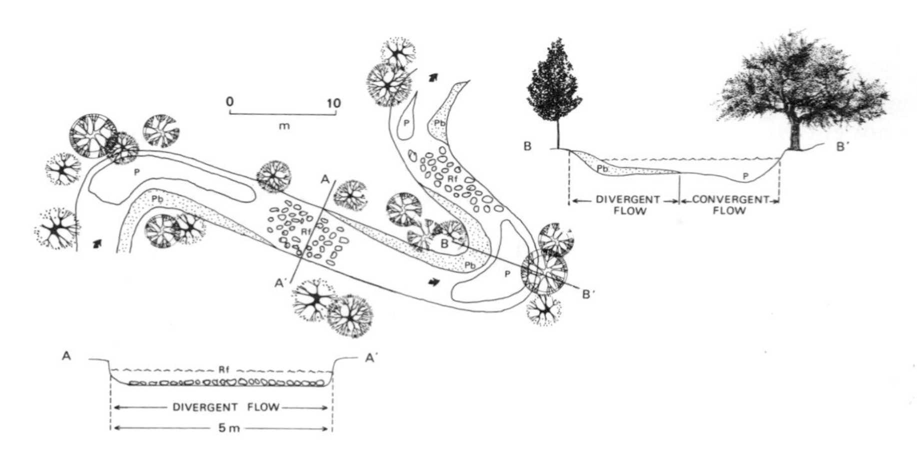

Figure 54. Idealized natural channel.

Morphological Diversity

In natural channels stream flow varies considerably over short distances and temporally over the range from low to high discharge. Flow is characteristically unsteady and non-uniform and nowhere are streamlines parallel to each other or to the stream banks. Leliavsky (1955) proposed a convergence–divergence criterion whereby at very high discharges convergence of stream flow is associated with scour, and divergence of stream flow corresponds to deposition. This criterion may be used to explain the presence of bedforms in stream channels which may result from the interaction of flowing water and mobile sediments. In a sinuous river with a gravel bed there are permanent morphological irregularities, which remain stable over a range of flows and are self-maintaining (Figure 54). The following features are important components of a natural alluvial channel:

(1) *Pool*. A pool is a topographically low area created by scour and corresponding to convergent flow at high discharges. It is generally located immediately downstream from the axis of a bend and is characterized by relatively deep, slow-moving water at low flow. The bed material is usually composed of fine-grained sand.
(2) *Riffle*. This is a topographically high area corresponding to divergent flow at high discharges. The bed material is composed of a concentration of larger rock sizes, often gravel. At low discharges the flow is fast and the water surface gradient steep. The cross-section is typically symmetrical.
(3) *Point bar*. The inner side of a bend is typically an area of deposition in contrast to the erosion of an outer bank. Accumulation of material forms a point bar adjacent to a pool, producing an asymmetrical profile.
(4) *Floodplain*. A natural river channel and adjacent floodplain are parts of a single-system. Rivers overflow their banks on average about once per year and this process may be important in building a floodplain by deposition. Streams with sufficient power erode laterally across the floodplain which is created and continuously modified by the processes of erosion and deposition. Floodplains also include the flat 'bottomlands' adjacent to the banks of rivers in the USA.
(5) *Bank vegetation*. Bank vegetation can be added to this idealized model, which provides shade to the channel and bank stability (Zimmerman *et al*., 1967) and organic debris (Mosley, 1981; Keller and Swanson, 1979) (Figure 54). The preference of fish for stream areas with protective cover has been recognized for a long time (Tarzwell, 1932, 1937; Boussu, 1954). Bankside vegetation prevents excessive illumination and water temperatures. Boussu (1954) demonstrated the relationship between trout populations and cover on Trout Creek in Gallatin County

Table XL. Aspects of the fluvial system which should be incorporated in alternative designs (after Keller, 1976)

(A) The channel and floodplain together form an open system tending towards a dynamic or quasi-equilibrium in which there is an approximate balance between the load imposed and the work done such that channel form and fluvial processes are interdependent.

(B) Flow in natural streams is characterized by alternating convergence and divergence of flow that facilitates morphologic stability, development of pools and riffles and channel maintenance.

(C) Geomorphic thresholds exist which partially control erosion, deposition and channel patterns.

(D) Complex relations between erosion, deposition and sediment concentration influence channel stability.

in Montana. By applying artificial brush cover the total biomass of fish increased, whilst biomass decreased in those sections where the natural brush cover (sedge and willow) was removed. Table XL summarizes the key aspects of the fluvial system which should be considered when planning and designing alternative procedures.

Recommendations which Minimize the Adverse Effects of Conventional Engineering Procedures

The adverse effects of conventional channelization methods may be minimized by careful selection of options at the planning stage, or by limiting the degree to which a channel is modified or maintained.

Option Selection

The selection of engineering options may depend on the type of river channel for which a solution is sought. Unfortunately few river management classification systems have been developed. One exception is the 'Bauer geohydraulic river zones' classification developed for rivers in Washington and Oregon (Palmer, 1976). A number of zones are recognized which have characteristic combinations of valley cross-section, channel pattern, gradient, and bedload size. The boulder zone is a river environment adjusted to transport water and sediment through a steep, resistant 'V'-shaped valley. The river is self-maintaining and intrusions such as landslides and rockfalls are soon flushed out. By contrast the floodway zone is more problematical since high energy and low bank resistance combine to make a dynamic channel. Efforts to straighten the channel will intensify the velocities and create adverse effects. In the pastoral zone the channel is more stable because of the cohesive strength of the finer sediment load and a gentler gradient. The rate of meander migration in this zone is considerably less.

In many countries the selection of a particular engineering option is determined principally by economic, social and political constraints but increasingly with regard to environmental impacts (Chapter III). Consultation procedures between engineers and environmental interests have been in existence for several years (e.g. Wessex Water Authority, 1976; Water Space Amenity Commission, 1980a,b). Interdisciplinary appraisal of engineering proposals and environmental impact assessments are now routine in many countries (Thames Water Authority, 1988b; Brookes and Gregory, 1988). Table XLI lists the conventional engineering practices most commonly used on alluvial rivers, tentatively ranked in order of increasing environmental impact. These guidelines were promulgated by the US Soil Conservation Service and the US Fish and Wildlife Service to guide personnel in complying with the National Environmental Policy Act of 1969 (Soil Conservation Service, 1977b). Embanking may have the least impact where banks are constructed of imported materials and the original channel is left intact. An engineering option should be chosen only after an examination of local channel characteristics and drainage basin conditions. By using data obtained from case studies it is possible to give an approximate indication of the occurrence and nature of morphological adjustment to be expected from channelization in particular regions of the World (Schumm *et al.*, 1984; Brookes, 1984, 1987a,c).

Table XLI. Types of channel modification listed in ascending order of impact on fish and wildlife resources (after Soil Conservation Service, 1977b)

1. Riprapping (placement of rock as bank protection)
2. Selective snagging (selective removal of objects such as fallen trees)
3. Clearing and snagging (removal of debris such as shoals and vegetation)
4. Widening (enlargement of channel by widening)
5. Deepening (enlargement of channel by deepening)
6. Realignment (construction of a new channel)
7. Lining (placement of non-vegetative smooth lining)

REVISED CONSTRUCTION PROCEDURES

Realignment

The biological impact of realignment can be reduced if guidelines relating to the design, construction and 'clean-up' phase are followed (McClellan, 1974). The Highways Administration in the USA produced a series of guidelines in 1979, describing the restoration of fish habitat in relocated streams (Federal Highways Administration, 1979). At the design stage there should be minimal reduction of channel length, the amount of excavation and fill should be controlled and equipment which minimizes destruction of bank and streamside growth should be used. Banks should be replanted wherever possible and riprap placed such that the growth of vegetation close to the stream edge is not impeded. During construction, access by vehicles should be strictly controlled and minimal disruption to the stream bed and banks can be attained by educating foremen and specifying the types of equipment that can be used in particular areas. Finally, in the 'clean-up' phase it is recommended that gravel and larger rocks are replaced in the stream bed to approximate conditions existing prior to construction and to restore stability. Replanting and/or reseeding of banks with native trees, plants or grass provides shelter and cover for wildlife. Table XLII summarizes recommendations for reducing the impact of realignment.

Table XLII. Recommendations to reduce the impact of realignment—based on a procedure carried out by the Oregon State Highways Department (after McClellan, 1974) and the guidance provided by Yearke (1971) and Brice (1981)

1. *Feasibility/planning*
   The existing channel for $\frac{3}{4}$ km or more upstream should be examined with particular attention given to the material within the channel and its susceptibility to erosion, in order to attempt prediction of the reaction after construction. This provides information on the need for erosion-control measures. Aerial photographs can be used to determine if the channel prior to relocation is unstable (e.g. slumped banks, fallen trees and wide point bars). Vertical instability can be assessed at structures such as bridges.

2. *Design*
   The engineer should attempt to duplicate the hydraulic properties of the natural stream in the relocated section. Since significant steepening of the channel grade will usually result in degradation, every effort should be made to approximate the slope of the natural channel through the channel relocation. In long relocations this may be achieved by the introduction of artificial channel meanders.
   (a) minimize the reduction of channel length; as a rule of thumb erosion may occur if relocation is over a distance greater than 250 channel widths.
   (b) carefully control the amount of excavation and fill.
   (c) the width change factor (ratio of bottom widths of relocated to natural channels) is important—a factor of 1 to 1.4 will be adequate to offset any increase of slope.
   (d) a trapezoidal cross-section of slopes less than 2 : 1 can be expected to contribute to stability.
   (e) employ construction methods and equipment that will minimize the destruction of bank and streamside growth—make provision in the contract for replanting trees.
   (f) place riprap such that the growth of vegetation close to the stream edge is not retarded.
   (g) avoid bends at the junctions of the relocated and natural segments.

3. *Construction*
   (a) control equipment so that least disruption to the channel bed and bank occurs.
   (b) specify the types of equipment that may be used.
   (c) educate the foremen regarding the need for minimal damage to the stream.

4. *Clean-up Phase*
   (a) replace large rock and gravel in the stream bed to restore stability; install in positions existing prior to construction.
   (b) replant or reseed banks with native trees, plants and grass to provide cover for wildlife.

5. *Maintenance*
   Routine maintenance is required to control overhanging trees; bank collapses etc. which reduce the conveyance of the channel.

Van Bendegom (1973) described conceptually how shortening may be undertaken with minimum geomorphological change. A straightened river which has not been dredged or protected is free to degrade its bed and banks. Excess energy may be expended by incision, which occurs over a vertical distance Δz, where Δz approximates the length by which the river is shortened times the original slope. To avoid deep incision followed by re-meandering a cutoff can be dredged so as to lower the bed immediately by an amount Δz over the entire reach. Degradation can also be limited by the addition of a sufficient amount of coarse material to the channel.

A method of preventing further incision of a straightened reach of the River Sur in Upper Bavaria, West Germany, utilized 'gravel ramps' at selected points in the channel (Geiger and Schröter, 1983). These ramps were cheaper than using complicated weir structures and they prevented downcutting with the result that the banks became stabilized and were colonized by nesting birds.

Meander Preservation

Where the location of a channel has to be moved then meandering alignments may be more expensive to construct than straight channels because of increased excavation costs. However, environmental benefits and reduced maintenance costs may offset increased construction costs over the life of the project. Well-designed meandering channels are more stable, provide a greater variety of flow conditions and aquatic habitat diversity, and are aesthetically more pleasing (Keller and Brookes, 1984). To design a suitable alignment it is recommended that the existing meandering geometry and slope be used as a guide, based on field surveys, maps or aerial photographs. The size, shape, meander geometry, slope and bed roughness should be similar to the old channel.

Channel Enlargement

Enlargement of channels by modifying only one bank, and leaving the opposite bank almost entirely untouched, is now a commonly used practice in many countries (Shields, 1982a,b). Vegetation on the opposite bank is disturbed as little as possible, although potential obstructions to flow such as individual trees may be removed. The bank from which the work is undertaken can be designated on the basis of habitat value of the vegetation, aesthetics, shade and bank stability. If work is alternated from one bank to the other the aesthetic appearance may be improved and this enables avoidance of sensitive habitats. Retention of tall vegetation will shade out aquatic vegetation and thereby reduce maintenance costs. If the channel is widened then clearly all the vegetation on the working bank will be lost. However, where widening is not significant then it is beneficial to retain as much vegetation as practical. Damage can be minimized by using small equipment and disturbed areas should be revegetated. The use of waterborne machinery ensures minimal removal and damage to trees (e.g. Weeks, 1982).

The impact of excavation on the aquatic vegetation has been shown to be minimized by avoiding the creation of very deep pools which may serve as silt traps or preclude light from reaching the channel bed (Brookes, 1987d). It is also recommended that excessive widening is avoided since this is likely to reduce the depth of water in a channel for a given discharge and thus limit the space for growth of vegetation.

It is suggested that pools and riffles should be preserved wherever possible. Griswold *et al.* (1978) reported that a 35 km channelized section of the Little Auglaize River was completely dry for 2 months, and one recommendation was that unaltered areas within sizeable channelization projects should be retained to serve as biological refuges during periods of drought.

The preservation of the substrate is critical for the macroinvertebrate fauna. Alternatively the original substrate can be stockpiled and reinstated after excavation has been completed. For the Afon Gwyrfai in Wales (UK) recolonization of a gravel bed was a gradual process, taking about 1 year to complete (Brooker, 1982). The stability of the retained gravels is important: if the gravel is too fine, and therefore unstable, then species diversity and abundance will be less. For maximum diversity of species it is necessary to preserve or recreate morphological diversity, the variations of converging and divergent flows associated with a pool and riffle sequence, and a variety of substrates. Retention of stands of aquatic plants along the margins of a channel will be beneficial to the diversity and stability of macroinvertebrate habitats. Limiting the removal of bankside vegetation, including trees, will allow organic litter input to the stream, which is an important food source. Preserving bankside vegetation will also avoid excessive water temperatures and the luxuriant growth of aquatic vegetation which may otherwise limit habitat diversity. Since drift appears to be an important source of recolonization of animals dredging upstream is likely to have less impact than working in a downstream direction. The timing of operations is very important: the least damaging time for dredging (in the UK) is probably in late spring/early summer because recolonization can occur during July/August with breeding shortly afterwards (Pearson and Jones, 1975).

A number of recommendations have been formulated for minimizing the impact of excavation by widening and/or deepening on the aquatic vegetation growing within a channel (Table XLIII). The preservation/recovery of vegetation may be desirable for ecological or aesthetic purposes.

Embankments

Embankments often lack ecological, recreational and aesthetic values, except for incidental fish and wildlife benefits that are derived from borrow pits (Nunnally *et al.*, 1987). Typically embankments have a uniform and monotonous appearance and maintenance guidelines specify that grass

Table XLIII. Preliminary recommendations for minimizing the impact of channelization on the aquatic vegetation

1. Where possible try to recreate morphological characteristics similar to the channel before excavation, especially the substrate. This can be achieved by minimizing, deepening or restoring the substrate.
2. Where deepening is necessary avoid the creation of very deep pools which preclude the growth of vegetation or may function as traps for silt passing from upstream.
3. Do not create an excessively wide channel since this will reduce the depth of water in a channel for a given discharge and thus limit the space available for the growth of vegetation.
4. Avoid the use of artificial linings to the bed wherever possible.
5. Removal of trees for machinery access may encourage the abundant growth of plants, and it is therefore recommended that trees be selectively removed. Retention of trees may also promote bank stability. Replanting of trees following completion of works may further shade the channel and limit plant growth.

From Brookes (1987d). Reproduced by permission of Academic Press, Inc. (London) Ltd.

should be closely mowed. It is only recently that embankments, particularly those of the US Army Corps of Engineers, have been designed, constructed and maintained with environmental considerations in mind (Hynson, 1985). Table XLIV lists some guidelines for environmentally sensitive flood bank construction as used on larger rivers by the Corps.

A list of the principal guidelines adopted on smaller watercourses in Britain is presented in Table XLV.

Lined Channels

The lining or paving of channels creates a very unnatural bed and usually accompanies channel enlargement and/or straightening. Measures may be taken to minimize the adverse impacts including the placement of suitable bed materials and the retention or planting of vegetation. A curved alignment could be used. Concrete channels have been alternated with short lengths of natural channel which provide an acceptable habitat for fish. In Hawaii these have been termed as 'rest stops' and permit fish migration in an otherwise impossible situation (Parrish *et al*., 1978). A general review of environmental considerations of bank protection is provided by Henderson and Shields (1984) and sensitive methods have been applied to a number of areas (e.g. Platts, 1981; Schnick *et al*., 1982). Instream habitat devices can be installed to improve hydrologic/morphologic variability (see Chapter VIII, p.210).

The choice of materials for bank protection is important from the environmental viewpoint. Rigid linings such as reinforced concrete, grouted riprap, bagged cement, and filled mats and membranes, have perhaps the most detrimental effect on the aquatic habitat. By contrast, riprap of cobble

Table XLIV. Environmental considerations during the planning and design of US Army Corps of Engineers levees (after Nunnally *et al.*, 1987)

Feature	Procedure	Advantages
Selectively preserve trees	old or unusually sized or shaped trees and trees with unique wildlife value for food, resting or nesting should be preserved	maintains scenic and ecological values; breaks up monotony of embankments; need for revegetation is minimized
Overdesign of internal drainage ditches	by carefully selecting an appropriate value for hydraulic resistance	reduces the need for frequent maintenance and supports wildlife habitat
Borrow pits from which spoil is obtained	preserve; manipulate depth to either foster or discourage plant growth. Plant trees; retain islands; recreate marshes	useful as a habitat for birds; aesthetically pleasing; use as fish ponds
Seeding and planting	careful choice so that roots do not affect the structural integrity of the bank	provides habitat
Maintenance	selective management of vegetation	increased diversity

Table XLV. Guidelines for environmentally sensitive flood bank construction (as used on watercourses in Britain)

1. Flood bank should be set back from the immediate edge of the channel, especially if the river bank is to be protected

2. It may be desirable in other circumstances to create a shallow wet marginal shelf for reeds, etc. Where the bank material is cohesive, excavate the edge of the bank to provide spoil for the embankment

3. Retain trees and shrubs on the inside of a new flood bank

4. Enhance habitats for wildlife and visual appearance by planting reed beds or trees and shrubs between the river and embankment

5. Where annual maintenance is required use should be made of herb mixes and slow-growing grasses; trees and shrubs would be inappropriate

or rubble, gabions, gravel armouring, grasses and woody vegetation are more desirable (Mifkovic and Petersen, 1975; Klingeman and Bradley, 1976; Maynard, 1978; Simons *et al*., 1979). Although these linings do not provide a habitat for the same benthic community as the natural channel, they are much better than concrete and can be just as effective in preventing erosion.

Several authors have described experiences of using vegetation as a means of bank protection (e.g. Parsons, 1965; Whitehead, 1976; Bowie, 1982). For example, Keown *et al*. (1977) demonstrated that grass can reduce velocities at a boundary layer by as much as 90%. Vegetation can be used alone or in conjunction with structural protection but it is imperative that it becomes established before the next flood (see Chapter VIII, Biotechnical engineering, p.229). Experience with enhancement techniques, particularly in North America, has revealed the advantages of riprap, composed of natural rock or quarry stone, as opposed to more conventional stabilization methods (Keown *et al*., 1977; Fernholz, 1978; Nunnally and Keller, 1979; Keown, 1981). Vegetation can become established between the stones, particularly where a soil cover has been applied. It may also provide a stable substrate for benthic invertebrates and the weathering of stones may produce gravel beds suitable for fish spawning (Francis *et al*., 1979). Increased primary productivity and abundance of invertebrates have been associated with riprap (e.g. Hansen, 1971a,b). Design procedures for riprap are listed by a number of authors (Anderson *et al*., 1970; Myers and Ulmer, 1975; Richardson *et al*., 1975).

The length of bank protection measures should be limited to the absolute minimum required for stabilization. One simple technique is the 'Principle of Velocity Reflection' of Leliavsky (1955) which has been applied to determine the likely locations of scour in meanders (e.g. Nunnally and Keller, 1979; Brookes, 1987b). Examples of the reflection of velocities at different-sized bends, producing scour and deposition, are shown in Figure 55.

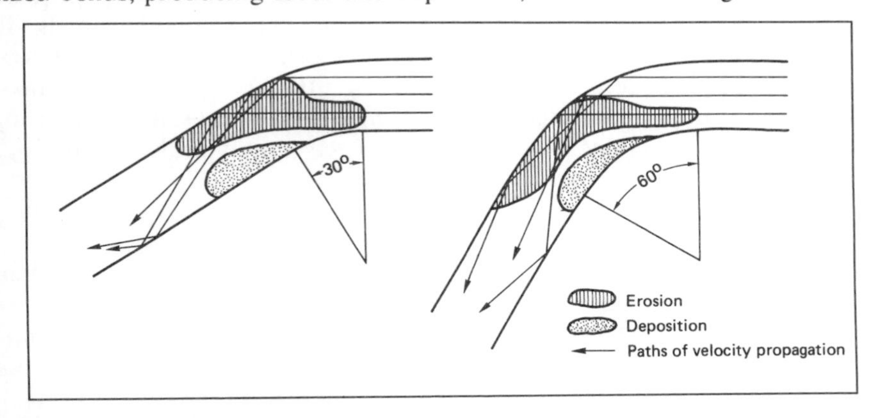

Figure 55. Use of Leliavsky's principle of velocity reflection to determine where riprap protection is required. (After Nunnally, 1978a. Reproduced by permission of Springer-Verlag, New York Inc.)

Dike Fields

Environmental guidelines for dike fields are available (e.g. Shields, 1983b). Whilst the major objective of dikes has been to stabilize long lengths of rivers such as the Missouri and Mississippi, recent studies have shown that dikes can provide an extremely valuable habitat for fish and macroinvertebrates (e.g. Bingham, 1982; Burress *et al.*, 1982). The main problem is to design dike fields which do not fill with sediment. This can be achieved by varying the length and height, but constriction gaps or notches in dikes are the most widely employed environmental feature at present (Shields, 1983b). These allow water to flow through the dike at intermediate stages and prevent sediment accretion by scouring. Detailed notch design affects the performance (Shields, 1983b). A variety of notch widths, shapes and depths are recommended through a reach to provide spatial and temporal habitat diversity. Notches should be wide enough to develop the desired habitat, yet not so wide that erosional damage occurs. Notch depth similarly affects the extent of scour downstream. Shields (1983b) advocated the monitoring of dikes with environmental features over long periods (10–20 years) to refine the designs.

Revised Maintenance Procedures

Given the literature on the morphological adjustments of engineered reaches in a wide variety of environments (Chapter IV) it should be possible to tentatively forecast the type of maintenance required at a particular site. Therefore maintenance implications and the possible adverse effects on the recovery of a channel could influence the initial design of a channelization project.

Environmental impacts may be considerably reduced by partially maintaining channels. The results from studies of selective channel enlargement, described above, apply equally to partial dredging. Recovery of channels following major works may occur in the absence of maintenance, whilst conventional maintenance practices, which disrupt the channel bed and banks, may be substituted by less detrimental techniques such as selective clearing and snagging and shading. Careful timing of the activity is also important.

Biological Recovery

Biological recovery from traditional channelization practices is dependent upon an improvement in the habitat, which may occur in the absence of maintenance (see also Chapter V, pp. 140–144). The potential environmental consequences of maintenance following construction are rarely considered at the design stage of major river works. One exception is a project on Badger Creek in Colorado which was intentionally designed, using gabion control structures, to allow some adjustment of the channel bed following excavation (Jackson and Van Haveren, 1984).

A study of 42 sites of channelization in North America revealed how maintenance destroys the habitats of various organisms by removing the bankside vegetation, encouraging bank instability and preventing the development of a stable substrate (Little, 1973). Recovery without the intervention of man may be a very slow process, taking between 50 and 100 years (Corning, 1975; Arner *et al.* 1975). Arner *et al.* (1975) found that there had been relatively little recovery of fish productivity even 43 years after channelization. Tarplee *et al.* (1971) suggested that species diversity increased with corresponding increases of cover, and that recovery may be achieved in 15 years provided that no further alterations of the stream bed, bank, forest canopy or aquatic vegetation occur. After 30 years a resectioned reach of the Chariton River in Missouri had reverted from a uniform width and depth back to natural conditions with meanders and pools, whilst the deposition of logs increased the instream cover (Congdon, 1971). Channelized rivers in Iowa of between 10 and 15 years age had as many fish as natural sections where brush piles accumulated in the channel in the absence of maintenance (Bulkley, 1975a,b). Seasonal peaks in species diversity of fish may attain levels typical of unmodified streams, but the stability of a fish community may be lower because seasonal changes in stream quality are high in disturbed reaches (Gorman and Karr, 1978). The amount of recovery of 18 realigned channels in Oregon depended on climate, topography and time elapsed since construction (McClellan, 1974).

Selective Clearing and Snagging

Clearing and snagging may be defined as the removal of woody vegetation and debris from stream channels and banks to preserve flood capacity, minimize erosion or maintain navigation. Traditionally all vegetation has been removed to the detriment of the environment and hence a number of alternative procedures have been developed (McCall and Knox, 1978; Gregory and Stokoe, 1980; Shields and Nunnally, 1984). These alternatives focus on limiting the amount of vegetation which is removed, or on using techniques which minimize disruption to the environment. It has been recommended that clearing and snagging should only occur at localized points where significant blockages occur (Table XLVI) (McConnell *et al*., 1980). Logs should be removed only if they obstruct flow, causing upstream ponding and/or sediment deposition, or if they are free. Logs which are rooted, embedded or waterlogged in the channel or floodplain and not causing blockage problems should not be removed. Rooted trees should be removed only if they are leaning over the channel at an angle of 30 ° or more, or are dead and are likely to fall into the channel within one year and create a blockage to flows. Very small accumulations of debris or sediment need not be removed, unless they obstruct flow to any significant degree.

Hand removal or hand-operated equipment (e.g. a power saw) is preferred for the removal of logs; alternatively a small water-based crane or small crawler tractor with winch can be used. Access routes should be carefully selected to avoid damage to floodplain vegetation. Rooted trees should be cut well above the base, leaving the roots and stumps undisturbed. Cut logs should be cleared away from the floodplain to avoid the risk of them re-entering the channel through over-bank flows.

Table XLVI. Summary of clearing and snagging guidelines for the Wolf River System

I.	General guidance	Minimum clearance should only be at locations where absolutely necessary.
II.	Materials to be removed	Only log jams which cause a sufficient blockage should be removed. Affixed logs should be retained unless impeding the flow; free logs can be removed; rooted trees should not be touched unless they have been severely undermined. Debris accumulations, sediments and soils should only be removed if they cause a significant blockage.
III.	Work procedure and equipment	Hand-operated machinery is preferred; water-based equipment (e.g. crane on barge); small crawler tractor with winch; access routes should be selected for minimal disturbance to the floodplain (i.e. no trees to be cut). Stumps of rooted trees adjacent to the channel should be left. Logs should be disposed well away from the floodway to prevent re-entrainment. Burning is desirable. Sediment blockages: careful use of machinery; spoil banks should not be continuous.
IV.	Reclamation	All disturbed areas should be reseeded or replanted.

Adapted from McConnell *et al.* (1980). Reproduced by permission of the Soil and Water Conservation Society.

The 'George Palmiter River Restoration Techniques' are based on the experience of Mr Palmiter of Montpelier, Ohio (Herbkersman, 1981; Willeke, 1981). The techniques were developed principally to make small rivers navigable for canoes and small boats by removing log jams by either power-driven or hand-held saws, protecting eroding banks by either brush piles or logs anchored to the bank with rope or wire and the removal of accumulated sand and gravel. Where the removal of trees is necessary then the stump should be left intact to preserve bank stability. A vital component of the work is careful management which indicates a survey of river problems, preparation of a work plan, supervision and documentation of the work and preservation of good public relations.

The George Palmiter techniques are presented as an information package, designed to help understanding and application. The package contains a training manual, three tape/slide programmes and a videotape from the Institute of Environmental Sciences of Miami University, Ohio. The first programme, 'Let the River Do the Work', is an overview which can be used at training sessions or public meetings, whilst the second programme 'The Palmiter Techniques of River Restoration', explains in detail the steps involved. A third programme, 'How to Manage a Palmiter River Restoration Project', describes the management duties.

Revised Dredging Practice

Working from one bank is important to preserve wildlife habitat by retaining trees on the opposite bank. However, selective or sensitive maintenance is desirable to limit the impact on the aquatic flora and fauna.

Very large numbers of crayfish can be removed during routine maintenance which involves the dredging of silt and marginal plant species. The destruction of their habitat also has a drastic effect on the crayfish populations, and recolonization of such areas can be slow. To overcome these problems, Hogger and Lowery (1982) recommended that the animals be washed from the dredged material by submerging the excavator bucket in the river. On the River Beane near Hertford (UK) this was shown to reduce the number of crayfish removed by about 35%. It was also suggested that if only one bank is excavated then those animals returning to the river have only a short distance to migrate to an undisturbed habitat.

Selective Weed Cutting

Since tall plants are used as shelter by birds or mammals cutting should not be undertaken during critical periods (e.g. during the nesting period) and the cutting should be done along selective reaches so that the animals can find cover. The presence of macrophytes substantially increases the habitat and cover for macroinvertebrates and fish. The greater the variety of macrophytes, the greater the diversity of habitats. The wholesale removal of macrophytes is traumatic and often destabilizes the system (Dawson, 1986). Frequently plants which are non-invasive in their habitat have been cut unnecessarily and this should be avoided. Sensitive management of marginal vegetation involves preserving existing stands of plants where possible. Kern-Hansen and Holm (1982) recommended cutting along the thalweg of a channel, leaving bars along the banks, which can be used as a food source and shelter for macroinvertebrate and fish communities. The roughness effect is greatest when vegetation is centrally positioned and therefore retention of fringes of marginal vegetation may be more desirable from the hydraulic viewpoint. Table XLVII summarizes some of the alternative weed cutting procedures.

The effects of weed cutting have been studied on chalk streams in Britain. It has been suggested that a spring cut of *Ranunculus* sp. during the growing season may in fact stimulate further growth rather than control it (Westlake, 1968a; Westlake and Dawson, 1982; Dawson, 1986) whilst the vigorous

Table XLVII. Weed cutting recommendations

Method	Details	Comments
Partial cutting	25% of the cover should be left; cutting can be in the centre of the channel with marginal beds left untouched	conserves plants, invertebrates and fish (Swales, 1982c)
	suggested 50% of biomass should be preserved as bars across the channel	could be detrimental to drainage interests (e.g. Kern-Hansen, 1978; Nielsen, 1985)
Selective clearance	filamentous algae cleared in preference to ecologically benefical species	Swales (1982c)
Timing	spawning of phytophilous fish species may coincide with cutting in May/June. Suggested that spawning areas should be avoided	Mills (1981)
Alternative control	use grass carp *Ctenopharyngidon idella* but there may be effect on native fish	Stott and Robson (1970); Stott (1977)
Alternative control	control nutrient levels responsible for growth (e.g. fertilizers and sewage)	Mitchell (1974)
Alternative control	shade out plants	Krause (1977); Dawson (1978)

regrowth of vegetation following a first cut has been noted on the River Lambourn in Berkshire (UK) (Ham *et al.*, 1981). Cutting on a Hampshire (UK) chalk stream changed the percentage cover of each species compared with the previous year and the growth pattern of the most dominant species. The cutting of *Callitriche* sp. in the spring was very effective in limiting its regrowth during the following growing period.

There is a considerable amount of unpublished experience on how to cut stream weeds, particular methods being suited to specific rivers and preferred by individual river managers. Birch (1976) emphasizes the importance of aquatic weeds as an integral part of a fishery. Weeds provide cover in trout streams and a habitat for water-bred flies. From a fisheries management viewpoint weed cutting should be restricted to thinning and topping to allow the unhindered passage of a dry fly, and the removal of a complete stand of plants when silting has occurred (Birch, 1976). Complete removal by chain scythes can ruin a fishery and selective removal only is advisable. Further research is required to quantify the effects of different types of vegetation, in different arrangements, and under a variety of flow regimes, on the discharge capacity.

Shading of Watercourses

An alternative to conventional methods of mechanical cutting and chemical control of aquatic plants is the addition of overhanging marginal vegetation which provides a shading effect. Mitchell (1974) suggested that it was environmentally more acceptable to effect control of vegetation by indirect means such as shading. The shading technique has been practised for several decades throughout Europe (Lohmeyer and Krause, 1974; Leentvaar, 1978; Hermens, 1978; Seymour, 1978).

However, it is only recently that studies have been undertaken to quantify the degree of shading required in various situations (Pitlo, 1978; Dawson, 1978, 1979, 1981a; Dawson and Kern-Hansen, 1978, 1979; Dawson and Haslam, 1983; Dawson, 1986). Dawson and Kern-Hansen (1979) reported the effects of natural and artificial shade on the macrophytes of lowland streams in mid-Jutland in Denmark. Table XLVIII shows that the biomass in naturally shaded sections was significantly less compared to adjacent open sections. However, this was not always directly proportional to the light available at the stream surface. The effect of natural shade is governed by a number of factors, including the relationship between the position and size of the marginal vegetation and the general stream morphology and orientation. For example, vertical banks create more shade than gently sloping banks of the same height. The light available also depends on the degree of overhang of tree branches and the orientation of the stream relative to the course of the sun is fundamental.

Light should be reduced by about half of that in open sections, thereby reducing the standing crop of macrophytes by about a half. Dawson and Kern-Hansen (1979) recommended that shading to levels substantially below a half should not be undertaken, because this will lead to the disappearance of aquatic plants together with associated loss of habitats and hiding places for fish and invertebrates. Optimium shading conditions for lowland streams are summarized in Table XLIX.

For streams in lowland England it has been recommended that 50 m stands of trees are planted such that 20 m gaps are left every 70 m on the south bank of rivers of about 20 m width, flowing east to west (Dawson, 1978). These gaps allow the growth of plants in bars which provide cover for fish and invertebrates, whilst retaining tree leaves in a well-distributed manner for breakdown. Shading is particularly valuable because the submerged plant biomass eventually stabilizes and allows a diversity of habitats. Or-

Table XLVIII. A comparison of the submerged plant biomasses of *Ranunculus peltatus* and *Sparganium emersum* with the light available at the stream surface of the Gjern stream in Denmark in 1977 and 1978. Other species occasionally present in these sections are included in the totals

Site	Light (relative units)	Maximum biomass ($g\ m^{-2}$ dry wt)					
		1977			1978		
		R. pel.	*Sparg.*	Total	*R. pel.*	*Sparg.*	Total
1. Open grazed banks	1.00	249	12	271	123	20	219
2. South, bank shade (little overhang)	0.55–0.60	66	32	98	3	24	27
3. Dense overhang. NW. *Alnus*	0.24–0.40	30	4	34	-	-	-
4. Wood	0.02–0.03	1	4	5	4	4	4

Note: suitably calibrated photocells were used to measure the irradiance available at the stream surface.

From Dawson and Kern–Hansen (1979). Reproduced by permission of Akademie-Verlag.

Table XLIX. Optimum shade for lowland streams

1. On very small streams (2 m) stream banks should be left ungrazed by livestock, allowing herbaceous plants and marginal grasses to grow. Management required every 3–5 years during the autumn to limit plant development. One bush every 30 m is acceptable to provide variety.
2. Larger streams (3–8 m) require bushes or small trees on the south bank or larger trees on the north bank if the flow is from east to west.
3. Rivers greater than 15 m wide may require mature trees on the south bank (Dawson, 1978). Note: not all rivers flow from east to west, therefore each site needs to be considered individually to provide optimum shade.

From Dawson and Kern–Hansen (1979). Reproduced by permission of Akademie–Verlag.

ganic material derived from terrestrial leaves shed into the stream during autumn, and from the natural decay and die-back of submerged plants, provides an important source of food for invertebrates. During the summer months overhanging trees and grasses are likely to provide an important input of terrestrial food organisms (Seymour, 1978; Dawson, 1981b). Shade also has many other positive effects, including the avoidance of excessive water temperatures detrimental to the behaviour of fish such as brown trout. Trees and bushes need to be occasionally cut to prevent excessive shading. Shading is clearly a long-term method of management because of the time needed for trees to become established (Dawson, 1986). To determine the success of the shading technique in other countries further large-scale trials are required.

Other techniques have been used to control submerged macrophytes, including dyes (Bartsch, 1954; Buglewicz and Hergenrader, 1977; Dawson, 1981a) which are really only successful in very slow-flowing waters or static water bodies. Opaque matting or sheeting has also been used (e.g. Whitely, 1964; Mayer, 1978; Dawson and Hallows, 1983) but is logistically difficult to apply in flowing waters.

Basin-Based Approaches

Controls on Channel Adjustments

Many river channels may not be in equilibrium since they have been affected to some extent by changes occurring within the basin upstream, and frequently these changes are induced by man. It is increasingly necessary to envision channel management within the context of the drainage basin as a whole (Nunnally, 1985; Brookes and Gregory, 1988). The need for basin-based management of water resources has been accepted for some years but it is necessary for river channel management to be placed in a basin context. Failure to consider a river channel in the context of the basin when planning a flood control project frequently leads to environmental problems. Channelization may itself be part of a more widespread programme involving several changes in the basin upstream. Both the hydraulic geometry and regime approaches to river channels are essentially empirical and consider streams as equilibrium systems (Chapter II, pp. 40–46). However, it is important to consider the effects that changes in a basin will have on channelization projects. In turn channel adjustments may trigger changes in the ecosystem. Furthermore adjustments will require substantial maintenance which will further disrupt a channel and offset biological recovery.

A large number of studies have shown how river channels respond to changes in water and sediment discharges resulting from land use changes such as the conversion of forest to agricultural land and urbanization, and from the building of dams (Chapter I, pp. 3–5). These changes have been responsible for alterations in the magnitude and frequency of water and sediment discharge, which have in turn initiated adjustments of channel morphology. Without a complete set of deterministic equations it is not possible to precisely predict the morphological response to drainage basin changes. However, channel behaviour can be deduced to a certain extent by integrating responses of stream channels that have already been altered. Lane recognized the major importance of fluvial morphology in hydraulic engineering in 1955 and advocated its wider application (Lane, 1955b). To this end Lane produced a general expression for the qualitative analysis of the problems of stream morphology:

$$Qs\ d \approx Qw\ s$$

where: Qs is the quantity of sediment
d is the particle diameter or size of sediment
Qw is the water discharge
s is the slope of the stream

If any of the four variables are altered then changes in one or more of the others can be anticipated to restore equilibrium. For example, if sediment load in a stream is decreased, equilibrium can be restored if the water discharge or slope are decreased sufficiently or if the diameter of the sediment is increased.

Schumm (1969, 1977) produced a series of qualitative process–response equations which given an estimation of changes to be expected as a result of changes in discharge and bedload (Table L). These indicate that the dependent morphological variables should increase (+) or decrease (-) for given combinations of changing water discharge (Qw) and bedload yield (Qs). These relationships may be tentatively used in planning and design to give an indication of the direction of channel adjustment to be expected from drainage basin changes (Nunnally and Shields, 1985). For example, increased sediment discharge resulting from intensive cropping of land may produce a decrease of median sediment size and cause a wider, shallower channel to be formed. By contrast the creation of impervious areas by urbanization will restrict soil erosion but increase the magnitude and frequency of flooding. Urban streams often respond by eroding their bed, thereby increasing the size of the channel, or by forming a segregated gravel layer (Wolman, 1967; Hammer, 1972).

General application of such relationships should be treated with caution. Thresholds exist which control erosion, deposition and channel patterns (Schumm, 1977). For example, changes occurring within the upstream drainage basin may be insufficient to cause the pattern to change from meandering to braiding or vice-versa. Furthermore there is the problem of indeterminacy whereby channels may adjust to change in numerous ways. Gregory (1976) identified nine categories of potential adjustment. Adjustments of size, shape and composition can occur in the channel cross-section, pattern or drainage network. Adjustment of any particular river channel may involve changes in any one of the nine categories. For example, an increase in peak discharge due to the effects of urbanizing a catchment could lead to an increase in channel cross-section, an alteration of channel shape, an increase in the size of meanders and a metamorphosis of planform from a single-thread to multi-thread. There still remains a need to understand how several potential adjustments interrelate.

Table L. Qualitative models of channel change, illustrating the direction of morphological response for particular combinations of changing discharge and sediment load (after Schumm, 1969)

1a. *Increase of discharge alone*
e.g. diversion of water into river

$Qw^{+} \sim b^{+}\ d^{+}\ L^{+}\ s^{-}\ F^{+}$

1b. *Decrease of discharge alone*
e.g. extraction of water from river

$Qw^{-} \sim b^{-}\ d^{-}\ L^{-}\ s^{+}\ F^{-}$

2a. *Increase in bedload*
e.g. deforestation or intense farming

$Qs^{+} \sim b^{+}\ d^{-}\ L^{+}\ s^{+}\ P^{-}\ F^{+}$

2b. *Decrease in bedload*
e.g. improved land use or soil conservation

$Qs^{-} \sim b^{-}\ d^{+}\ L^{-}\ s^{-}\ P^{+}\ F^{-}$

3. *Discharge and bedload increase together*
e.g. during construction of an urban area

$Qw^{+}\ Qs^{+} \sim b^{+}\ d^{\pm}\ L^{+}\ s^{\pm}\ P^{-}\ F^{+}$

4. *Discharge and bedload decrease together*
e.g. as a result of dam construction

$Qs^{-}\ Qs^{-} \sim b^{-}\ d^{\pm}\ L^{-}\ s^{\pm}\ P^{+}\ F^{-}$

5. *Discharge increases as bedload decreases*
e.g. following urbanization when sediment source areas have been paved over

$Qw^{+}\ Qs^{+} \sim b^{\pm}\ d^{+}\ L^{\pm}\ s^{-}\ P^{+}\ F^{-}$

6. *Discharge decreases as bedload increases*
e.g. resulting from increased water and land use

$Qw^{-}\ Qs^{+} \sim b^{\pm}\ d^{-}\ L^{\pm}\ s^{+}\ P^{-}\ F^{+}$

A plus or a minus exponent is used to indicate how various aspects of channel morphology change with an increase or decrease in water discharge or bedload.

Qw = water discharge (e.g. mean annual discharge or mean annual flood); Qs = bedload (expressed as a percentage of total load); b = width; d = depth; L = meander wavelength; s = channel slope; P = sinuosity; F = width–depth ratio.

CHAPTER VIII

Mitigation, Enhancement and Restoration Techniques

INTRODUCTION

Recommendations which aim to modify conventional channelization procedures have been discussed in the previous chapter (Chapter VII) and these relate mainly to the degree, extent and timing of construction. The following discussion considers: (1) procedures which have been developed to mitigate the impact or enhance planned or existing channelization projects; (2) unique designs from particular areas of the world which incorporate an understanding of fluvial processes; (3) projects which attempt to restore river channels to their pre-channelization state; and finally (4) floodplain alternatives, such as the corridor concept, which recognize rivers as open systems. In contemplating the use of these alternative techniques reference should be made to the basin approach described at the end of the previous chapter. In basins with a high sediment load it may be necessary to install a sediment trap upstream to eliminate sedimentation within the modified reach.

METHODS OF MITIGATION AND ENHANCEMENT

It is possible to mitigate or enhance a channel against the adverse effects of channelization by installing structures or taking measures at some time after construction. These measures can be used in the planning, design, construction or maintenance of flood control channels and they improve the net environmental effect. Many of these methods are similar to those structures which have been used for several decades to improve unaltered watercourses. Habitat structures have been used successfully to accelerate biological recovery of channelized streams.

Instream Devices

Instream structures or devices have been widely used to increase the diversity of habitat by altering the flow, channel morphology or substrate, or by providing cover. These have been classified according to their effect: namely whether they impound or modify river flow, provide direct cover, or improve spawning areas (Swales and O'Hara, 1980). Structures may be installed immediately but have been applied at dates ranging up to 100 years after channelization. A number of mitigation techniques have been applied to various areas, including Michigan (Hubbs *et al.*, 1932), Maine (Warner and Porter, 1960) and the Upper Mississippi River System (Schnick *et al.*, 1982).

Deflectors function either to direct flow and eliminate accumulated sediments, or to narrow a channel, thereby increasing the velocity and creating a scour pool with a corresponding riffle downstream (Spillett, 1981). Experiments have been carried out on three lowland streams in North Jutland, Denmark, using various designs and combinations of stream deflectors (Holm, 1984). Figure 56 illustrates the bed contours before and 1 year after installation of a single deflector on the River Voer å, a deep hole having formed immediately downstream from the deflector. They may be installed on alternate banks to produce a meandering thalweg. A spacing of 5 to 7 channel widths has been recommended for deflectors, corresponding to the pool–riffle spacing found in natural streams (Borovicka, 1968; Everhart *et al.*, 1975; White, 1968, 1975). One of the largest studies of current deflectors has been undertaken on Lawrence Creek in Wisconsin (Hunt, 1969, 1971, 1976; Hunt and Graham, 1972). Hunt showed that the mean annual biomass of trout, mean annual number of legal-sized trout and annual production, increased markedly in the first 3 years, reaching a maximum development after 5 years.

Small weirs or sills diversify the habitat by impounding a greater depth of water above the structure, and by increasing the velocity downstream to erode a scour pool (Figure 57) (Gard, 1961; Barton and Winger, 1973a,b; Winger *et al.*, 1976). These structures extend across the entire channel, although some have a notch to concentrate flows locally. They have a minimal backwater effect. Material removed from the downstream scour pool may be deposited as a riffle a short distance downstream. Weirs can be constructed from logs, rocks, gabions, sheet piles or concrete, and must be keyed-in to the bed and banks to prevent flanking. The technique has been successfully applied to increase fish populations within relatively short periods of time in the United States and in West Germany (Januszewski and Range, 1983; Geiger and Schröter, 1983). Gard (1961) showed how brook trout which were introduced into a Californian stream with low flow dams grew rapidly and reproduced over the period of the 4-year study. No trout were able to survive prior to mitigation. Structures may fail in high flows and fish populations may not be enhanced because of increased sediment loads (Keown, 1981; May, 1975). However, many sills have been shown to provide a superior habitat compared with the modified channel (McCall and Knox, 1978; Carline and Kloslewski, 1981).

Devices which provide direct cover may either be fixed to the bed or banks of a channel or allowed to float and adjust their level with varying discharge (Figure 58) (White and Brynildson, 1967; Cooper and Wesche, 1976; White,

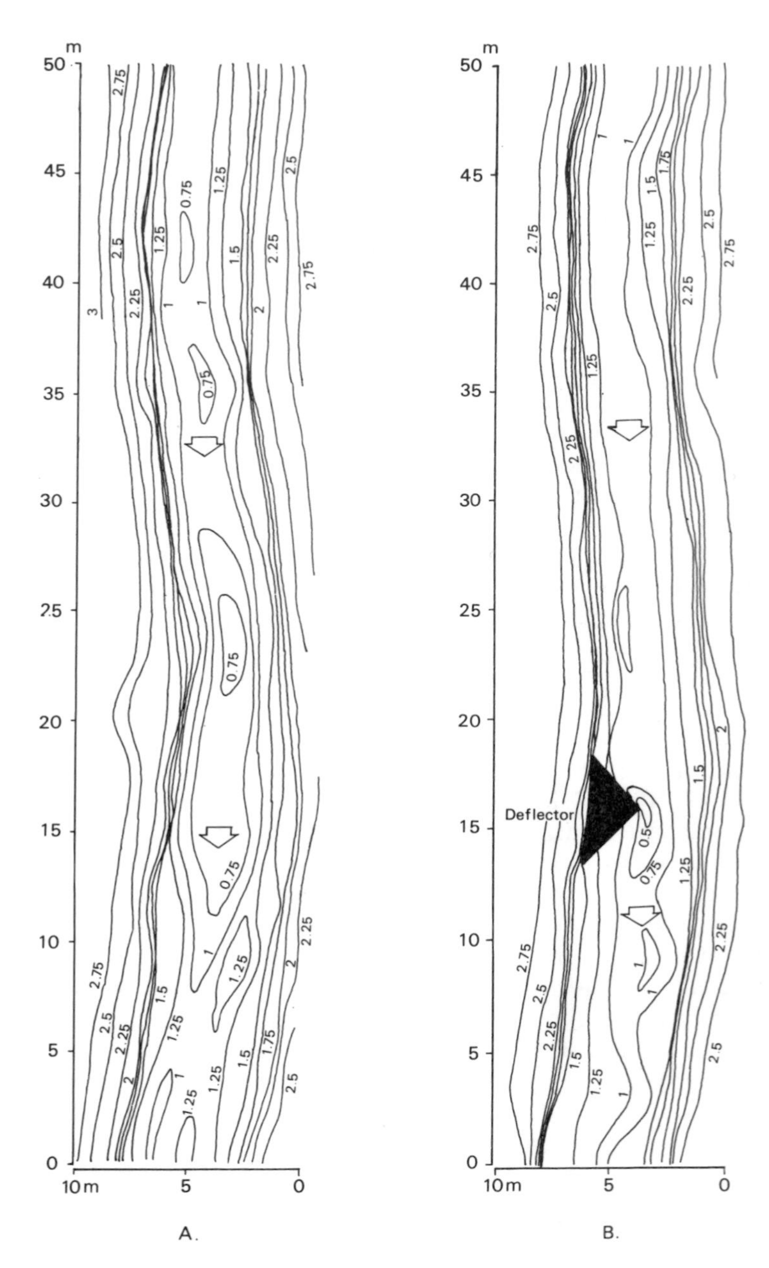

Figure 56. The impact of a deflector on the bed topography of a small stream (River Voer å) in Denmark. (A) Before installation and (B) formation of pool one year after installation. (From Kern-Hansen *et al.*, 1983. Reproduced by permission of the National Agency of Environmental Protection, Denmark.)

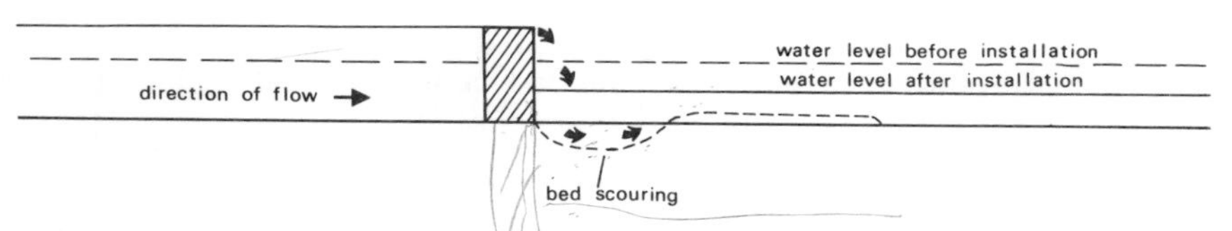

Figure 57. The effects of a low dam on current flow and channel morphology. (After Swales, 1982b. Reproduced by permission of Academic Press Inc. London.)

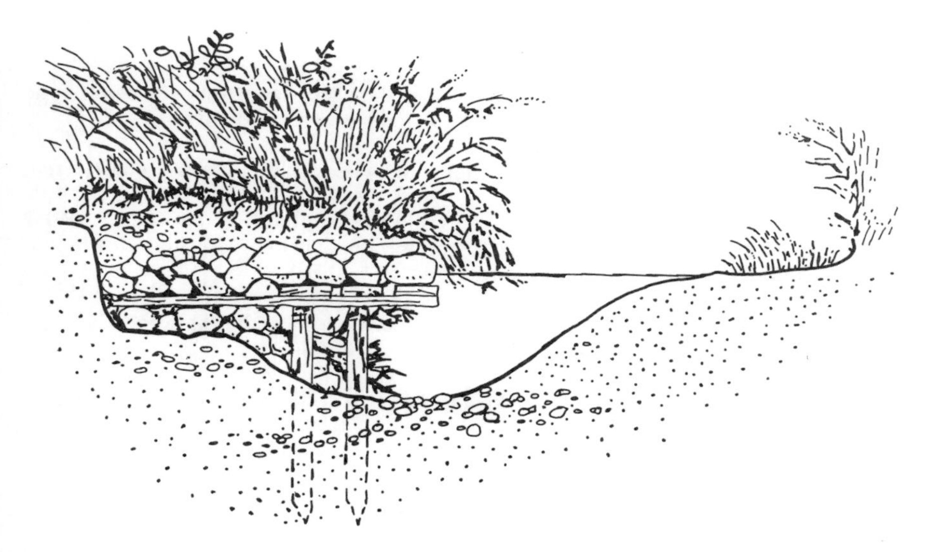

Figure 58. Cross-section showing deflector and bank cover device. (From White, 1968. Reproduced by permission of Verlag Paul Parey.)

1968, 1972). White (1975) reported how fixed cover on a stream in Wisconsin increased the number of trout by over 100% in 3 years. Compared to four reaches of the Big Roche-a-Cri Creek left unmanaged, the youngest trout numbers were up to 11% higher, whilst older or larger trout were 200% more numerous in the spring and autumn in the managed channel. Application of interwoven willow branches to a stream in Montana caused a marked increase in the numbers and biomass of trout (Boussu, 1954). Rock-filled gabions and granite rubble may provide suitably sized holes for successful colonization by macroinvertebrates. For example, Hogger and Lowery (1982) found that for part of a flood scheme on the River Ware near Hertford (UK) colonisation by crayfish had occurred only 18 months after construction was completed in 1976. Rocks may be placed randomly in channels to produce small scour holes and areas of reduced velocity.

Table LI. Mitigation of habitat losses due to channelization along high gradient trout streams

1. Alter original stream channels only when absolutely necessary and then keep alterations to a minimum. Maintain the original stream length and meander the new stream channel as much as possible to correct the ditch-like appearance.

2. Trees and shrubs should be retained to provide bank stability and shade. When topsoil and vegetation are lost banks should be sloped and topsoil replaced and seeded down to the high-water mark. Trees and shrubs can be planted to stabilize stream banks.

3. When riprap is used it should be covered with subsoil and topsoil down to the high-water mark and then re-vegetated with grass, trees and shrubs.

4. Instream devices to create pools (e.g. deflectors) should be properly engineered.

5. In gravel-bed streams mitigating devices should be placed at between 5 and 7 channel widths, alternating from each streambank. This corresponds with pool–riffle sequences found in unaltered sections.

6. Mitigating structures should be placed in currents close to the thalweg to be most effective in providing trout habitat.

7. Further random placement of rocks can be used to create larger mid-channel pools.

Reproduced with permission from Lund (1976).

Table LII. Design guidelines for habitat structures (after Shields, 1983a)

1. *Feasibilty study*
   Determine projected habitat characteristics and assess stability of channel (ie. erosion or deposition may be detrimental) must be adequate for desired fish species and related organisms.

2. *Design*
   (a) Determine habitat deficiencies
   Desired depths, velocities and substrates should be determined using a method such as the Instream Flow Methodology (Stalnaker, 1980). Alternatively the morphology and shade characteristics of the unaltered stream or unaltered neighbouring reaches/streams can be used as a basis for the habitat structure plan.

Table LII. (*continued*)

(b) Locating structures
Detailed location to avoid drop structures, etc. Number and frequency of structures should be based on the habitat requirements. Can be placed at irregular intervals. Barton and Cron (1979) suggest random rock placement at a density of one boulder per 27 m^2. Kanaly (1975) recommends one boulder of 2 m diameter per 70 m^2.

(c) Types of structure
Depends on the habitat requirements, the physical effects, velocity regime, bed sediments, economics, availability of materials, constructability and aesthetics, e.g. deflectors are ineffective in low velocity streams of 0.6—0.9 m sec^{-1} (e.g. Nelson *et al.*, 1978). However, a notched sill will concentrate flows to create scour.

(d) Size the structures
Structures should have a maximum effect at low flows and negligible effect on high flows. Sill should be no more than one-third bankfull depth (Barton and Cron, 1979). Spacing of a minimum of 3 pool lengths is recommended (Winger *et al.*, 1976) to allow intervening riffles. Lowest point of a sill should be in the centre to avoid flanking.
Size of deflectors depends on extent of desired depositional zone and the acceptability of erosion of the opposite bank. White and Brynildson (1967) recommended deflectors at one-third stream width. If bank protection is placed opposite a deeper channel may result.
Random rocks should be no more than one-fifth channel width at normal flow (Barton and Cron, 1979). Rocks need to withstand given velocities.

(e) Investigate hydraulic effects of structures.
Use a mathematical model (e.g. backwater). Structures may be modelled as roughness elements, low weirs or contractions. Scour holes should also be considered, and their predicted size based on experience. Increases in channel roughness due to placement of rocks are analysed by Barton and Cron (1979).

(f) Consider effects on sediment transport
Type and size should be selected with the sediment load and grain sizes in mind.

(g) Select materials and design
Materials include logs, rock, riprap, gabions, sheet piling and concrete (e.g. Maccaferri, 1968). Local materials salvaged from channelization may be used. Ehlers (1956) showed that after 18 years log sills proved superior to rock structures. Rock should be sized according to velocity.

(h) Construction
Biologist + engineer should be able to make minor adjustments on site during construction. Close supervision is required.

(i) Follow-up
Regular monitoring, adjustments and maintenance should be undertaken.

Instream habitat devices have been installed successfully over a range of physical environments. Structures are not effective in unstable channels that experience rapid aggradation because sediment can fill pools and holes induced by structures and cover rocky surfaces important as a substrate for macroinvertebrates (Harvey and Watson, 1988). This is especially true for many streams in Denmark which have low slopes (Holm, 1984). Trade-offs between the restoration of the aquatic habitat and channel capacity may be required; structures increase the hydraulic roughness. Measures are required to prevent undercutting, flanking and undesirable bed and bank erosion (Shields, 1982a,b). Many channels recover their morphology naturally with time and in those circumstances habitat structures may be regarded as an interim measure between completion of a scheme and recovery.

A combination of procedures was used to mitigate against the adverse effects of channelization of the St. Regis River in Montana (Table LI) (Lund, 1976). An evaluation of the fishery resources between 1973 and 1975 revealed that mitigation structures had been effective in providing fish habitat comparable to the unaltered sections. Game fish populations in altered sections with mitigations recovered from construction work in about one year. By contrast trout populations in altered sections of more than 50 years old failed to recover to levels found in unaltered sections. The number of pools per section as measured by pool–riffle periodicity provided the closest correlation to trout population changes. Problems associated with channelization which were not mitigated included the loss of streambank vegetation, the destruction of natural stream aesthetics and the loss of stream length.

The majority of the available literature focuses on the use of devices in natural streams, but a preliminary basis for design based on documented experiences is available (Table LII) (Shields, 1983a).

The design procedure for structures should also consider processes of erosion occurring in the upstream watershed. Many of the stream improvement methods used successfully in North America are difficult to apply to the Pacific Coast of British Columbia in Canada because the streams have a greater range of flow and higher gradient than eastern streams (Government of Canada, Fisheries and Oceans, 1980). The results of stream improvement projects are therefore less predictable, and the useful life of structures and improvements within the stream will tend to be lower than experienced in central and eastern North America. The following combination of measures have been suggested to improve the habitat of anadromous salmonids in British Columbia. These measures recognize the need to control both the processes operating both within a channel and in the basin upstream: (1) forest harvesting plans must include measures to protect streams and the adjacent land area; strips of trees are left to provide shade; protect the banks from erosion; cause undercut banks favourable for juvenille salmonids. Trees, particularly deciduous species, are important for the production of aquatic insects and organic material. (2) Eroding banks provide a continuing source of gravel for the stream. High stream flows may shift gravel bars, cleanse spawning beds and form or deepen pools, all to the benefit of spawning and rearing of salmonids. The erosion of finer materials (silt and sand) can degrade the environment and therefore these banks must be stabilized. Debris jams often aggravate this erosion. Gabions, rock or logs can be used to protect the bank; wing deflectors to direct current away. Spaces between rocks and gabions will provide holding and rearing places for juvenile fish. Trees can be planted to stabilize banks; elsewhere fast-growing grass and legume mixtures. Fences can be erected to prevent access by cattle. (3) Control of upland erosion is recommended to prevent silt from entering the channel. Revegetation of unstable slopes, such as road cuts, is essential. Overburden waste of mining operations should be terraced and then hydroseeded with mulch, fertilizer and selected grasses. Settling ponds can be used to collect stormwater draining from mines. Erosion of access roads should be minimized by providing rock-filled ditches to divert flow across the road at intervals.

Pools and Riffles

The procedure for the design of pools and riffles depends on whether the pools and riffles are to be constructed in lined or unlined channels, whether additional material is to be added to the channel or whether the bedforms are to be allowed to reform naturally in the absence of maintenance (Table LIII) (Nunnally and Shields, 1985). As a preliminary step it is necessary to assess the flow characteristics and channel morphology to determine if pools and riffles are appropriate instream habitat features. The main criterion is usually the stream's ability to support a fishing resource. Pools and riffles are not usually installed on ephemeral streams, in channels with a steep gradient, where there is a high sediment transport, or where the banks are unstable. Pools and riffles have rarely been constructed where the bed material is too coarse to be moved under present hydrological conditions. However, this latter condition does not adversely affect channel stability.

Table LIII. Guidelines for the design of pools and riffles

Three principal steps should be followed:
(1) An assessment of the flow characteristics and channel morphology to determine if pools and riffles are appropriate instream habitat features.
(2) Determination of the spacing and sizing of pools and riffles for the particular channel type.
(3) Calculating the size of bed material used to construct the riffles.

The spacing of pools and riffles is not critical in lined channels. For unlined channels spacing can be determined from neighbouring streams with similar characteristics, or from other reaches of the same watercourse. Generally an average of 5–7 channel widths has been found to be sufficient to emulate natural conditions (Stuart, 1959; Keller, 1975, 1978). In Britain the range for the natural pool–riffle sequence is 3–10 channel widths. Regular spacing should be avoided. A meandering alignment should be incorporated, the riffles being located in straight reaches and pools at the bends. Proper spacing facilitates self-maintenance. Stuart (1959) placed gravel at regular intervals along the length of a Scottish stream and this was dispersed by winter floods during the following three months.

Pool and riffle dimensions are not critical and may be varied to suit habitat requirements. However, pools which are too wide, too deep, or excessively long, may trap silt and require periodic maintenance. Experience shows that pools should have a minimum low-water depth of 0.3 m, and riffles should not project from the bed by more than 0.3–0.5 m. Individual pools and riffles should not be longer than 3 channel widths or shorter than 1.

In those cases where the riffles are to be dynamic and self-maintaining, they should be constructed from natural stream gravels with a size-distribution typical of the existing bed material (Stuart, 1959). Otherwise they can be constructed from gabions, cobbles, or riprap sized to withstand discharges up to a selected return interval (Edwards *et al*., 1984). Keller (1978) showed how planned manipulation of the channel form can be used to induce the development of pools and riffles at desired locations.

Pools and riffles have been successfully reconstructed in channels which are maintained by flows with a recurrence interval of between 1 and 2 years. Keller (1975) suggested that designing a channel for flood control is more involved because such projects require a channel which can carry the 25- or even the 100-year flood. Pools and riffles could not be successfully installed in a 100-year channel and maintained by the 2-year flood; such a channel would probably braid and become choked with migrating sand bars. Keller (1975) therefore recommended that a pilot channel maintained by a 2-year flood should be excavated within the larger flood channel. This pilot channel would have pools and riffles and would provide a fish habitat during low and normal flows. The larger channel would be vegetated.

From studies of seven streams in North Carolina, California and Indiana, Keller (1975) reported that pools and riffles reformed naturally on slopes ranging from about 0.0015 to 0.005 in periods of less than 25 years. Thus it was recommended that a gravel-bed stream of initial slope 0.0025 should not be shortened more than 50% if pools and riffles are desired. Pool–riffle sequences are very sensitive to channel width, and channelization that changes the width will alter the pool–riffle environment.

Macroinvertebrate abundance, diversity indices, standing stock in the benthos and drift were significantly higher in a channelized reach of the Olentangy River in Ohio which had been mitigated with pools and riffles, and approached values found in natural streams (Figure 59) (Edwards *et al*., 1984). The diversity and abundance of game fish were also higher in the mitigated reach. However, certain non-game species were relatively more abundant in the mitigated area when compared to the natural area. An earlier study had shown that, 24 years following channelization, the number of species of fish compared with a natural reach was 22% less in a conventional channelized reach but only 5% less in a reach where artificial riffles had been constructed (Edwards *et al.*, 1975; Edwards, 1977).

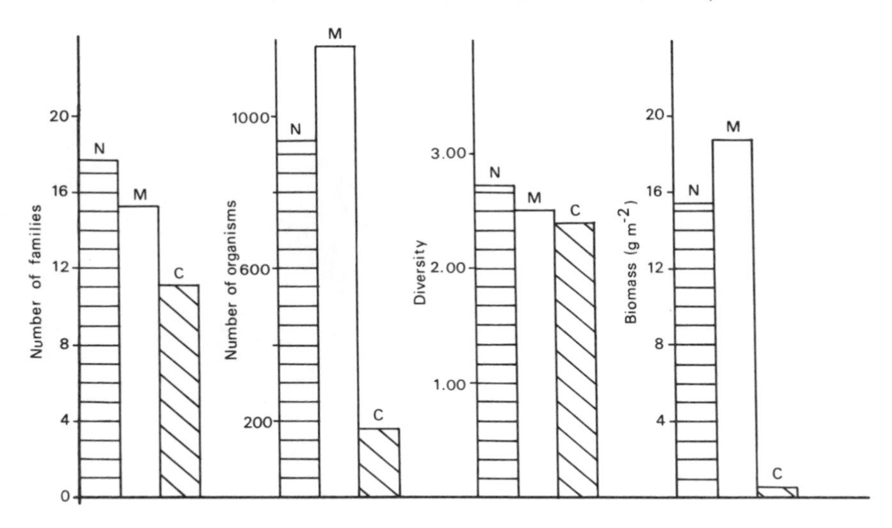

Figure 59. Number of benthic families and organisms, diversity and biomass for the natural channel (N), mitigated channelized reach (M, channelized in 1970) and unmitigated channelized reach (C, channelized in 1950) of the Olentangy River, Ohio (average of 13 sampling visits between June 1974 and October 1976). (Based on data in Edwards *et al.*, 1984. Reproduced by permission of American Fisheries Society.)

Substrate Reinstatement

The replacement of natural bed sediments following completion of a channelization scheme may speed recovery (cf. Gore and Johnson, 1980). This is most successful where well-sorted gravels and cobbles replace unsorted sediments or bedrock (eg. Hjorth and Tryk, 1984). Placement of artificial materials such as crushed rock may also improve the habitat for fish and for macroinvertebrates (Stuart, 1953; Spillett and Armstrong, 1984). For the River Ock in Oxfordshire (UK) Spillett and Armstrong (1984) introduced crushed limestone and flint gravels over a bare clay bed. Surveys indicated a significant increase in invertebrate populations and improvement in the biological quality at intervals of 4, 10 and 20 weeks after reinstatement (Table LIV). The limestone substrate supported a much higher invertebrate density than the clay bedrock, and higher abundance and diversity as expressed by the Chandler Biotic Score (Chandler, 1970). The rate of recolonization was much faster in the limestone area. In the Afon Gwyrfai (UK) recolonization of a reinstated gravel bed was a gradual process, taking about a year

Figure 60. Reinstatement of gravels to a small Danish stream, using a wheelbarrow. (Photograph by Ulrich Kern-Hansen.)

Table LIV. Number of invertebrates (range in parentheses) and Chandler Biotic Score* on newly reinstated artificial substrates (after Spillett and Armstrong, 1984)

Time after instatement	Crushed limestone		Clay bedrock	
	No. m^{-2}	Biotic score	No. m^{-2}	Biotic score
4 weeks	6913	878 (746–903)	2770	324 (266–443)
10 weeks	9732	877 (786–974)	3388	650 (429–1028)
20 weeks	10065	1039 (719–1490)	3047	874 (531–1217)

*Chandler, (1970)

to complete (Brooker, 1982). The stability of retained or reinstated gravels is important: if the gravel is too fine and therefore unstable, then species diversity and abundance will be less. The methods for the replacement of suitably sized spawning gravel are outlined by various authors (Figure 60) (e.g. Platts *et al.*, 1979; Shirazi and Seim, 1979).

Where sediment loads are high due to upstream land use changes or channel works then the restoration of gravel for spawning may be unsuccessful.

The majority of reinstated spawning beds in Denmark have become covered with sand and silt, thus reducing intra-gravel water flows and depriving fish eggs of dissolved oxygen (e.g. Hermansen and Krog, 1984). Sand traps placed immediately upstream have been straightened as a means of overcoming this problem. By widening and/or deepening a stream's cross-section the velocities are reduced, inducing the deposition of sediment particles. High-velocity water jets have also been used to flush fine material from gravel before the silt-laden water is pumped into a separator (Mih, 1978, 1979; Mih and Bailey, 1981). The silt slurry is then jetted onto the bank. The effects of gravel cleaning on the bottom organisms have been assessed for three streams in Alaska (Meehan, 1971).

ALTERNATIVES FROM SPECIFIC AREAS

Whilst the alternatives discussed so far have perhaps a more universal application, there are a number of techniques which have been developed and tried for particular geographic and hydrologic areas.

Non-uniform Channel Geometries

Research undertaken in West Germany has advocated the preservation or construction of meanders and asymmetrical cross-sections to emulate the morphology of natural channels by directing the flow and inducing the development of pools and riffles (Seibert, 1960). Similar techniques have been applied in Bavaria to produce a diverse range of habitats, thus creating channels which are biologically productive (Binder and Grobmaier, 1978).

Experiments were commenced in 1974 on Gum Branch in Charlotte, North Carolina to determine if manipulation of the cross-sectional channel morphology could cause a stream to converge and diverge the flow as in a natural stream, and thus induce the stream to develop a series of point bars in desired locations along 130 m of channel (Keller, 1978). This was accomplished by varying the inclination of the channel bank from 2 : 1 to 3 : 1. The asymmetric cross-section should converge the high-flow water and cause scour near the bank with the 2 : 1 slope, whilst inducing deposition of a point bar adjacent to the bank with the 3 : 1 slope. The symmetrical cross-profile with both channel banks at 2 : 1 is designed to diverge flow. Thus the stream should be induced to construct a series of point bars and scour areas, similar to that found in natural streams. In practice bars emerged following the first above-normal flow after construction was completed. The bars formed adjacent to the bank with the 3 : 1 slope, where planned. During the period from summer 1974 to autumn 1975 these bars remained stable despite four overbank flows.

Conventional and Unconventional Two-stage Designs

Two-stage channels confine the normal range of flows to the original channel whilst flood-flows are contained within a larger channel constructed above bankfull by widening out the floodplain (Figure 61) (Dobbie *et al.*, 1971; Hinge and Hollis, 1980). Low-flow channels prevent excessive sediment deposition in the enlarged channel and enable the migration of fish and avoid the excessive water temperatures associated with conventional flood channels. A limited wetland flora may develop on the berm areas. They comprise either the original natural channel, a newly excavated channel, or a combination of the two (Shields, 1982a).

When constructing a new low-flow channel it is desirable to emulate the morphological characteristics of the original channel, thereby providing stability, and biological and aesthetic diversity. When the existing natural channel is retained as a low-flow channel the original substrate, bedforms and alignment are preserved. If the original stream has sufficient power to erode its channel then it may be necessary to restrain the low-flow channel from migration.

A more unconventional 'flexible' two-stage design is that of the River Roding in Essex, England (Figure 62). The original channel was left relatively undisturbed and the design is flexible because excavation of the second-stage alternates from bank to bank (Wojcik, 1981; Paynting, 1982; Weeks, 1982). The flood berms require periodic maintenance to restrict plant growth. Surveys of the fish populations undertaken before and immediately after construction showed that the scheme had not resulted in any serious changes to the type or overall biomass of fish populations (Dearsley and Colclough, 1982). The average biomass was 23.9 g m^{-2} before construction and 25.8 g m^{-2} after. This was attributed to the fact that the flood-berming techniques involved minimal disruption to the existing dry weather channel. However, it was also recognized that surveillance would have to be undertaken for several years until the berms became stabilized and potential siltation in the dry weather channel had been assessed. It was also felt that increased solar radiation may result from the reduced cover and therefore encourage macrophyte growth.

For a 5 km length of two-stage channel on the River Roding the aquatic vegetation growing under, or emerging from, the low-flow water level was monitored during the period 1979–82 (Raven, 1985, 1986). Following construction it was found that the plant communities remained stable in the absence of river engineering. The in-channel vegetation cover, species richness and diversity were maintained when the undisturbed lowermost portion of river was incorporated into a two-stage channel. The vegetation cover increased for over two years after the excavation of the flood berm. Species richness increased in the short term, but subsequently declined as invasive species, particularly *Sparganium erectum*, proliferated. Increased light and reduced scour along the channel edge created an ideal habitat

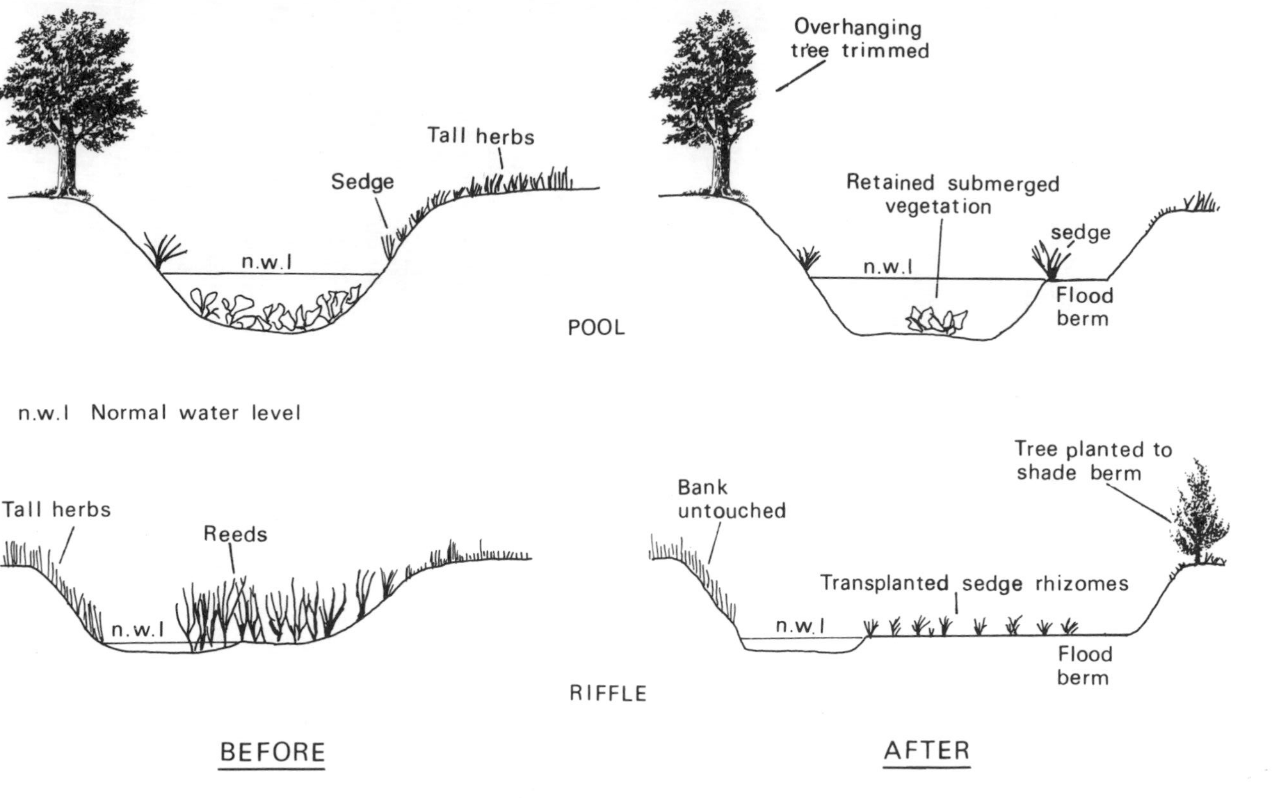

Figure 61. Conventional two-stage channel design, proposed for the River Ray near Bicester, Oxfordshire (UK). (Based on Hinge and Hollis, 1980.)

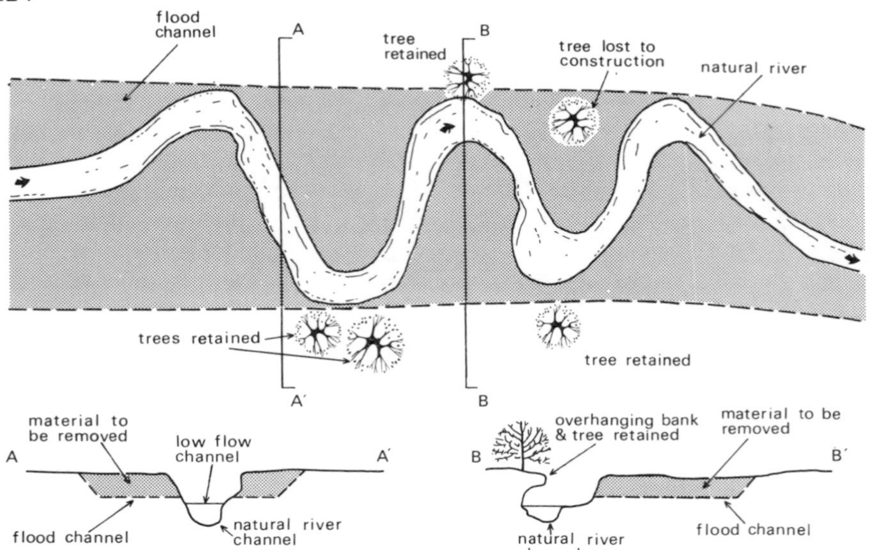

Figure 62. Flexible two-stage design devised for the River Roding, England. (Based on Keller and Brookes, 1984. Reproduced by permission of the American Society of Civil Engineers.)

for tall emergent vegetation and *S. erectum, Phragmites australis* and *Typha latifolia* proliferated along reaches bordered by low-level flood berms. Although two-stage channel construction was considered to be ecologically preferable to other forms of channelization, it is recommended that appropriate measures should be taken to counter excessive vegetative growth, through tree shading or by using self-scouring channel profiles if the short-term hydraulic efficiency is to be maintained into the longer term (Raven, 1986).

The River Roding catchment has a low sediment yield and the channel banks are composed of resistant clay. The design would probably not work for rivers with a high sediment load and erodible banks (Keller and Brookes, 1984). This is due to the reduced capacity of the meandering low-flow channel which favours frequent overbank deposition in the flood channel, thereby reducing the capacity and the level of flood protection.

Further work is required to improve the understanding of the hydraulics of compound channels and a five year programme is currently under-way in the UK using models such as the Science and Engineering Research Council Flood Channel Facility and field examples (Knight and Sellin, 1987). Research will concentrate on straight channels (1986–89), meanders (1989–1992) and sediment movement (1992–1995). Preliminary results from an experimental channel indicate that the processes involved in two-stage channels are extremely complex and involve boundary shear stresses, secondary flows and momentum transfer between the main channel and berm (Figure 63) (Knight, 1981; Knight and Demetriou, 1983; Knight and Hamed, 1984).

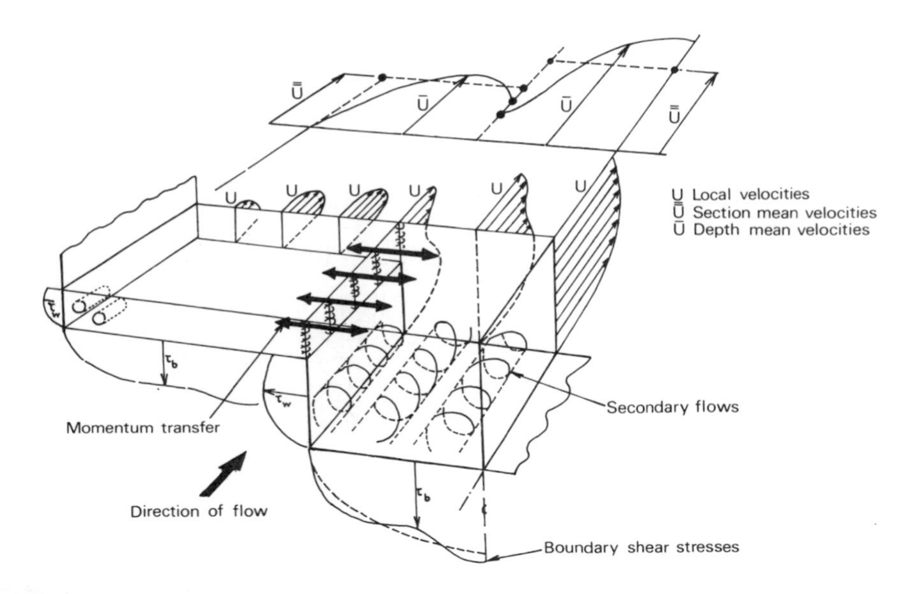

Figure 63. Hydraulic aspects of overbank flow. (From Knight and Hamed, 1984. Reproduced by permission of the American Society of Civil Engineers.)

Partial Restoration

The use of fluvial morphology in the design of river projects to re-establish or simulate natural conditions has been demonstrated by a number of authors (e.g. MacBroom, 1980). Möller and Wefers (1983) described the restoration of two cutoffs on the River Krückau, Penneberg, West Germany, which had been abandoned and backfilled several years previously. The old channels were excavated to form backwaters and were planted with indigenous species of reeds, shrubs and trees.

The Water Resources Research Institute of the State University of North Carolina has developed a procedure for restoring urban streams which reproduced some of the characteristics of natural streams but at the same time achieved the engineering objective of flood prevention (Table LV) (Keller and Hoffman, 1976, 1977; Nunnally, 1978a,b; Nunnally and Keller, 1979). Stream restoration is a technique used in urban streams and involves the removal of trash, extensive growth of small trees and bush, and removal of large trees which have fallen into the stream channel, creating a drainage problem (Figures 64 and 65). The first guideline for stream restoration is to minimize channel straightening and the consequent increase of slope. This is intended to avoid the undesirable adjustments arising from straightening, which eventually lead to a less efficient channel for discharging floodwaters. It was recognized that even stable alluvial channels tend to change position as a result of selective erosion and deposition within the channel, and that

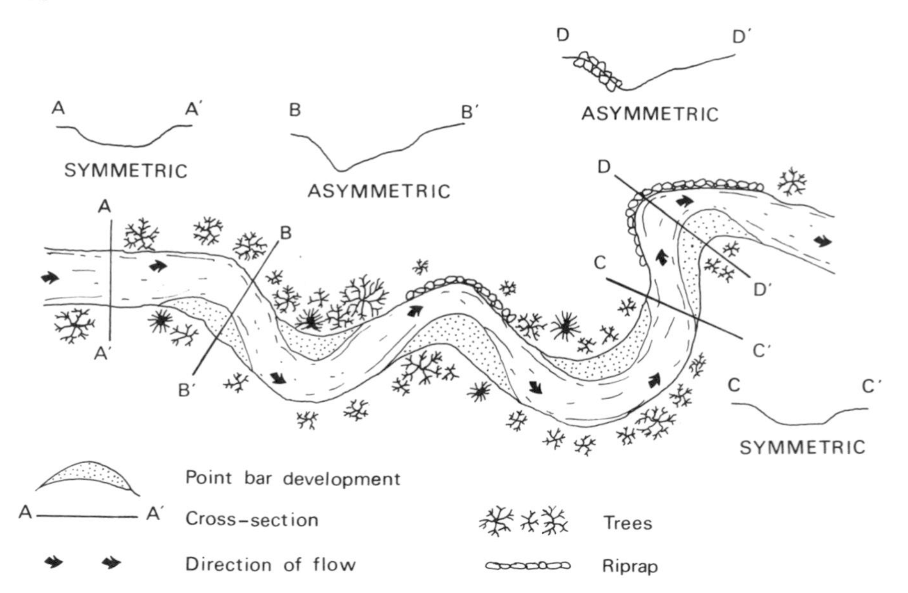

Figure 64. The concept of channel restoration. (From Keller and Brookes, 1984. Reproduced by permission of the American Society of Civil Engineers.)

in urban areas where structures may be threatened, then bank protection measures may be required. The second design principle is therefore to promote bank stability. This is achieved by several methods, including retaining as many trees as possible during construction. Only trees which obstruct the flow, or are in immediate danger of falling into the channel, or which restrict the access of equipment, are removed. The removal of trees can be further minimized by using the smallest equipment feasible for the work. Where trees have to be cut then stumps should be left in place for bank stability. Figure 65a shows the selection of trees for retention prior to restoration of Briar Creek in North Carolina in the summer of 1975. In rapidly meandering streams then riprap can be used as local protection for the outsides of bends where the flow impinges. To avoid aesthetic impact, only minimal use

Table LV. Summary of the design principles used in stream renovation (after Nunnally, 1978a)

1. Straighten the channel and increase the slope as little as possible. This minimizes undesirable adjustments.
2. Promote bank stability by retaining as many trees as possible; minimizing channel reshaping; reseeding disturbed areas promptly; careful placement of riprap.
3. Emulate nature in designing channel form. Use asymmetrical cross-sections at bends; symmetrical in straight reaches; avoid uniform trapezoidal sections and steep slopes.

Figure 65. Briar Creek, Charlotte, North Carolina, restoration project (a) selecting trees for retention; (b) after completion. (Photographs reproduced by permission of E.A. Keller.)

of riprap is recommended, but this can be supplemented at future dates. Disturbed areas should be quickly seeded with grasses that grow quickly and develop extensive root systems.

The third guideline for stream restoration is to emulate nature in designing channel form rather than using uniform trapezoidal cross-sections. Modified channels can be designed with a more natural asymmetrical cross-section at a bend, using a 2 : 1 slope on the inside bank and a 3 : 1 slope on the outside bank. This facilitates the development of a point bar (sand bar) on the inside of each bend as found in natural channels. Figure 65b shows Briar Creek after completion of the restoration project, incorporating the three design

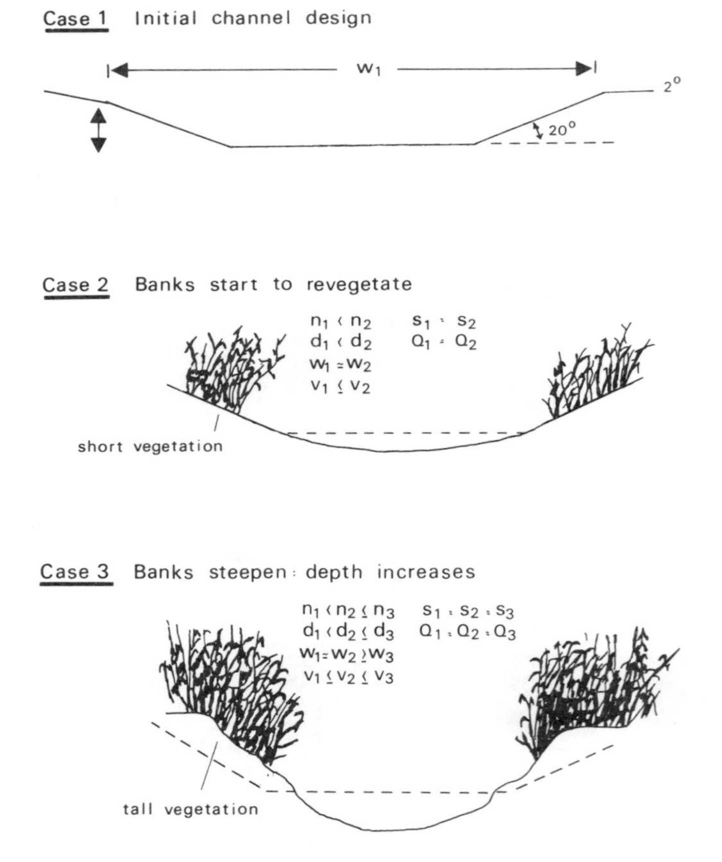

Figure 66. Expected evolution of the rehabilitated channel of Badger Creek from Case 1 (initially constricted) to Case 2 (initial vegetative stabilization of banks) to Case 3 (final regime channel). (From Jackson and Van Haveren, 1984. Reproduced by permission of the American Water Resources Association.)

guidelines. This procedure trades off a small loss of flow efficiency for a more stable and biologically productive morphology.

The objective of restoring Badger Creek (Drainage area: 569 km^2) was to duplicate the characteristics of a more natural channel and obtain a more stable channel (Jackson and Van Haveren, 1984). Overgrazing, deforestation and road building in the watershed had increased peak flows and sediment delivery, causing severe erosion of the channel. Sizing was achieved by the Manning equation and slope measurements. An initial trapezoidal design was obtained (Figure 66) and critical shear stress equations were used to check bed and bank stability. From this initial shallow and wide section it was anticipated that revegetation in approximately 2–5 years would cause the banks to become more stable and that the channel will become narrower, inducing increased velocities and bed erosion. Gabions were also recommended at intervals along the channel to prevent lateral migration and excessive bed erosion. This would eventually recreate a channel with a regime geometry adjusted to current watershed conditions, the stream transporting the sediment load supplied from upstream. The channel was designed to approximately the 1 : 2 year flood (8.5 $m^3\ sec^{-1}$). The floodplain was therefore an integral part of the project, being used to dissipate flood flow energies and encourage the deposition of silt.

Many of these restoration techniques are cheaper on a per kilometre cost basis compared to constructed immobile channels. However, there is a clear risk of project failure, particularly in the short term before vegetation becomes established. Channel inspections and maintenance are important aspects of restoration. Without periodic inspection and maintenance to remove trash, fallen trees and overhanging vegetation, restoration would only be short-lived and moderately successful.

Biotechnical Engineering

In several countries there has been extensive research into the use of living vegetation rather than artificial materials as a means of bank protection, but particularly in West Germany and Austria the application of such methods has been apparent for the past 40 years (Bittmann, 1957; Olschowy, 1957; Hautum, 1957; Miers, 1977, 1979; Pfeffer, 1978; Binder *et al*., 1983; Wentz, 1983; Januszewski and Range, 1983). Living plants as a method of bank protection can be used in conjunction with inanimate construction materials such as steel sheet piles, wooden fascines, stone revetment, concrete paving or gabions. It is necessary to establish for each stream, even for each part of a stream, the method which is hydraulically and biologically correct. Seibert (1968) summarized how vegetation can be used to stabilize the channel bed, banks, floodplain and embankments (Figures 67 and 68). A zone of 'aquatic plants', permanently submerged, protects the channel bed from erosion by reducing the velocity and therefore the power of erosion. Species include watercrowfoot, pondweed and white water-lily. There may be a need to limit

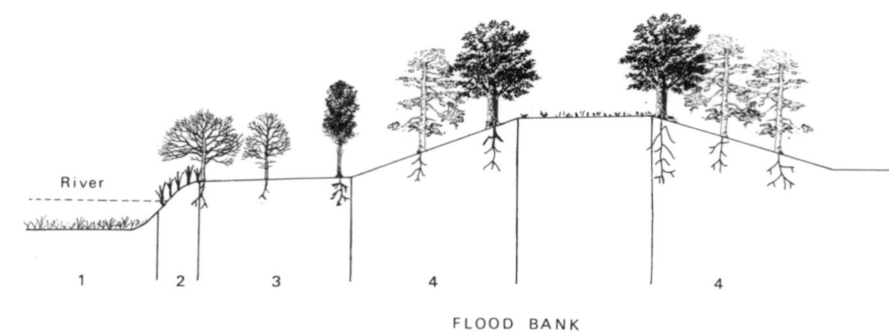

Figure 67. Zones of vegetation applied to Bavarian Rivers. (From Keller and Brookes, 1984. Reproduced by permission of the American Society of Civil Engineers.)

Figure 68. Use of vegetation to stabilize the River Vils, Bavaria, West Germany (photographed in 1983).

luxurious growth by cutting. However, the artificial implantation of aquatic plants is difficult and has so far seldom been tried. Emergent plant species implanted in the 'reed bank zone' at the margins of a channel restrict bank erosion by waves and currents. Species include bulrushes, sedges and reed grass; the lower part of the zone is permanently submerged for about half the year only. Roots, rhizomes and shoots bind the soil under the water whilst parts of plants above the soil surface protect against the currents and waves by friction. Experimental work on the River Thames in Oxfordshire (UK) has shown the effectiveness of marginal plants for protecting against bank erosion by boat wash (Bonham, 1980). The common reed (*Phragmites communis*) provided the maximum degree of protection; they are very robust plants and the shoots and rhizomes are deeply and strongly rooted and densely intertwinned. They bind the soil more firmly than any other reed and protection continued through the winter. They are capable of regeneration and proliferation and can be replanted. All types of reeds and sedge may be planted in clumps. The aerial organs of the reed colony are scythed and then square clumps are cut out, which are placed in a pit prepared in advance on the chosen site, at a depth in which they are submerged to two thirds of their height at the most.

Individual rhizomes and shoots of common reed, reed mace, bulrush, sweet flag, yellow flag and reed grass can be planted. Here too plants are taken from existing beds during the inactive season, after the aerial organs have been cut. Rhizomes and shoots are placed in holes or narrow trenches along the line of the average summer water level, so that only the aerial sprouts are showing above the soil.

In many cases, these methods do not consolidate the banks sufficiently during the period immediately after planting. Combined structures have therefore been designed, in which the bank is initially protected by inanimate materials. Reed rolls have been used for this purpose. A 40 cm trench is dug behind a row of stakes; wire netting is then stretched from both sides between upright planks; on this netting fill material such as coarse gravel is then dumped, and then covered with reed-clumps until the two edges of the netting can just be held together with wire. The upper edge of the roll should not be more than 5 cm above the level of the water. Finally the planks are taken out, and any gaps along the sides of the roll are filled in with earth and fragments of reed-clump. Reeds can also be planted in conjunction with paving or stone-facing by planting clumps, rhizomes or shoots in the crevices and gaps along the line of the average summer water level.

The roots of softwood trees such as willows and alders which are planted immediately adjacent to a channel may also stabilize banks 'zone of softwoods' (Grant and Fenton, 1948). Tree roots help to prevent erosion and are the deepest penetrating form of protection. Where there is a thick network of roots the bank is seldom breached. Trees collapse only if they are growing high on a bank and the soil below the roots is washed away. During periods of high-water, shrubs and bushes reduce the water velocity and thereby the erosive force of the water. It is essential that the wooded growth on banks must be continuous, especially around a bend where there is a risk of attack from high flows.

All softwood species have a high regenerative capacity; all species throw up suckers, and willows can also develop secondary roots on cut branches. The willows can be put into the soil, either individually as cuttings, slips or stems, or bound together to ensure immediate bank protection. They can be wattled or wired together to create fascines. Willow cuttings should be sunk into the soil to a depth of at least 20 cm. Fascines and fascine rolls are usually 4–20 m long and between 10 and 40 cm in diameter. These should contain brushwood sticks if possible, tightly wired. The parts which are intended to take root should touch the ground above the water level where they can get sufficient moisture. Covering with earth improves contact with the ground. Fascine rolls should be filled with coarse gravel or rubble.

The degree of protection can be increased by using barriers, mattresses or packed fascines. The barriers are made of willow switches 2–3 years old and 1.5–2.0 m long which are placed in line at intervals of 1–5 cm, perpendicular to the current and sloping downstream. They are set in a trench 15 cm deep, which is later filled in. The willow switches are held in place by wire, fascines or willow hurdling. Before the switches are set out, the stakes (0.6–1.0 m long) needed for the wire and for anchoring the hurdles to the ground, are driven in so that 10–20 cm are still showing. After setting up the switches the stakes are bound together with galvanized wire, then driven home until the switches are held firmly to the ground. If fascines or willow hurdles are used for consolidation they are spaced at intervals of 1.0–1.2 m. The whole barrier is lightly covered with earth, so that the branches are set in earth but not completely covered up.

A 'zone of hardwoods' may be established on the floodplain to protect embankments from erosion. Although trees have formed a valuable means of protecting alluvial river banks in various parts of the world, experience has shown that serious erosion and increased flooding can be caused by uncontrolled growth. In New South Wales, trees growing on sand and gravel deposits on the stream bed can cause severe erosion of the banks by directing and concentrating the flow (Rankin, 1980). Tree roots growing on the top of the bank may not penetrate to the toe, leading to undercutting of the trees, which then fall into the channel. Elsewhere the application of biotechnical engineering techniques may be restricted by the need for a large amount of land (Miers, 1977).

Reconversion

This category includes those instances where a channelized reach is reconverted to the original natural channel. Synonymous terms include renaturation or rehabilitation. Clearly this can be undertaken only where an enlarged or modified channel is no longer required to meet an engineering objective (Glitz, 1983; Brookes, 1987b). High-energy channels which have been straightened may regain their original size and sinuosity in the absence of maintenance. Several authors have examined the conditions under which straight alluvial channels develop meandering thalwegs (e.g. Callander, 1969; Ponce and Mahmood, 1977; Brookes, 1987c). However, to achieve reconversion it is often necessary to recreate the original channel with the aid of machinery.

There is evidence to suggest that, prior to logging earlier this century, free-flowing streams drained areas of the coastal plain of North Carolina that are now characterized by sluggish streams choked with debris that have created man-made swamps. In Onslow County channel work is carried out to eliminate these swamps and therefore mosquitoes. Flood control is not therefore a primary concern and engineers have found that if a meandering channel is excavated in black organic-rich soils which probably formed along the original course prior to logging, bank erosion is much reduced (Keller, 1976). Unfortunately this experience is not sufficiently documented in developing design criteria.

Sizing Channels

Stream channels in the coastal plain of North Carolina have experienced aggradation due to land use changes, such as farming, improved roads and ditches and logging operations, causing overbank flooding and constituting health hazards (Nelson and Weaver, 1981). They are restored to their former capacities by sizing neighbouring natural channels unaffected by sedimentation, and the procedure is summarized in Table LVI. The Manning equation is used to determine the discharge of the natural channel cross-sections, which are then related to the drainage area. The capacity of the aggraded channel is then determined and compared to that of the idealized natural channel for the same drainage area. The degree of excavation can then be estimated. The purpose of channelization was to remove accumulated sediment which constituted a health hazard by inducing inundation of the floodplain at greater frequencies and the pooling of water increased the mosquito breeding potential. Consequently the newly excavated channels were designed to accommodate the 2-year flood and this relatively simple technique was not recommended where significant impedance of flow downstream could affect the water surface profile. Although it is suggested that the procedure may be applicable to other southeast States with similar geographic and climatic characteristics, for other regions new formulae need to be developed.

Table LVI. Criteria for sizing North Carolina's coastal plain stream channels (after Nelson and Weaver, 1981)

Step 1. Estimate the discharge of the two year flood (Q_2) for each section from $Q_2 = 59.3(A)^{0.664}$, where A is the drainage area.

Step 2. Determine the existing capacity of each section using the Manning equation and assuming bankfull.

Step 3. Compare the existing capacity with the Q_2 to give the reduction of capacity. The discharge capacity can be increased by excavating the channel depth, the new dimensions being determined from the Manning equation.

Restoring the Sinuosity

The former sinuosity, cross-sectional dimensions, slope and substrate of a small straightened stream channel in southern Denmark were recreated in 1984/5 (Figures 69 and 70); (Brookes, 1987b). The Stensbæk stream has a drainage area of only 6 km^2. The new channel was intended to replace an 800 m length of straightened channel which was severely degrading. The original course was determined from historical maps, from a field reconnaisance of depressions in the floodplain and from comparison of naturally sinuous streams in neighbouring catchments with similar physical characteristics. A series of trenches excavated in the floodplain enabled the approximate cross-sectional dimensions of the old channel to be determined and these were checked by reference to natural reaches, with a similar drainage area, elsewhere in the catchment and from neighbouring basins. The trenches also provided information on the longitudinal oscillations associated with a pool–riffle sequence and on the substrate type. Reconstruction of the detailed variations in the long profile was left until the construction phase. A combination of grasses, for quick initial stabilization, and woody species for longer-term stability were planted. Native species were obtained from the straightened channel before it was backfilled. The recreated course required stabilization by riprap on the outsides of bends before vegetation became established, and there was a ready supply of local assorted stones derived from glacial deposits. Stones were also stockpiled for later use to counter localized scour, particularly evident after high flows. Gravels were placed on the stream bed to recreate a naturally occurring layer, sufficiently stable to restrict renewed downcutting. The new course restored morphologic and hydrologic diversity and the channel became successfully colonized by a variety of fauna and flora.

The technique assumes that no major watershed changes have occurred since initial straightening, which would disrupt the equilibrium of the restored channel through altered sediment and water discharges. The technique has been recommended for a number of lowland streams in Denmark, but clearly each case needs to be considered individually. These are relatively low-energy streams which under natural conditions do not actively change their positions. Further studies are required in a range of physical environments (Brookes, 1987b). Application to actively migrating channels should be treated with caution.

Glitz (1983) recreated the former meandering course of the Wandse in Hamburg-Rahlstedt, West Germany, a lowland river about 1.5 m in width. This was done in 1982 over a length of about 1 km. However, the stream profile was not recreated in detail because it was believed that the stream would adjust naturally. However, a survey carried out in the summer of 1985 indicated that only limited reformation of pools and riffles had taken place, probably because of the extremely low energy of the stream. One novel aspect of the project was to leave as many of the abandoned

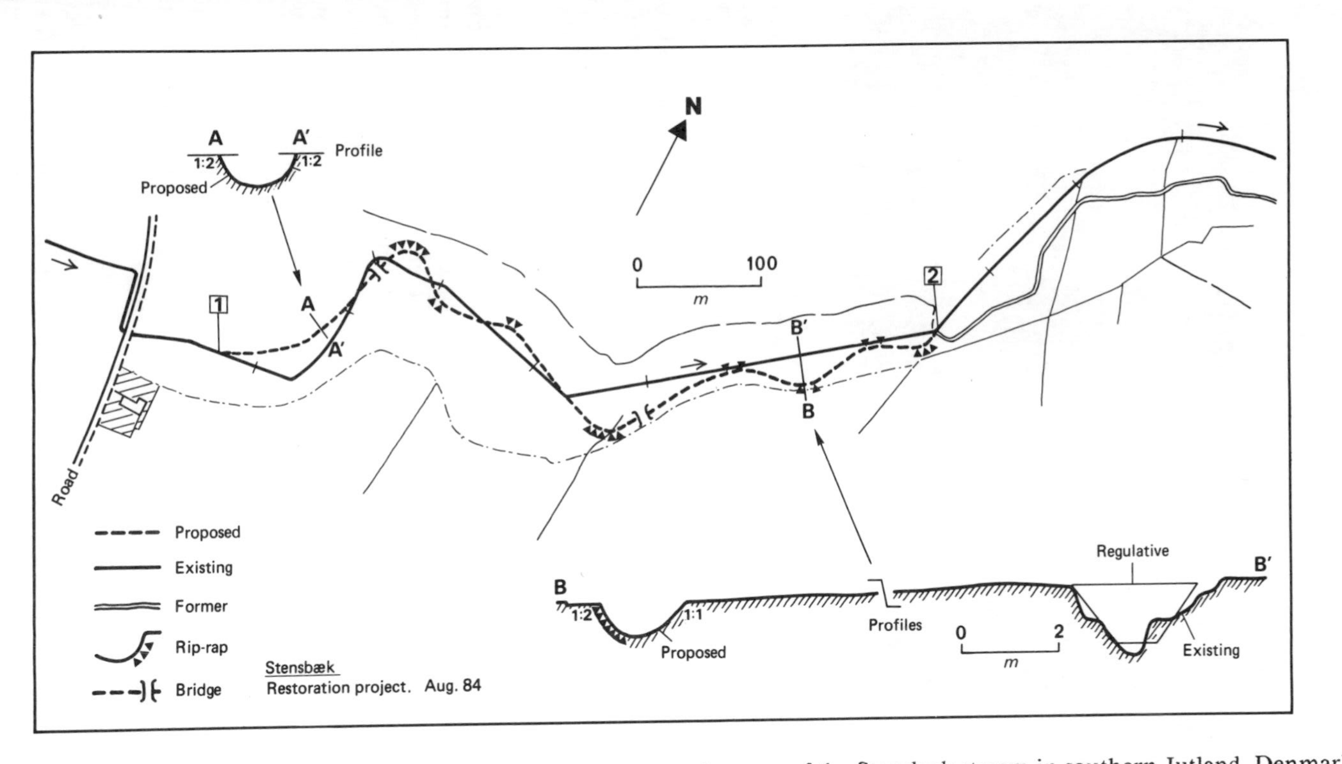

Figure 69. Plan showing the position and dimensions of the restored course of the Stensbæk stream in southern Jutland, Denmark. (From Brookes, 1987b., Reproduced by permission of Springer-Verlag, New York.)

Figure 70. Completed course of the Stensbæk in summer 1985, looking downstream from near the footbridge. Species of grass have grown from seed planted along the banks to provide stability and a natural appearance. (From Brookes, 1987b. Reproduced by permission of Springer-Verlag, New York.)

straightened sections as possible as backwaters, to allow rapid colonization of the new course by plants and animals. The intention has been to monitor the biological changes every two years.

Roughness Elements

In streams in northern California which were cleared in the late 18th and early 19th century for navigation, the habitat for salmonids has been improved by reintroducing large-scale roughness elements (Lisle, 1981). These include boulders and woody debris which change the distribution of hydraulic forces over a stream bed and thereby cause scour and the sorting of fine sediment from gravel. Figure 71 shows the predicted effect of placing large organic debris on the right bank of Jacoby Creek in California.

Rehabilitation of Incised Channels

A geomorphic approach to rehabilitation of incised channels, which differs markedly from the engineering approach of total structural control, has been described for northern Mississippi (Harvey *et al*., 1982; Schumm *et al*., 1984). A review of historical records and intensive fieldwork enabled the equilibrium status of a channel to be assessed. If the channel is in disequilibrium then the magnitude of the problem can be determined at low cost by making a *Reconnaisance level* study involving photographs, maps, field

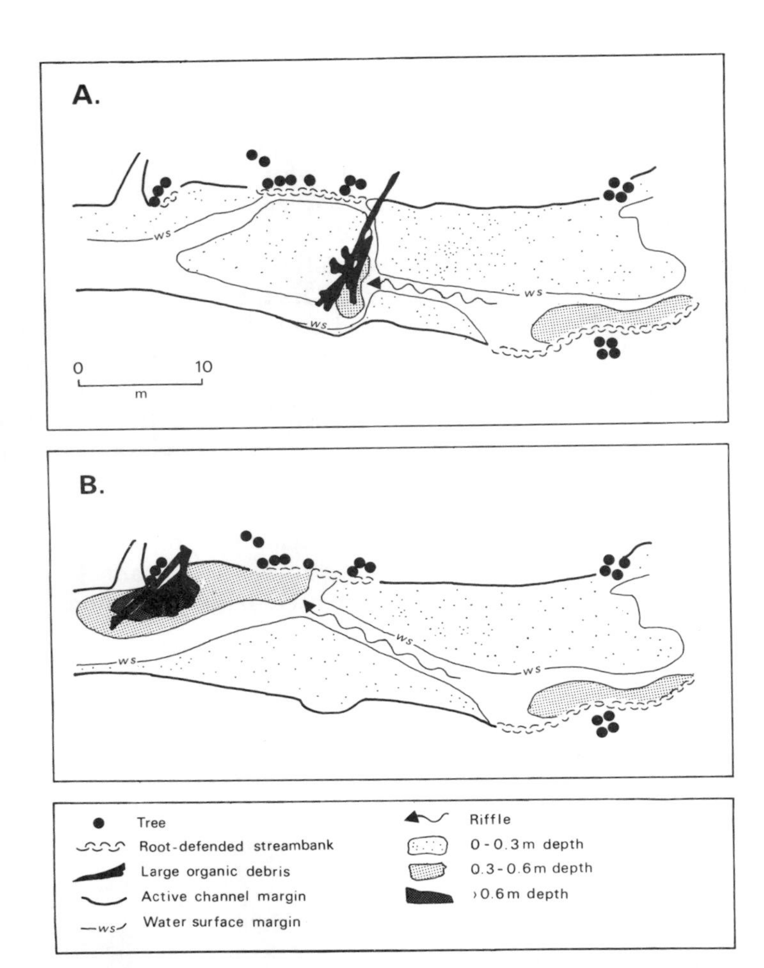

Figure 71. Survey of Jacoby Creek at low flow, showing expected results of moving and stabilizing root wad and associated debris to a new position on the right bank (August–September 1981). (After Lisle, 1981. Reproduced by permission of Thomas J. Hassler.)

inspection and calculation of drainage areas and channel width. At the *Survey level* allowable channel slopes and discharges can be determined without detailed hydrologic and hydraulic studies. The exact locations and numbers of grade-control and flow-control structures can then be determined at the *Design stage*.

Floodplain Approaches

This category includes those management procedures which utilize the floodplain for flood control as an alternative to modifying the existing channel. Such approaches enable existing morphologic and biologic characteristics of the channel to be preserved.

Corridors

For rivers which are actively changing their courses it has been suggested that management must provide a sufficient corridor to allow continued change (Figure 72) (Palmer, 1976). This 'floodway zone' develops a valley floor by the river channel shifting across the floodplain. In this zone the supply of sand and gravel is balanced by the ability of the stream to transport sediment and a moderately steep slope provides the energy for this transport. Channel features such as pools, riffles and point bars, have a high value for wildlife habitat. Although a river might shift over the entire floodplain in a period of several hundred years, migration of individual bends is usually accomplished in a few tens of years. Thus the meander belt may be confined to a fixed position on a floodplain over a period of time for planning and management. Geomorphologists can attempt to determine the width and location of a corridor that can be maintained for planning objectives (50–100 years). Whilst river response cannot be accurately predicted the corridor concept may provide a valid alternative to confining a channel to a single position by either straightening and/or extensive bank protection.

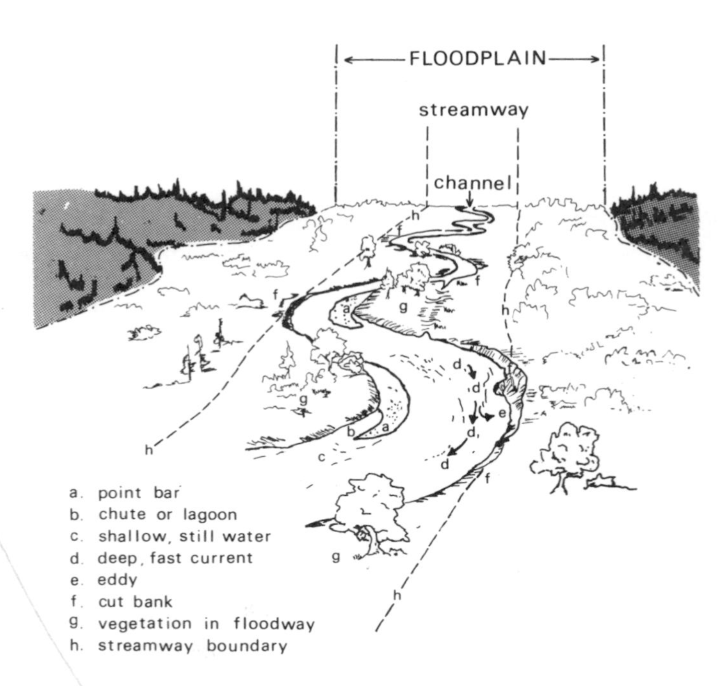

Figure 72. The streamway concept. (Modified from Palmer, 1976. Reproduced by permission of George Allen & Unwin.)

Binder *et al.* (1983) recommended the preservation of the dynamics of meandering channels based on experiences in Bavaria, West Germany. In certain cases it is possible to leave rivers to their own devices. Within this belt meanders, braided reaches and pools and riffles are preserved. This maintains natural habitat conditions. Where space is restricted then bank stabilization may be necessary to limit the migration of rivers which would otherwise take up the whole valley floor. The preservation of old meander loops and land adjacent to the channel provide habitats for a variety of fauna and flora (Table LVII).

Table LVII. Recommendations for the care of watercourses (after Binder and Grobmaier, 1978)

1. Acquisition of the watercourse, together with a 5 m wide strip on both sides of the channel. Acquire abandoned meander loops and connect to river to provide flow of water.
2. Acquire stands of trees on the former meander loop.
3. Preserve stands of polars and other indigenous species.
4. Acquire reed bank in the former meander loop.
5. Acquire the land forming the slip-off slope of the meander; plant with trees and protect with plants.
6. Acquire land with the remnants of the water meadow forest trees growing on the floodplain.
7. Plant the river banks with indigenous species, particularly to protect the outer bank.
8. Maintain existing trees growing along the watercourse.

For artificially straightened channels in Denmark which were regaining their former sinuous courses in the absence of maintenance, it has been proposed that a streamway corridor be delineated within which the stream can continue to migrate through the processes of erosion and deposition (Brookes, 1984, 1987c). This can be a far more economic solution than using extensive engineering methods to constrain a stream's course. Corridors preserve the instream habitat and avoid erratic responses associated with more conventional channelization procedures. Clearly a corridor cannot be created in those circumstances where continued migration of a channel threatens property. A further limiting factor, highlighted by the examples from Bavaria, is the requirement for purchase of large tracts of land by the local government. However, education of owners in the subject of river processes may ultimately help to overcome unnecessary concern (Palmer, 1976).

Floodways

Floodways or bypass channels are separate channels constructed on floodplains into which high flows are diverted (Nunnally and Shields, 1985). The existing natural channel is left unaffected.

Research Needs

Whilst some alternatives such as mitigation and enhancement techniques have general application to a broad range of river types, the more specialized procedures have been developed for specific areas with particular geographic and hydrologic characteristics. Table LVIII attempts to summarize the suitability of alternatives in relation to channel characteristics and basin conditions, based on experience which is documented in the published literature. Application of a particular technique may be described either as successful, problematical or unsuccessful. It is apparent that the majority of available techniques are not applicable to high energy, braided river channels. There are many environments where alternatives have not been tried and there is a need for further published studies. For example, Miers (1977, 1979) realized the need for research into the application of biotechnical solutions to the UK. In particular, studies are needed to ascertain the effect of vegetation on channel capacity, the use of vegetation to restrict bank erosion, the effects of trees and shrubs on the stability of floodbanks, the use of non-uniform channel geometries and methods of maintaining such watercourses. It is likely that there are many examples of alternatives which have been formulated and implemented by those organizations responsible for river management, but not published in a readily available form. Studies and handbooks concentrate mainly on the successful application of alternatives, and it is difficult to find documented failures.

It is also clear that whilst alternatives have been shown to be successful in diversifying the channel morphology, biological improvements have not been adequately quantified. Scientific evidence of the biological advantages of adopting alternative approaches is needed to ensure continued application by engineers. Monitoring of biological populations before, during and after installation is required. There is also a need to assess the aesthetic merits of alternative practices, compared to conventional channelization practices (Chapter III). Although techniques for the evaluation of the aesthetic quality of rivers were advocated in the late 1960s and early 1970s (McHarg, 1969; Leopold and Marchand, 1968; Melhorn *et al.*, 1975) little has since been done to develop and apply such techniques.

Whilst wildlife and conservation interests increasingly demand alternative designs which treat rivers sympathetically (Water Space Amenity Commission, 1980a,b; Newbold *et al.*, 1983; Lewis and Williams, 1984), it should be appreciated that engineers involved in river channel management do not yet have a full understanding of the hydraulics of such unconventional designs

Table LVIII. Suitability of alternative designs according to physical conditions

	IMPACT MINIMISATION						MITIGATION / ENHANCEMENT		UNIQUE DESIGNS				RECONVERSION			FLOODPLAIN	
	REALIGNMENT	MEANDERING ALIGNMENT	ENLARGEMENT	READJUSTMENT	SELECTIVE CLEARING & SNAGGING	SHADING	INSTREAM HABITAT DEVICES	POOLS AND RIFFLES	ASYMMETRICAL SECTION	TWO-STAGE CHANNELS	PARTIAL RESTORATION	BIOTECHNICAL ENGINEERING	SIZING	SINUOSITY	ROUGHNESS	CORRIDORS	FLOODWAYS
CHANNEL CHARACTERISTICS																	
CHANNEL PATTERN																	
BRAIDED	○	○					○	○	○	○	○		○	○	○	○	○
MEANDERING	●	●		●					●	□	●			●		●	●
STRAIGHT	○	○								●							
BANKFULL DISCHARGE																	
LOW							●										
MEDIUM							□								●	●	
HIGH							○										
SLOPE																	
LOW							□						●	●		□	
MODERATE	●	●					●				●					●	
HIGH	□	□		●			○			□	□					○	
SEDIMENT LOAD																	
HIGH BEDLOAD		□					○	□	□	○	○					□	
HIGH SUSPENDED LOAD							□										
LOW LOAD							●		●	●			●	●			
CHANNEL STABILITY																	
STABLE				○									●	●	●		
UNSTABLE	○	□						□	□	□					□	□	
SUBSTRATE																	
BEDROCK				○			□		□		○				●	□	□
GRAVEL							●								●		
SAND				□			□	○		□	○			□		●	
SILT										●			●			●	
BANK COHESIVENESS																	
HIGH				□						●							
LOW	□	□						□		□	□		●	●	□	□	
POOL & RIFFLE SEQUENCE																	
WELL DEVELOPED								●									
POORLY DEVELOPED								□									
BASIN CONDITIONS																	
UNDISTURBED							●						●	●	●		
DISTURBED				□			□	□	□	□	□		○	○			

● SUCCESSFUL APPLICATION

□ PROBLEMATICAL

○ UNSUCCESSFUL

(Newson, 1986). Many of the techniques have a limited influence on the hydraulics of a channel. For example, reinstated pools and riffles or other instream habitat devices may drown-out at higher flows and have a minimal impact on the flood conveyance of a channel. However, studies have shown how the growth of thick vegetation such as willows and shrubs in channelized reaches can reduce the discharge capacity by up to 50% in less than a year (e.g. Wilson, 1973). Further work is required to improve the understanding of flood channel hydraulics, particularly non-straight channels, non-parallel floodplains and the seasonal management of floodplain vegetation. This deficit is gradually being rectified by research throughout the world, and a five-year programme is currently under way in the UK to investigate the hydraulic behaviour of compound channels, including two-stage designs. Until a better understanding of the hydraulics of certain alternatives is available it will be difficult for design engineers in several countries to move away from traditional engineering solutions.

Hydraulics Research (UK) is currently proposing research into environmentally sensitive scheme design in conjunction with several UK water authorities. The first stage will be to produce a review of available literature and results of the hydraulic effects of alternative designs such as the influence of vegetation on Manning's '*n*'.

Postscript and Prospects

CHAPTER IX

Postscript and Prospects

For more than a century studies have been undertaken with the objective of revealing the character and extent of changes in the environment produced by human action (e.g. Marsh, 1864) and more recently humans have been identified as 'the most important geomorphological factor' (Demek, 1973). During the last two decades throughout the developed world there has been an accelerated growth of awareness relating to environmental impacts such as deforestation, cultivation, reclamation, urbanization and pollution. Alterations to river channels are required to achieve the objectives of flood protection, drainage, navigation and the prevention of erosion. In view of the proliferation of channelization to achieve these objectives the extent of impact on river channels should not be underestimated.

River engineering solutions involve a conflict between man and nature and, whilst the undesirable consequences have been recognized in the United States for at least two decades, it is only recently that environmental issues have come to the fore in other countries. In 1977 the Committee of Ministers of the Council of Europe adopted a resolution for the protection of lakes and river banks. A meeting held in Athens in 1984, attended by members of the Council, discussed the planning and management of the water's edge. The European Information Centre for Nature Conservation mounted a campaign under the title 'The Water's Edge', which includes some of the most endangered habitats in the natural environment. After centuries of destruction it was recognized that the last vestiges of natural habitat of the water's edge including river banks should be preserved. An issue of *Naturopa* in 1984 described rivers and streams (Broggi, 1984). Whilst recognizing that in some European countries government directives ensure that river engineering is consistent with nature conservation, Broggi recommended the following action: (1) make inventories of the biological condition of watercourses; (2) make provision for the preservation of all parts of watercourses which are in need of protection; (3) prohibit further rectification and canalization of watercourses; (4) use alternative procedures wherever possible (e.g. biotechnical measures and diverse morphology); (5) maintain and enhance natural riparian vegetation (e.g. sow wild grass); (6) limit and control regulation of recreational traffic. 'The European Year of the Environment' from 21 March 1987 to 20 March 1988 was an example of the concern for the environment and a number of water organizations throughout Europe undertook to restore watercourses throughout this period.

Continued Pressure for Floodplain Development

Human occupancy of floodplains will continue and the pressures for development will remain. At a time of food surpluses the demand for drainage associated with more intense agricultural practices may be temporarily abated. However, roads will still be built and improved. Urban areas will continue to expand, and when they do expand they will accelerate runoff, thereby exacerbating flooding and leading to the demand for flood control measures. A major dilemma in the United States is that in spite of increasing annual expenditures for flood control, losses from flooding continue to rise. This results from an overdependence on federally supported structural solutions to flood problems (Costa, 1978b). The construction of a flood scheme gives a false sense of security to floodplain dwellers and both reduces their vigilance and makes the refusal of planning permission for new buildings in the protected area difficult (Hollis, 1976). The design discharges of schemes will be exceeded at some time in the future. A reduction in the frequency of flooding can be achieved but this will only increase the enormity of floods when they do occur in towns and cities. For the United States, Costa (1978) advocated that risk and responsibility should be given to the local government and that development should be more strictly controlled, otherwise the only alternative would be to modify every river and stream in the country. Because these pressures will continue there is a need for improved understanding of the environmental effects of channelization, for environmental impact assessments, for interdisciplinary approaches to channel management and for the development of alternative strategies.

Channelization and the Environment: Antithesis ?

> The engineer who alters natural equilibrium relations by ... channel improvement measures will often find that he has the bull by the tail and is unable to let go—as he continues to correct or suppress undesirable phases of the chain reaction of the stream to the 'initial stress' he will necessarily place increasing emphasis on study of the genetic aspects of equilibrium in order that he may work with rivers, rather than merely on them.
>
> (J. Hoover Mackin, 1948)

The dramatic and adverse effects that channelization practices can have were first appreciated in the United States of America. Since channelization involves manipulation of one or more of the dependent hydraulic variables of slope, depth, width and roughness, then feedback effects may be instituted which proceed to promote a new state of equilibrium. Channelization projects, particularly if their design is based on inadequate background data, may induce instability not only in the improved channel reach but also upstream and/or downstream of the reach. In each section erosion or deposition may take place. Perhaps the most dramatic adjustments occur in response to slope changes associated with channel straightening or regrading, or extended bottom widths. However, maintenance works such as clearing and snagging can also adversely affect the channel morphology.

Although these effects can be quite severe in some cases, the environmental impacts associated with channelization do not occur with equal frequency or severity. Considerable regional contrasts in the consequences can be found throughout North America (Hill, 1976) and clearly there is a requirement for many more studies which document the responses to channelization in a variety of fluvial environments (Ritter, 1979). It is also unfair to suggest that all channelization projects result in a deteriorated environment (Ruhe, 1970) and channel works are often undertaken to maintain rivers which have become choked with sediment and debris arising from land use changes in the watershed upstream (Hay and Stall, 1974; Richardson and Christian, 1976; Keller, 1976). Streams carrying a high sediment load can have a detrimental effect on the stream ecosystem because sediment reduces the growth of valuable waterfowl food plants (Grizzell, 1976).

However, the manipulation of hydraulic variables by channelization, and subsequent adjustments, may substantially alter the channel morphology with corresponding dramatic consequences for the aquatic ecosystem. Loss of morphologic variability, erosion and sedimentation affect all forms of aquatic life. The most documented effects are those relating to fish and macroinvertebrates. Perhaps the most commonly stated reason for change in fish populations is the loss of a natural pool–riffle sequence which provides a variety of low-flow conditions suitable as cover for fish and for the organisms on which fish feed. Shelter areas are also required at high flows to protect fish from abnormally high water velocities, and these conditions are absent where a meandering stream is replaced by an artificially straight channel, composed mainly of riffle. Disturbance of the natural substrate may also adversely affect the bottom-dwelling organisms on which fish feed. Clearance of bankside and instream vegetation during construction will eliminate the valuable cover and shade required by fish. The synergistic action of altering these parameters may be a dramatic reduction in the numbers of native species and diversity, and colonization by less desirable species and smaller fish.

> Channelization of natural streams and the productivity of fish and wetland wildlife ecosystems are unequivocally antithetical. Channelization severely degrades the functional and structural properties of natural stream and riverine communities by destroying habitat, species diversity, stability, and the productive capacity of the systems. Channelization accelerates and makes possible the drainage, conversion and loss of wetland habitat and bottomland hardwood forest ecosystems. Chan-

> nel maintenance, when practiced, arrests recovery of stream and wetland ecosystems. Lack of maintenance permits a progressive recovery, but never approaching natural stream productivity or usefulness as an aquatic-wildlife resource.
>
> (Philadelphia Academy of Natural Science, quoted in Committee on Government Operations, 1973b, p.10)

The degree of impact may also depend on the size of a project. Long-reach channelization (greater than 1 km length) may have the most dramatic impact, whilst the effects of short-reach channelization (up to 0.5 km) may be relatively insignificant, although difficult to assess (Bulkley *et al*., 1976).

However, once modified, morphological and biological recovery may be very slow. Straightened channels in particular environments can develop a series of pools and riffles and may ultimately regain their sinuosity, thereby becoming more biologically productive. Where a channel design is based on inadequate data then there may be serious problems of erosion and sedimentation, requiring costly maintenance which further offsets biological recovery.

AN INCREASING ROLE FOR THE ENVIRONMENTAL SCIENTIST

Assessing the Impacts

Jean-Pierre Ribaut suggested in 1982 that:

> Further research and study is universally required to provide the data essential for qualitative and quantitative preservation of our natural heritage (or what remains of it).

Few studies have identified and quantified the environmental variables which are altered by channelization and that affect the distribution and abundance of organisms. For some groups of organisms and environmental situations biological measurements are totally lacking. Information is also required from a variety of environments on the nature and rate of biological recovery. There is not yet a sufficient number of case studies available to make accurate predictions about recovery following channelization. Monitoring a range of channel works before, during and after construction to assess the physical, hydrological and biological responses would provide valuable databases for future decision-making. Further data could be compiled by universities as part of a research programme or as contract research, by consultants, or by the scientists working in those organizations responsible for river management. Post-monitoring could be done in conjunction with an assessment of other aspects of channel works, including economic performance and recreational interest. There is also a need for a classification of the relationships between morphological variables and biological parameters which will provide information for improving engineering design to minimize the biological impact.

Interdisciplinary Management

The alteration of natural river channels must be carefully planned and executed if a stable and environmentally acceptable channel is to be achieved (MacBroom, 1980). River engineering schemes have often been described as 'horses for courses', and problems have often been solved by a combination of good judgement and past experience rather than resorting to empirically-based equations. Traditionally, works have been appraised principally in terms of economic costs and benefits (Handmer and Penning-Rowsell, 1987). The preliminary appraisal or design of channel modifications should be an interdisciplinary effort in order to ensure consideration of all types of existing and potential projects (Shields, 1982a,b). This should include the developer, engineer, geomorphologist, biologist, conservationist and other local interest groups. Such an approach, based on environmental impact assessment, has already been adopted in several countries, including the United States and West Germany. Environmental impact statements are required before, during and after channelization to enable more accurate predictions of the likely impacts.

Geomorphologists can contribute to the planning stage by identifying those sites where instability might be expected following construction. Coates (1976b) argued that the 'geomorphic engineer' is complementary to the civil engineer *per se*. Whereas the engineer is concerned with the design, siting and construction of particular structures, the geomorphic approach entails background knowledge of the function of structures, together with an understanding of the local and regional effects. The participation of the geomorphologist in river engineering should therefore be sought at the planning stage and not as an after-thought once problems have arisen. Such participation will avoid failure of structures, unexpected extra costs due to maintenance and loss of time. The biological response of a given channel type to a given set of physical changes can also be anticipated at the planning stage. The environmental scientist can further contribute to the design, construction and maintenance stages. Environmental objectives are best served when channel design and construction are painstaking processes with meticulous attention being given to the details of a site, rather than an overall analysis of the structural and hydraulic considerations (Shields, 1982a,b).

Anticipating the Consequences

The effects of river engineering projects are very difficult to predict. This is because some phenomena in the formation and deformation of river channels are still not yet fully understood and models cannot yet give totally reliable results (Jansen *et al.*, 1979). Simple problems such as the prediction of the final state of equilibrium of a river bed can be solved using a mathematical model. It is also possible to predict the intermediate stages although the accuracy of these predictions can be questioned. Scale models, which are often used to investigate complex river bed changes, are also very limited because of scale effects. Therefore inaccuracy in predictions should be allowed for at the design stage by incorporating flexibility.

Unfortunately without a set of deterministic equations it is not possible for the geomorphologist to precisely predict the mophological response to alterations of width, depth, slope, roughness or channel planform caused by channelization. However, behaviour can be deduced to a certain extent by observing responses of stream channels that have already been altered. For example, if a sufficient number of case studies are used then increases of channel slope, bankfull discharge, and therefore of stream power, following channelization delineate the general magnitude of potential post-construction adjustments. This allows objective decisions to be made on the most appropriate engineering solution to a particular river problem. Ultimately it may be possible to predict more precisely the future morphology of river channels at the planning stage of an engineering scheme. Predictions of the morphological response to river engineering can be used as an input to wildlife impact methodologies.

It also necessary to envision channel management within the context of the drainage basin as a whole (Brookes and Gregory, 1988). This is one of the major bridges that needs to be built between geomorphology and engineering. River channelization may not be necessary when floods are abated through the reduction of runoff by small headwater dams located in the basin upstream or by land treatment. Failure to consider the river as an integral part of basin management when planning and designing a flood control project has frequently led to environmental problems. Both the hydraulic geometry and regime approaches to river channels are essentially empirical and consider streams as equilibrium systems. However, most channels are in a transient state as a result of human activities either having a direct impact on a channel at the scale of a reach, or indirectly on catchment processes, altering the sediment and water discharges. It is important to consider at the design stage of a scheme the effects that changes in a basin will have, and a number of functional relationships have been produced which describe the nature and direction of channel adjustment (Schumm, 1969, 1977). In turn channel adjustments may trigger changes in the ecosystem. The results of research investigations from a number of areas in the world emphasize how much change has occurred and how diverse the changes can be. These studies embrace a range of channel adjustments including changes of channel cross-section (capacity, width, depth, hydraulic radius and width–depth ratio) and of channel reaches (volume, sinuosity, meander wavelength, channel slope, channel length, channel island area, surface area and floodplain area).

The particular combination of parameters or degrees of freedom which adjust in any particular area is a response to drainage basin and local channel controls. Although it is not easy to predict what will change at a particular location, the further adoption of a basin approach will assist understanding and facilitate prediction. As a result of connectivity in the fluvial system repercussions of channelization can be transmitted over a wide area, especially downstream. Downstream consequences which should be anticipated include increased flooding, morphological effects, water quality impacts and ecological changes.

Improved Designs

It is still not possible to predict and explain the 3-D mutual adjustment of all the degrees of freedom of self-formed alluvial channels. For this reason the design of artificial channels using traditional design techniques remains restricted to uniform trapezoidal sections in straight reaches and only in a narrow range of environments. Alternative strategies have been developed as a means of 'working with the river rather than against it' (Winkley, 1982) and are a substantial departure from the conventional notion that 'technology can fix it' (Leopold, 1977). Geomorphological understanding is an essential component of alternative channelization designs which emulate the natural characteristics of river channels. These alternatives include the preservation of natural channels, minimization of the impact of conventional engineering practices, enhancement of existing engineering projects and the design or re-creation of natural channels. Those aspects of fluvial geomorphology which are helpful in design have been listed in Chapter VII, p. 191 as: (1) the concept that streams are open systems in which channel form and process can evolve in harmony, (2) utilization of the convergent–divergent criterion, (3) identification of geomorphic thresholds for stream channels, and (4) recognition of the complex relationships between deposition, erosion and sediment concentration (Keller, 1976).

It is necessary to establish for each stream the method which is hydraulically and biologically correct. Some indication of the physical constraints which have determined the success or failure of alternative designs can be obtained from published results. Many of the alternatives are new concepts developed for specific areas and therefore their application to other sites, particularly in different physical environments, should be treated with caution. Although there is no absolute guarantee that proposed stream restoration projects will be entirely successful in the longer term, many of the schemes implemented so far demonstrate that alternative techniques based on an understanding of the form and process of natural channels can be successful when their design is tailored to local conditions. Many of the projects described in Chapter VIII must be carefully evaluated over longer time periods to evaluate hydraulic and environmental performances. In particular there is a need for scientific studies which quantify the biological effects and for cost studies which determine the effectiveness of alternatives. Implementation of environmental measures requires close supervision and inspection of construction and maintenance.

Further advances are clearly necessary in hydraulic engineering, and work is currently under way to assess environmentally sensitive design techniques at the US Army Waterways Experiment Station in Vicksburg, Mississippi, and at Hydraulics Research Ltd, in England. Research topics include two-stage designs, the stability of reinstated habitat features over the longer term, and the impact of regrown or replanted vegetation on the discharge capacity of a channel. Considerable advances in research are required before all adverse environmental effects can be minimized. The research programme at the Waterways Experiment Station has so far achieved much to fill the gaps in knowledge and to link research with action (e.g. Nunnally and Shields, 1985). Preliminary design guidelines for future channel projects that reduce environmental impacts were compiled mainly from the early experiences of conservation agencies.

Without improved and fuller understanding the authors of practical handbooks of river management will continue either to omit the hydraulic aspects or to substitute the gaps in knowledge with generalized descriptions which may be misleading. Decision-making becomes less subjectively based as this information becomes available and ultimately the environmental scientist will be better placed to advise on channel changes and management techniques.

THE FUTURE

Alternative strategies may continue to be developed and applied throughout the world as governments and organizations responsible for river management become aware of the increasing need for reconciling engineering objectives with nature conservation. For example, the imminent development of environmental impact legislation in the European Community may lead to detailed scrutiny of the application of existing technologies and the development of alternative methods. Whilst it cannot be denied that many alternatives have been environmentally successful, and should continue to be applied, there remains the need for further investment in hydraulic and geomorphic research to convince the practising river manager.

The advancement of scientific knowledge, summaries of which have been the focus of this book, is not enough in itself to ensure improved environmental management. It is only when scientists become involved in the formulation of public policy and the production of practical guidelines and handbooks that environmental research and action are linked. The Council of Europe is an international organization with the responsibility of informing, liaising and educating. Since the European Committee for the Conservation of Nature and Natural Resources was set up in 1961 it has been aware of the need to collate scientific data and to make specific recommendations on conservation, development and management (e.g. Seibert, 1968).

Advances are well demonstrated by the studies carried out under contract by various research organizations throughout the United States for the Stream Alteration Programme of the Fish and Wildlife Service. Within the framework of the National Agency of Environmental Protection in Denmark researchers have contributed to public policy by producing a series of guidelines which are used by those organizations responsible for river management (e.g. Kern-Hansen *et al*, 1983; Holm, 1984; Brookes, 1984). Within the UK a number of scientists have contributed to, and are increasingly being asked to give advice on, alternative strategies for the management of rivers (e.g. Lewis and Williams, 1984; Holmes, 1986; Thames Water Authority, 1988b; Countryside Commission, 1987, 1988). It is through these practical handbooks that scientific knowledge has become more rapidly disseminated and has reached the managers who make key decisions.

Appendix—Notation

The following symbols are used in this book

Symbol	Quantity	SI units
w or b	channel width	m
d	hydraulic mean depth	m
A	cross-sectional area	m^2
p	wetted perimeter	m
F	width–depth ratio	
R	hydraulic radius	
s or S	bed slope or energy gradient	
n	Manning's roughness coefficient	
f_L	Lacey's silt factor	
D_{50}	median size of bed sediments	mm
ϕ	mean grain size of sediment	mm
P	sinuosity	
d	particle diameter	
L	meander wavelength	
v	velocity	$m\ s^{-1}$
U	local velocities	$m\ s^{-1}$
$\bar{\bar{U}}$	section mean depth velocities	$m\ s^{-1}$
$\bar{U}$	depth mean velocities	$m\ s^{-1}$
Q or Q_w	water discharge or bankfull discharge	$m^3\ s^{-1}$
Q_s	bed material discharge	
g	gravity	$m\ s^{-2}$
p	density of water	$kg\ m^{-3}$
w	specific weight of water	$kN\ m^{-3}$
ω	power expenditure per unit area	$W\ m^{-2}$
τ	shear stress (force per unit area)	$N\ m^{-2}$
τ_0	unit tractive force	$N\ m^{-2}$

τ_b	maximum tractive force	N m^{-2}
τ_c	critical tractive force	N m^{-2}
τ_s	permissible shear	N m^{-2}
$\varnothing$	angle of repose	
θ	angle of channel side slope	
k	coefficient	
z	ratio of horizontal to vertical distance for bank slope	

References

Acheson, A.R. (1968). *River Control and Drainage in New Zealand*. Ministry of Works, Wellington North, New Zealand.

Adkin, W.B. (1933). *Land Drainage in Britain*. Estates Gazette Limited, London.

Agostini, R., and Papetti, A. (1978). *Flexible Linings for Canals and Canalized Water Courses*. Officine Maccaferri S.P.A, Bologna, Italy.

Agostini, R., Conte, A., Malaguti, G., and Papetti, A. (1985). *Flexible Linings in Reno Mattresses and Gabions for Canals and Canalized Water courses*. Officine Maccaferri S.P.A, Bologna, Italy.

Agostini, R., Bizzarri, A., and Masetti, M. (1981). *Flexible Gabion Structures in River and Stream Training Works. Section one: weirs for river training and water supply*. Officine Maccaferri S.P.A, Bologna, Italy.

Ames, T. (1970). From swamps to rich farms. *Illinois History*, **24**, 19–20.

Andersen, T. (1977). Danish legislation on the use and protection of freshwater areas. In *Danish Limnology: Reviews and Perspectives* (ed. Hunding, C.), *Folia Limnologica Scandinavica*, **17**, 125–30.

Anderson, A.G., Paintal, A.S. and Davenport, J.T. (1970). Tentative design procedure for riprap lined channels. *National Cooperative Highway Research Program Report*, 108, National Academy of Sciences.

Anon (1972). Whiskey Creek improved looks and flow. *Michigan Contractor and Builder*, October 14.

Anon (1977). *River improvement*? Environmental Awareness Publication No. 2, Conservation Council of Victoria, Victoria, Australia.

Apman, R.P., and Otis, M.B. (1965). Sedimentation and stream improvement. *New York Fish and Game Journal*, **12**, 117–26.

Arner, D.H., Robinette, H.R., Frasier, J.E. and Gray, M.H. (1975). Report on effects of channelization modification on the Luxapalila River. In *Symposium on Stream Channel Modification*. Harrisonburg, VA, pp. 77–96.

Arner, D.H., Robinette, H.R., Frasier, J.E., and Gray, M.H. (1976). *Effects of Channelization of the Luxapalila River on Fish, Aquatic Invertebrates, Water Quality and Fur Bearers*. Report No. FWS/OBS-76/08, Office of Biological Services, Fish and Wildlife Service, US Department of the Interior Washington DC.

Ash, C.G. (1973). Channelization of the United States Corps of Engineers. In Panel on stream channelization and trout fisheries (ed. White, R.J.). *Trout Magazine*, **14**, 24–25.

Aubert, J. (1949). *Barrages et Canalisation*. Dunod, Paris.

Aus, P.B. (1969). What is happening to the wetlands? *Transactions of the North American Wildlife Conference*, **34**, pp. 315–323.

Averyanov, S.F., Minayeua, E.N., and Timoshkina, V.A. (1971). Increasing agricultural productivity through irrigation and drainage. In *Natural Resources of the Soviet Union: their use and renewal* (eds Gerasimov, I.P., Armand, D.L., and Yefron, K.M.). W.H. Freeman & Co., San Francisco, pp. 131–160.

Ayers, O.C. (1939). *Land Drainage and Reclamation*. McGraw-Hill, New York.

Bagby, G.T. (1969). Our ruined rivers. *Georgia Game and Fish*, **4**, 2–16.

Bailey, A.D. and Bree, T. (1981). Effect of improved land drainage on river flood flows. In *Flood Studies Report: 5 years on*. Institute of Civil Engineers, London, pp. 134–142.

Baker, M, Jr, Inc. (1973). *Evaluation of the Environmental Impact to Appalachian Pennsylvania Waters of the 1972 Flood and Subsequent Stream Channelization with Future Policy Recommendations*. Report No. ARC-73–185–2562, Pennsylvania Fish Commission under contract No. FC-12453 and for the Pennsylvania Department of Environmental Resources as study EER-128 with funding from the Appalachian Regional Commission.

Baldock, D., Hermans, B., Kelly, P., and Hermet, L. (1984). *Wetland Drainage in Europe: the effects of agricultural policies in four EEC countries*. Institute for European Environmental Policy, International Institute for Environment and Development, London.

Barclay, J.S. (1980). *Impact of Stream Alterations on Riparian Communities in South-central Oklahoma*. Report No. FWS/OBS-80/17, Office of Biological Services, Fish and Wildlife Service, US Department of the Interior, Washington, DC.

Barnard, R.S. (1977). Morphology and morphometry of a channelized stream: the case history of the Big Pine Creek Ditch, Benton County, Indiana. *Studies in Fluvial Geomorphology*, Technical Report No. 9, Water Resources Research Center, Purdue University, West Lafayette, IN.

Barnard, R.S., and Melhorn, W.N. (1982). Morphologic and morphometric response to channelization : the case history of the Big Pine Creek Ditch, Benton County, Indiana. In *Applied Geomorphology* (ed. Craig, R.G. and Craft, J.L.). George Allen & Unwin, London, pp. 224–239.

Barrett, P.R.F. (1978). Aquatic weed control—necessity and methods. *Fish Management*, **9**, 93–101.

Barstow, C.J. (1970). Impact of channelization on wetland habitat in the Obion-Forked Deer Basin, Tennessee. *Transactions of the 36th North American Wildlife Conference*, pp. 362–375.

Barstow, C.J. (1971). Impact of channelization on wetland habitat in the Obion-Forked Deer Basin, Tennessee. In *Stream Channelization—a Symposium* (eds Schneberger, F. and Funk, J.L.). Special Publication. No. 2, North Central Division, American Fisheries Society, Bethesda, MD, pp. 20–30.

Barton, B.A. (1977). Short term effects of highway construction on the limnology of a small stream in Southern Ontario. *Journal of Freshwater Biology*, **7**, 99–108.

Barton, J.R. and Cron, F.W. (1979). *Restoration of Fish Habitat in Relocated Streams*, Report No. FHWA-IP-79–3, Federal Highways Administration, US Department of Transportation, Washington, DC.

Barton, J.R., and Winger, P.V. (1973a). *A Study of Channnelization of the Weber River, Summit County, Utah*. Final Report submitted to the Utah Division of Wildlife Resources and Utah State Department of Highways, Provo, Utah.

Barton, J.R., and Winger, P.V. (1973b). Rehabilitation of a channelized river in Utah. Hydraulic Engineering and the Environment. *Proceedings of the 21st Annual Hydraulic Division Speciality Conference*, Montana State University, Bozeman, MT, pp. 1–10.

Barton, J.R., and Winger, P.V. (1974). *Stream Rehabilitation Concepts*, Research Report submitted to Utah State Department of Highways, Provo, Utah.

Barton, J.R., Peters, E.J., White, D.A. and Winger, P.V. (1972). *Bibliography of the Physical Alteration of the Aquatic Habitat (Channelization) and Stream Improvement*. Brigham Young University Publications, Provo, Utah.

Barton, N.J. (1962). *The Lost Rivers of London—a study of their effects upon London and Londoners, and the effects of London and Londoners upon them*. Leicester

University Press, Phoenix House, London.
Bartsch, A.F. (1954). *Practical Methods for Control of Algae and Water Weeds*, US Public Health Service, Public Health Report, **69**, 749–757.
Bauer, E.A. and East, B. (1970). The gravediggers. *Outdoor Life*, July 1970.
Bayless, J., and Smith, W.B. (1967). The effects of channelization upon the fish population of lotic waters in eastern North Carolina. *Proceedings of the Annual Conference South-east Association Game and Fish Commission*, **18**, 230–38.
Beaumont, P. (1978). Man's impact on river systems: a worldwide review. *Area*, **10**, 38–41.
Beckett, D.C., Bingham, C.R., and Sanders, L.G. (1983) Benthic macroinvertebrates of selected habitats of the lower Mississippi river. *Journal of Freshwater Ecology*, **2**, 247–261.
Beland, R.D. (1953). The effect of channelisation on the fishery of the lower Colorado River. *California Fish and Game*, **39**, 137–139.
Belusz, L. (1970). Missouri Department of Conservation memorandum to Lee Redmond, dated 29 September, 1970. *Outdoor America*, **36**, 10.
Benson, N.G., and Weithman, S.A. (1980). *A Summary of Seven US Fish and Wildlife Service Stream Channelization Studies*. Office of Biological Services, Fish and Wildlife Service, US Department of the Interior, Kearneysville, W.VA.
Beschta, R.L., and Jackson, W.L. (1979). The intrusion of fine sediments into a stable gravel bed. *Journal of the Fisheries Research Board, Canada*, **36**, 204–210.
Bianchi, D.R. and Marcoux, R. (1975). The physical and biological effects of physical alteration on Montana trout streams and their political implications. In *Symposium on Stream Modification*, Harrisonburg, VA, pp. 50–60.
Bilby, R.E., and Likens, G.E. (1980). Importance of organic debris dams in the structure and function of stream ecosystems. *Ecology*, **61**, 1107–1113.
Binder, W. (1979). *Grundzuge der Gewasserpflege*. Landesamt fur Wasserwirtschaft, Munich, Germany.
Binder, W., and Grobmaier, W. (1978). Bach- und Flusslaufe-Ihre Gestalt und Pflege. *Garten Und Landschaft*, **1**, 25–30.
Binder, W., Jürging, P., and Karl. J. (1983). Natural river engineering—characteristics and limitations. *Garten und Landschaft*, **2**, 91–94.
Bingham, C.R. (1982). Benthic macroinvertebrate study of a stone dike. *Environmental and Water Quality Operational Studies Information Exchange Bulletin*, E-82–4, US Army Engineer Waterways Experiment Station, Vicksburg, MS.
Birch, E. (1976). Trout stream weeds: how to cut, and how not to, for fish and fly. *Salmon and Trout Magazine*, **206**, 49–53.
Bird, J.F. (1980). Geomorphological implications of flood control measures: Lang Lang River, Victoria. *Australian Geographical Studies*, **18**, 169–83.
Bittmann, E. (1957). Der biologische Uferschutz an den Bundeswassertraßen unter besonderer Berücksichtigung der Verwendung von Röhricht. *Naturnaher Ausbau von Wasserläufen* (eds Olschowy, G. and Köhler, H.). Hiltrup bei Münster, pp. 47–61.
Blackburn, B.B. (1969). Where conservation is a bad word. *Field and Stream*, December, 12–14, 58, 59.
Blair, D.A. (1973). The effects of stream channelization on fish populations. MS thesis, Pennsylvania State University, State College, Pennsylvania, PA.
Blench, T. (1952). Regime theory for self-formed sediment-bearing channels. *Transactions of the American Society of Civil Engineers*, **117**, 383–400; discussion 401–408.
Blench, T. (1957). *Regime Behaviour of Canals and Rivers*. Butterworth, London.
Boelter, D.H. (1972). Water table drawdown around an open ditch in organic soils. *Journal of Hydrology*, **15**, 329–340.
Bonham, A.J. (1980). Bank protection using emergent plants against boat wash in

rivers and canals. *Report IT 206 Hydraulics Research Station*, Wallingford, Oxfordshire, UK.

Borovicka, R.L. (1968). Stream preservation and improvement. *United States Bureau of Land Management Manual*, US Deparment of the Interior, Washington DC.

Bou, C. (1977). Conséquences écologiques de l'extraction des alluvions récentes dans le cours moyen du Tarn. *Bulletin écologiques*, **8**, 435–444.

Bouchard, B., Clavel, P., Hamon, Y., and Romaneix, C. (1979). Ecological incidences of dredging up alluvial materials and regulation of streams on the aquatic ecosystem. *Bulletin Français Pisciculture*, **273**, 137–56.

Boussu, M.F. (1954). Relationship between trout populations and cover on a small stream. *Journal of Wildlife Management*, **18**, 229–39.

Bovee, K.D. (1978). The incremental method of assessing habitat potential for coolwater species, with management implications, Special Publication No. 11, American Fisheries Society, Bethesda, MD, pp. 340–346.

Bovee, K.D., and Hardin, T.S. (1983). Use of the instream flow incremental methodology to design habitat improvement features: St. Vrain rehabilitation study. Presented at Stream Mechanics Course sponsored by Soil Conservation Service, US Department of Agriculture, at Colorado State University, Fort Collins, CO.

Bowie, A.J. (1982). Investigations of vegetation for stabilizing eroding streambanks. *Transactions of the American Society of Agricultural Engineers*, **25**, 1601–1611.

Bradt, P.T., and Wieland, G.E. III (1978). *The Impact of Stream Reconstruction and a Gabion Installation on the Biology and Chemistry of a Trout Stream*. Department of Biology, Lehigh University, Bethlehem, PA.

Brandon, T.W. (ed.) (1987). *River Engineering—Part I, Design Principles*. Published for the Institute of Water and Environmental Management by the IWES, London.

Bray, D.I., and Cullen, A.J. (1976). Study of artificial cutoffs on gravel bed rivers. *Proceedings of Conference Rivers '76*. American Society of Civil Engineers, pp. 1399–1417.

Bree, T., and Cunnane, C. (1980). The effect of arterial drainage on flood magnitude. In *Project 5.1 of the International Hydrological Programme. Casebook of methods of computation of quantitative changes in the hydrological regime of river basins due to human activities*. UNESCO, pp. 116–126.

Brice, J.C. (1981). *Stability of Relocated Stream Channels*. Technical Report No. FHWA/RD-80/158, Federal Highways Administration, US Department of Transportation, Washington, DC.

Brinson, M.M., Swift, B.L., Plantico, R.C., and Barclay, J.S. (1981). *Riparian Ecosystems: their ecology and status*, Report No. FWS/OBS-81/17, Office of Biological Services, Fish and Wildlife Service, US Department of the Interior, Kearneysville, W.VA.

British Waterways Board (1964). *The Future of Waterways*. Her Majesty's Stationery Office, London.

British Waterways Board (1973). *The Last Ten Years*. British Waterways Board, London.

Broggi, M.F. (1984). Rivers and streams: neglected habitats, *Naturopa*, No. 46-1984, pp. 11–13.

Brooker, M.P. (1976). The ecological effects of the use of Dalapon and 2,4-D for drainage channel management. I. Flora and chemistry. *Archives of Hydrobiology*, **78**, 396–412.

Brooker, M.P. (1981). The impact of impoundments on the downstream fisheries and general ecology of rivers. *Applied Biology*, **6**, 91–152.

Brooker, M.P. (1982). Biological investigations in the Welsh Water Authority: An assessment of macroinvertebrate colonisation of two reinstated salmonid spawning gravel beds in the Afon Gwyrfai, Snowdonia, Wales. Report by Welsh Water Authority, Brecon, Wales, UK.

Brooker, M.P. (1985). The ecological effects of channelization. *Geographical Journal*, **151**, 63–9.

Brookes, A. (1981). Channelization in England and Wales. Discussion Paper No.11, Department of Geography, University of Southampton, Southampton, UK.

Brookes, A. (1983). River channelization in England and Wales: downstream consequences for the channel morphology and aquatic vegetation. Unpublished PhD thesis, University of Southampton, Southampton, UK.

Brookes, A. (1984). *Recommendations Bearing on the Sinuosity of Danish Stream Channels—consequences of realignment, spatial extent of natural channels, processes and techniques of natural and induced restoration.* Technical Report No. 6., Freshwater Laboratory, National Agency of Environmental Protection, Silkeborg, Denmark.

Brookes, A. (1985a). River channelization: traditional engineering methods, physical consequences and alternative practices. *Progress in Physical Geography*, **9**, 44–73.

Brookes, A. (1985b). Downstream morphological consequences of river channelization in England and Wales. *Geographical Journal*, **151**, 57–62.

Brookes, A. (1986). Response of aquatic vegetation to sedimentation downstream from channelisation works in England and Wales. *Biological Conservation*, **38**, 351–367.

Brookes, A. (1987a). River channel adjustments downstream from channelization works in England and Wales. *Earth Surface Processes and Landforms*, **12**, 337–351.

Brookes, A. (1987b). Restoring the sinuosity of artificially straightened stream channels, *Environmental Geology and Water Science*, **10**, 33–41.

Brookes, A. (1987c). The distribution and management of channelized streams in Denmark. *Regulated Rivers*, **1**, 3–16.

Brookes, A. (1987d). Recovery and adjustment of aquatic vegetation within channelization works in England and Wales. *Journal of Environmental Management*, **24**, 365–382.

Brookes, A. (1988). Alternative channelization procedures. In *Alternatives in Regulated Flow Management* (eds Gore, J.A., and Petts, G. E.). CRC Press, Bocca Raton, Florida, in press.

Brookes, A., and Gregory, K.J. (1988). River channelization and public policy. In *Geomorphology and Environmental Planning* (ed. Hooke, J.). John Wiley & Sons, Chichester, pp. 145–167.

Brookes, A., and Penning-Rowsell, E.C. (1988). Integrating geomorphology and public policy: the reality of contributing to river management. Paper presented to the Institute of British Geographers, Loughborough, January 1988. Publication, Flood Hazard Research Centre, Middlesex Polytechnic, Middlesex, UK.

Brookes, A., Gregory, K.J., and Dawson, F.H. (1983). An assessment of river channelization in England and Wales. *Science of the Total Environment*, **27**, 97–112.

Brooks, A. (1976). *Waterways and Wetlands*. Published by the British Trust for Conservation Volunteers Ltd, Regents Park, London.

Brown, B.L. (1931). Straighten the Mississippi. *Civil Engineering*, **1**, 660.

Bryant, M.D. (1983). The role and management of woody debris in west coast salmonid nursery streams. *Journal of Fisheries Management*, **3**, 322–330.

Buckner, C.H. (1955a). Small mammal populations on a changing site, Canada Department of Agricultural Science, Service Division, *Forest Biology, Bi-monthly Progress Report*, **2**, 2.

Buckner, C.H. (1955b). Small mammals as predators of sawflies. *Canadian Entomologist*, **87**, 121–123.

Buglewicz, E.G., and Hergenrader, G.L. (1977). The impact of the artificial reduction of light on an entrophic farm pond. *Transactions of the Nebraska Academy of Science*, **4**, 23–34

Buie, E.C. (1973). The channelization program of the SCS. Panel on stream channelization and trout fisheries (ed. White, R.J.). *Trout Magazine*, **14**, pp. 26–27.

Bulkley, R.V. (1975). A study of the effects of stream channelization and bank stabilization on warm water sport fish in Iowa. *Subproject 1. Inventory of Major Stream Alterations in Iowa*, Project No FWS/OBS-76-11. Iowa Cooperative Fishery Research Unit, Ames, IO.

Bulkley, R.V., Bachman, R.W., Carlander, K.D., Fierstine, H.L., King, L.R., Menzel, B.W., Witten, A.L., and Zimmer, D.W. (1976). *Warmwater Stream Alterations in Iowa: extent, effects on habitat, fish and fish-food, and evaluation of stream improvement structures (Summary report)*. Report No. FWS/OBS-76/16, Office of Biological Services, Fish and Wildlife Service, US Department of the Interior, Washington, DC.

Bull, W.B., and Scott, K.M. (1974). Environmental monitoring: impact of mining gravel from urban stream beds in southwestern United States. *Geology*, **2**, 171–174.

Burke, T.D., and Robinson, J.W. (1979). River structure modifications to provide habitat diversity. In *The mitigation symposium: A national workshop on mitigating losses of fish and wildlife habitats* (ed. Swanson, F.J.), Technical Report RM-65, Forest Service US Department of Agriculture, Fort Collins, Colorado, pp. 556–561.

Burke, W. (1969). Drainage of blanket peat at Glenamoy. In *Transactions of the Second International Peat Congress, USSR*, pp. 809–817.

Burkham, D.E. (1976). *Hydraulic Effects of Changes in Bottom-land Vegetation on 3 Major Floods, Gila River in Southeastern Arizona*. Professional Paper 655-J, Geological Survey US Department of the Interior, Washington, DC.

Burkhard, W.T. (1967). *Effects of Channelization on the Trout Fishery of Tomichi Creek*. Federal Aid Project F-26-R-4, Job Completion report. Colorado Game, Fish and Parks Department.

Burns, J.W. (1972). Some effects of logging and associated road construction on northern California streams. *Transactions of the American Fisheries Society*, **101**, 1–17.

Burnside, K.R. (1967). *The Effects of Channelisation on Fish Populations on Boeuf River in Northeast Louisiana*. Unpublished MS thesis, N.E. LA State College, LA.

Burress, R.M., Krieger, D.A., and Pennington, C.H. (1982). *Aquatic Biota of Bank Stabilisation Structures on the Missouri River, North Dakota*. Technical Report No. E-82–6, US Army Engineer Waterways Experiment Station, Vicksburg, MS.

Burt, T.P., Donohoe, M.A., and Vann, A.R. (1983). The effect of forestry drainage operations on upland sediment yields: the results of a storm-based study. *Earth Surface Processes and Landforms*, **8**, 339–346.

Burt, T.P., Donohoe, M.A., and Vann, A.R. (1984). Changes in the yield of sediment from a small upland catchment following open ditching for forestry drainage. In *Channel Processes: Water, Sediment, Catchment Controls* (ed. Schick, A.P.). Catena supplement 5, Braunschweig, pp. 63–74.

Burwell, R.W., and Sugden, L.G. (1964). Potholes—going, going .... In *Waterfowl Tommorrow*. US Bureau of Sport Fisheries and Wildlife, Washington, DC, pp. 369–380.

Callander, R.A. (1969). Instability and river channels. *Journal of Fluid Mechanics*, **36**, 465–480.

Callison, C.H. (1971). *Questions of Senator Buckley's Committee Regarding Stream Channelization Directed at Charles H. Callison*. National Audubon Society, N.Y.

Campbell, K.L., Kumar, S., and Johnson, H.P. (1972). Stream straightening effects on flood runoff characteristics. *Transactions of the American Society of Agricultural Engineers*, **15**, 94–98.

Canterbury, J.H. (1972). Channelisation—the farmer's friend. *Soil Conservation*, **38**, 23–24.

Carey, W.C. (1966). Comprehensive river stabilisation. *Journal of the Waterways and Harbors Division, American Society of Civil Engineers*, **92**, 59–86.

Carline, R.F., and Kloslewski, S.P. (1981). *Reponses of Macroinvertebrates and Fish Populations to Channelization and Mitigation Structures in Chippewa Creek and River Styx, Ohio*. Final report on Project 761091/711102. The Ohio State University Research Foundation, Columbus, OH.

Carling, P.A. (1984). Deposition of fine and coarse sand in an open-work gravel bed. *Canadian Journal of Fisheries and Aquatic Science*, **41**, 263–270.

Carlston, C.W. (1965). The relation of free meander geometry to stream discharge and its geomorphic implications. *American Journal of Science*, **263**, 864–85.

Carothers, S.W., and Johnson, R.R. (1975). The effects of stream channel modification on birds in the southwestern United States. In *Symposium on Stream Channel Modification*, Harrisonburg, VA, pp. 60–70.

Chancellor, A.P. (1958). *The Control of Aquatic Weeds and Algae*. Her Majesty's Stationery Office, London.

Chandler, B. (1970). A biological approach to water quality management. *Water Pollution Control, London*, **69**, 415–422.

Chang, H.H. (1986). River channel changes: adjustments of equilibrium. *Journal of Hydraulic Engineering, American Society of Civil Engineers*, **H2**, 43–55.

Chapman, D.W. (1962). Effects of logging upon fish resources of the West Coast, *Journal of Forestry*, **60**, 533–537.

Chapman, D.W., and Knudsen, E. (1980). Channelization and livestock impacts on salmonid habitat and biomass in western Washington. *Transactions of the American Fisheries Society*, **109**, 357–363.

Charlton, F.G. (1980). *River Stabilisation and Training in Gravel Rivers*. Report, Hydraulics Research Station, Wallingford, Oxfordshire, UK.

Charlton, F.W., Brown, P.M., and Benson, R.W. (1978). *The Hydraulic Geometry of some Gravel Rivers in Britain*. Report, IT 180. Hydraulics Research Station, Wallingford, Oxfordshire, UK.

Ching-Shen, P., Shao-Chuan, S., and Wen-Chung, T. (1977). *A Study of Artificial Cutoffs of the Middle Yangtze (Peking)*. Foreign Language Press, Peking, China.

Choate, J.S. (1971). *Wetland Drainage in the Hawk Creek Pilot Watershed*. Special Publication No. 91, Minnesota Division of Game and Fish, MN.

Choate, J.S. (1972). Effects of stream channeling on wetlands in a Minnesota watershed. *Journal of Wildlife Management*, **36**, 940–43.

Chow, V.T. (1959). *Open-channel Hydraulics*. McGraw-Hill, New York.

Chutter, F.M. (1969). The effects of silt and sand on the invertebrate fauna of streams and rivers. *Hydrobiology*, **34**, 57–76.

Clark, C.C. (1944). The freshwater naiades of Auglaize County, Ohio. *Ohio Journal of Science*, **54**, 167–76.

Clavel, P., Cuinat, R., Hamon, Y., and Romaneix, C. (1978). Effects des extractions de materiaux alluvionnaires sur l'environment aquatique dans les cours superieurs de la Loire et de l'Allier. *Bulletin Français Pisciculture*, **268**, 121–154.

Clemens, G.R. (1936). Straightening the Father of Waters. *Engineering News Record*, February.

Coates, D.R. (ed.) (1972). *Environmental Geomorphology and Landscape Conservation*. Benchmark papers in geology. Dowden, Hutchinson & Ross.

Coates, D.R. (1976a). Geomorphology in legal affairs of the Binghamton New York Metropolitan Area. In *Urban Geomorphology* (ed. Coates, D.R.). Special Paper No. 174, Geological Society of America, Boulder, CO. pp. 111–148.

Coates, D.R. (ed.) (1976b). *Geomorphology and Engineering*. George Allen & Unwin, London.
Coates, D.R. (1984). Geomorphology and public policy. In *Developments and Applications of Geomorphology* (eds Costa, J.E., and Fleisher, P.J.). Springer-Verlag, Berlin, Heidelberg, pp. 97–132.
Coffey, A. (1982). Stream improvement: the Chicod Creek episode. *Journal of Soil and Water Conservation*, **37**, 80–82.
Cole, G. (1976). Land drainage in England and Wales. *Journal of the Institute of Water Engineers*, **30**, 354–361.
Committee on Government Operations (1973a). *Stream Channelization (Parts I–V)*. Hearings before a subcommittee of the Committee of Government Operations, House of Representatives, 92nd Congress. US Government Printing Office, Washington, DC.
Committee on Government Operations (1973b) *Fifth report: Stream Channelization. What federally financed draglines and bulldozers do to our nation's streams*. House Report, 93–530. US Government Printing Office, Washington, DC.
Congdon, J.C. (1971). Fish populations of channelized and unchannelized sections of the Chariton River, Missouri. In *Stream Channelization: A Symposium* (eds Schneberger, E., and Funk, J.L.). North Central Division, American Fish Society, Special Publication No. 2, Bethesda, MD, pp. 52–83.
Cooper, A.C. (1956). *A Study of the Horsefly River and the Effect of Placer Mining Operations on the Sockeye Spawning Grounds*. Publication No. 3. International Pacific Salmon Fisheries Commission.
Cooper, C.O., and Wesche, T.A. (1976). *Stream Channel Modifications to Enhance Trout Habitat under Low Flow Conditions*. Water Resources Research Institute, Wyoming University, WY.
Corbett, E.S., Lynch, J.A., and Sopper, W.E. (1978). Timber harvesting practices and water quality in the Eastern United States. *Journal of Forestry*, **76**, 484–488.
Cordone, A.J., and Kelley, D.W. (1961). The influence of inorganic sediment on aquatic life of streams. *California Fish and Game*, **47**, 189–228.
Corning, R.V. (1975). Channelization: shortcut to nowhere. *Virginia Wildlife*, **6**, 8.
Costa, J.E. (1978). The dilemma of flood control in the United States. *Environmental Management*, **2**, 313–322.
Countryside Commission (1987) *Changing River Landscapes: A Study of River Valley Landscapes*. Countryside Commission, Cheltenham, UK.
Countryside Commission (1988) *The Water Industry in the Countryside*. Countryside Commission, Cheltenham, UK.
Cowardin, L.M. (1982). Wetlands and deepwater habitats: A new classification. *Journal of Soil and Water Conservation*, **37**, 83–84.
Crisp, D.T., and Gledhill, T. (1970). A quantitative description of the recovery of the bottom fauna in a muddy reach of a mill stream in southern England after draining and dredging. *Archives of Hydrobiology*, **67**, 502–541.
Crocker, W.H. (1960). Land drainage and fisheries. *Salmon and Trout Magazine*, **158**, 63–73.
Cuinat, R. (1983). Consequences hydrobiologiques et piscoles de la chenalisation d'une petite riviere de limagne. In *Association Français de Limnologie: 27eme Congres National, Bordeaux 25–27 Mai 1982*, Association Français de Limnologie, Paris, pp. 63–68.
Cummins, K.W. (1974). Structure and function of stream ecosystems. *Bioscience*, **24**, 631–641
Cummins, K.W. (1979). The natural ecosystem. In *The Ecology of Regulated Streams* (eds Ward, J.W., and Stanford, J.A.). Plenum Press, New York, pp. 7–24.
Dale, E.E. (1975). *Environmental Evaluation Report on Various Completed Chan-*

nel Improvement Projects in Eastern Arkansas. Water Resources Research Centre, University of Arkansas, Fayetteville, AR.

Danels, P., Chapus, E., and Dhaille, R. (1960). Tetrapods and other precast blocks for breakwaters. *Journal of the Waterways and Harbors Division, American Society of Civil Engineers*, **86**, 1–14.

Daniels, R.B. (1960). Entrenchment of the Willow Creek Drainage Ditch, Harrison County, Iowa. *American Journal of Science*, **258**, 161–76.

Darby, H.C. (1956). *The Drainage of the Fens*. Cambridge University Press, Cambridge, UK.

Darnell, R.M., Pequegnat, W.E., James, B.M., Benson, F.J., and Defenbaugh, R.A. (1976). *Impacts of Construction Activities in Wetlands of the United States*. Report No. EPA-600/3–76–045, Office of Research and Development, US Environmental Protection Agency, Corvallis, Oregon.

Das, B.P. (1976). Channelisation of the River Bhargavi, India. In *Proceedings of Conference Rivers '76*, American Society of Civil Engineers, pp. 1700–1771.

Davidson, G. (1969). A vanishing part of Louisiana—it's streams, *Louisiana Conservation*, January–February, 1969.

Davies, G.H. (1982). Wildlife down the drain. *The Sunday Times*, 2 May, 1982.

Dawson, F.H. (1976). The organic contribution of stream edge forest litter fall to the chalk stream ecosystem. *Oikos*, **27**, 13–18.

Dawson, F.H. (1978). Aquatic plant management in semi-natural streams: the role of marginal vegetation. *Journal of Environmental Management*, **6**, 213–221.

Dawson, F.H. (1979). Ecological management of vegetation in flowing waters. In *Proceedings of Conference on Ecology and Design in Amenity Land Management* (eds Wright, S.E., and Buckley, G.P.). Wye College, 8–11 April 1979, Wye College, pp. 146–167.

Dawson, F.H. (1980). The origin, composition and downstream transport of plant material in a small chalk stream, *Freshwater Biology*, **10**, 415–435.

Dawson, F.H. (1981a). The reduction of light as a technique for the control of aquatic plants—an assessment. In *Proceedings of Conference on Aquatic Weeds and their Control*. Association of Applied Biologists, Oxford, UK. pp. 157–164.

Dawson, F.H. (1981b). The downstream transport of fine material and organic balance for a section of a small chalk stream in southern England. *Journal of Ecology*, **69**, 367–380.

Dawson, F.H. (1986). Light reduction techniques for aquatic plant control. In *Lake and Watershed Management. Proceedings of the 5th Annual Symposium, North American Lake Management Society*, November 1985, Geneva, Wisconsin. North American Lake Management Society, Washington DC, pp. 258–262.

Dawson, F.H., and Hallows, H.B. (1983). Practical applications of a shading material for macrophyte control in watercourses. *Aquatic Botany*, **17**, 301–308.

Dawson, F.H., and Haslam, S.M. (1983). The management of river vegetation with particular reference to shading effects of marginal vegetation. *Landscape Planning*, **10**, 147–169.

Dawson, F.H., and Kern-Hansen, U. (1978). Aquatic weed management in natural streams: the effect of shade by marginal vegetation. *Verhandlungen der internationalen Vereinigung für theoretische und angewandte Limnologie*, **20**, 1440–1445.

Dawson, F.H., and Kern-Hansen, U. (1979). The effect of natural and artificial shade on the macrophytes of lowland streams and the use of shade as a management technique. *International Revue Der Gesamten Hydrobiologie*, **64**, 437–455.

Dawson, F.H., Castellano, E., and Ladle, M. (1978). Concept of species succession in relation to river vegetation and management. *Verhandlungen der internationalen Vereinigung für theoretische und angewandte Limnologie*, **20**, 1429–1434.

De Vries, M. (1975). A morphological time scale for rivers. *Delft Hydraulics Laboratory Publication*, **147**.

Dearsley, A.F., and Colclough, S.R. (1982). The River Roding Flood Alleviation Scheme—Contract 4 (Abridge to Loughton): effects on the fish populations.*Internal Report, Directorate of Scientific Services, Thames Water Authority*, Reading, UK.

Dee, N., Baker, J.K., Drobny, N.L., Duke, K.M., and Fahringer, D.C. (1972). *Environmental Evaluation System for Water Resources Planning*. US Bureau of Reclamation, Battelle Memorial Institute, Columbus, Ohio.

Demek, J. (1973). Quaternary relief development and man. *Geoforum*, **15**, 68–71.

Department of the Environment (1977). *Assessment of Major Industrial Applications—A Manual*. DOE Research Report No. 13, prepared for the Scottish Development Department, Department of the Environment and the Welsh Office, UK.

Detwyler, T.R. (1971). *Man's Impact on the Environment*. McGraw-Hill, New York.

Diamant, R., Eugster, J.G., and Dverksen, C.J. (1984). *A Citizen's Guide to River Conservation*. The Conservation Foundation, Washington, DC.

Dobbie, C.H., Prus-Chacinski, T.M., and Bowen, H.C. (1971). Flood alleviation works. *Civil Engineering and Public Works Review*, 383–390.

Dodge, W.E., Possardt, E.E., Reed, R.J., and MacConnell, W.P. (1977). *Channelization Assessment. White River, Vermont: Remote sensing, benthos and wildlife*. Report No. FWS/OBS-76/07, Office of Biological Services, Fish and Wildlife Service, US Department of the Interior, Washington, DC.

Doran, W.E. (1959). The River Great Ouse Flood Protection Scheme. *River Boards Association Year Book*, pp. 36–44.

Drower, M.S. (1954). Water supply, irrigation and agriculture. In *History of Technology* (eds Singer, C., Holmyard, E.J., and Hall, A.R.). Clarendon Press, Oxford, pp. 520–557.

du Boys, P. (1879). Études due régime du Rhône et l'achan exercée par les eaux sur un lit à fond de graviers indéfiniment affoiullable. *Annales des ponts et chaisseés, ser. 5*, **18**, 141–195.

Duckworth, J.A., and Seed, G. (1969). Bowland Forest and Pendle Floods. *Association of River Authorities Year Book*, pp. 81-90.

Dunne, T., and Leopold, L.B., (1978). *Water in Environmental Planning*. W.H. Freeman, San Francisco, CA.

Duvel, W.A., Volkmar, R.D., Specht, W.L., and Johnson, F.W. (1976). Environmental impact of stream channelization. *Water Resources Bulletin*, **12**, 799–812.

East, B. (1973). Black day for mallards. *Outdoor Life*, February 1973, pp. 45–47 and 122–125.

Edwards, C.J. (1977). The effects of channelization and mitigation on the fish community and population structure in the Olentangy River, Ohio. PhD thesis, Ohio State University, Columbus, OH.

Edwards, C.J., Griswold, B.L., and White, G.C. (1975). An evaluation of stream modification in the Olentangy River, Ohio. In *Symposium on Stream Modification*, Harrisonburg, VA, pp. 34–49.

Edwards, C.J., Griswold, B.L., Tubb, R.A., Weber, E.C., and Woods, L.C., (1984). Mitigating effects of artificial riffles and pools on the fauna of a channelized warmwater stream. *North American Journal of Fisheries Management*, **4**, 194–203.

Edwards, D. (1969). Some effects of siltation upon aquatic macrophyte vegetation in rivers. *Hydrobiologia*, **34**, 29–37.

Ehlers, R. (1956). An evaluation of stream improvement devices constructed eighteen years ago. *California Fish and Game*, **42**, 203–217.

Eiserman, F., Dern, G., and Doyle, J. (1975). *Cold Water Stream Handbook for*

Wyoming. US Soil Conservation Service and Wyoming Game and Fish Department, Cheyenne, WY.

Elam, W.E. (1931). New Mississippi River considerations. *Civil Engineering*, **1**, 652.

Elam, W.E. (1946). *Speeding Floods to the Sea*. Hobson, New York.

Elliot, C.G. (1919). *Engineering for Land Drainage*, 3rd edn. John Wiley & Sons, New York.

Elliot, D.O. (1932). *The Improvement of the Lower Mississippi River for Flood Control and Navigation*. Mississippi River Commission, Vicksburg, MS.

Ellis, M.M. (1936). Erosion silt as a factor in aquatic environments. *Ecology*, **17**, 29–42.

Ellis, R.W. (1976). The impact of stream alteration on wildlife along tributaries of the Roanoke River, Charlotte County, Virginia. Unpublished MS thesis, Virginia Polytechnic Institute and State University, Blacksburg, VA.

Ellis, R.W., and Whelan, J.B. (1978). Impact of stream channelization on riparian small mammals and bird populations in Piedmont, Virginia. *Transactions of the Northeastern Section Wildlife Society*, **35**, 93–104.

Elser, A.A. (1967). Fish populations of a trout stream in relation to major habitat zones and channel alterations. Unpublished MS thesis, Montana State University, Bozeman, MT.

Elser, A.A. (1968). Fish populations of a trout stream in relation to major habitat zones and channel alterations. *Transactions of the American Fisheries Society*, **98**, 253–262.

Emerson, J.W. (1971). Channelization: a case study. *Science*, **173**, 325–326.

Engelhardt, W. (1951). Faunistisch-Ökologische Untersuchungen uber Wasserinsekten an den südlichen Zuflussen des Ammersees. *Mitteilungen Münchener Entomologischer Gesellschaft*, **41**, 1–135.

Erichsen, F.P. (1970). Let's take a look at channel improvement. In *Proceedings of Symposium on Interdisciplinary Aspects of Watershed Management*. American Society of Civil Engineers, Montana State University, Bozeman, MT, pp. 261–267.

Etcheverry, B.A. (1931). *Land Drainage and Flood Protection*. McGraw-Hill, New York.

Etnier, D.A. (1972). The effect of annual rechanneling on a stream fish population. *Transactions of the American Fisheries Society*, **101**, 372–375.

Everhart, W.H., Eipper, A.W., and Youngs, W.D. (1975). *Principles of Fishery Science*. Cornell University Press, Ithaca, N.Y.

Federal Highways Adminstration (1979). *Restoration of Fish Habitat in Relocated Streams*. Report No. FHWA-IP-79–3, Federal Highways Administration, US Department of Transportation, Washington, DC.

Fernholz, W.B. (1978). Determination and significance of wing dams and sand as a fishery habitat. In *Mississippi River Work Unit Annual Report 1977–78*. Wisconsin Department of Natural Resources, La Crosse, WI, pp. 14–30.

Forshage, A., and Carter, N.E. (1973). Effects of gravel dredging on the Brazos River. *Proceedings of the 27th Annual Conference, Southeastern Association Game and Fish Commission*, **24**, 695–708.

Found, W.C., Hill, A.R., and Spence, E.S. (1975). *Economic and Environmental Impacts of Land Drainage in Ontario*. Geographical Monograph No. 6, Atkinson College, York University, Canada.

Found, W.C., Hill, A.R., and Spence, E.S. (1976). Economic and Environmental impacts of agricultural land drainage in Ontario. *Journal of Soil and Water Conservation*, **31**, 20–24.

Fox, A.C. (1975). Guidelines for avoiding adverse impacts in modifying stream channels. In *Symposium on Stream Channel Modification*, Harrisonburg, VA, pp. 122–125.

Francis, G.R., Magnuson, J.J., Regier, H.A., and Talhelm, D.R. (eds) (1979). *Rehabilitating Great Lakes Ecosystems*. Technical Report No. 37, Great Lakes Fishery Commission.

Fraser, J.C. (1975). *Determining Fluvial Discharges for Fluvial Resources*. Fisheries Technical Paper No. 143, FAO.

Freeman, B.O. (1972). The problems of stream channelization in Mississippi. In *Forum on Environmental Quality*, Mississippi State College for Women, Columbus, MS, pp. 41–48.

Friedkin, J.F. (1945). *A Laboratory Study of Meandering Alluvial Rivers*. United States Army Corps of Engineers, Waterways Experimental Station, Vicksburg, MS.

Funk, J.L. (1968). *Missouri's Fishing Streams*. Fish Division, Missouri Department of Conservation, Jefferson City, MO.

Funk, J.L., and Robinson, J.W. (1974). Changes in the channel of the lower Missouri River and effects on fish and wildlife. *Aquatic Series II*, Missouri Department of Conservation, Jefferson City, MO.

Funk, J.L., and Ruhr, C.E. (1971). Stream channelization in the Midwest. In *Stream Channelization: A Symposium* (eds Schneberger, E., and Funk, J.L.). Special Publication No. 2, North Central Division, American Fisheries Society, Bethesda, MD, pp. 5–11.

Gammon, J.R. (1970). The effect of inorganic sediment on stream biota. *Water Pollution Control Research Series*, No. 18050, DWC 12/70, Water Quality Office, US Environmental Protection Agency, Washington, DC.

Gard, R. (1961). Creation of trout habitat by constructing dams. *Journal of Wildlife Management*, **25**, 384–390.

Garde, R.J., and Ranga Raju, K.G. (1977). *Mechanics of Sediment Transportation and Alluvial Stream Problems*. Wiley Eastern Limited, New Delhi.

Gardiner, J.L., Dearsley, A.F., and Woolnough, J.R. (1987). The appraisal of environmentally sensitive options for flood alleviation using mathematical modelling. *Journal of the Institution of Water and Environmental Management*, **1**, 171–184.

Gaufin, A.R. (1959). Production of bottom fauna in the Provo River, Utah. *Iowa State College Journal of Science*, **33**, 395–419.

Gebhards, S. (1970). The vanishing stream. *Wildlife Review*, March–April.

Gebhards, S. (1973). Effects of channelization on fish. In *Panel on Stream Channelization and Trout Fisheries* (ed White, R.J.). *Trout Magazine*, **14**, 23–24.

Geiger, H., and Schröter, E. (1983). Renaturation, predominantly through biologically based management : aims and methods of maintenance measures carried out on the Sur in Upper Bavaria. *Garten und Landschaft*, **2**, 114–116.

Gelroth, J.V., and Morzolif, G.R. (1978). Primary production and leaf litter decomposition in natural and channelized portions of a Kansas stream. *American Midland Naturalist*, **99**, 238–243.

George, M. (1976). Mechanical methods of weed control in watercourses—an ecologists view. In *Aquatic Herbicides*. British Crop Protection Council, Monograph No. 16, London, pp. 91–99.

Gibson, R.J., and Power, G. (1975). Selection by brook trout and juvenile salmon of shade related to water depth. *Journal of the Fisheries Research Board Canada*, **32**, 1652–1656.

Gillette, P. (1972a). Stream channelization: conflict between ditchers and conservationists. *Science*, **176**, 890–894.

Gillette, P. (1972b). The CEQ papers: secrecy is a sometime thing. *Science*, **177**, 38.

Glitz, D. (1983). Artificial channels—the 'ox-bow' lakes of tomorrow: the restoration of the course of the Wandse in Hamburg-Rahlstedt. *Garten und Landschaft*, **2**, 109–111.

Goddard, J.E. (1976). The nation's increasing vulnerability to flood catastrophe.

Journal of Soil and Water Conservation, **31**, 48–52.

Golden, M.F., and Twilley, C.E. (1976). Fisheries investigation of a channelised stream—Big Muddy Creek Watershed, Kentucky. *Transactions of Kansas Academy Science*, **37**, 85–90.

Goodman, B.W., and Staines, K.N. (1969). River Arun revetment—use of nylon mattresses. *The Association of River Authorities Year Book*, pp. 91–94.

Gore, J.A. (1978). A technique for predicting in-stream flow requirements of benthic macroinvertebrates. *Freshwater Biology*, **8**, 141–151.

Gore, J.A., and Johnson, L.S. (1980). *Establishment of Biotic and Hydrologic Stability in a Reclaimed Coal Strip Mined River Channel*. Institute of Energy and Environment, University of Wyoming, Laramie, WY.

Gorman, O.T., and Karr, J.R. (1978). Habitat structure and stream fish communities. *Ecology*, **59** , 507–515.

Goudie, A.S. (1981). *The Human Impact: man's role in environment change*. Blackwell, Oxford.

Government of Canada, Fisheries and Oceans (1980). *Stream Enhancement Guide*. Vancouver, British Columbia, Canada.

Graesser, N.W.C. (1979). How land improvement can damage Scottish salmon fisheries. *Salmon and Trout Magazine*, **215**, 39–43.

Graham, R. (1975). Physical and biological effects of alterations on Montana's trout streams. Unpublished paper presented at Symposium on Stream Channel Modification, Harrisonburg, VA.

Grant, A.P. (1948). Channel improvements in alluvial streams. *Proceedings of the New Zealand Institution of Engineers*, **34**, 231–279.

Grant, A.P., and Fenton, G.R. (1948). Willows and poplars for conservation and river works. *Bulletin of the Soil, Conservation and Rivers Control Council*. New Zealand.

Grantham, R.B. (1859). On arterial drainage and outfalls. *Minutes and Proceedings of the Institute of Civil Engineers*, **XIX**, 53–130.

Gray, J.K.A., and Edington, J.M. (1969). Effect of woodland clearance on stream temperature. *Journal of Fisheries Research Board, Canada*, **26**, 399–403.

Green, F.H.W. (1973). Aspects of the changing environment: some factors affecting the aquatic environment in recent years. *Journal of Environmental Management*, **1**, 377–391.

Green, F.H.W. (1979). Field under-drainage and the hydrological cycle. In *Man's Impact on the Hydrological Cycle in the UK* (ed. Hollis, G.E.). Geobooks, Norwich, UK, pp. 9–18.

Greenly, J.R. (1936). Progress of stream improvement. *Transactions of the American Fisheries Society*, **65**, 316–321.

Gregory, J.D., and Stokoe, J.L. (1980). Streambank management. *Proceedings of the Symposium on Fisheries Aspects of Warmwater Streams*. American Fisheries Society, pp. 276–281.

Gregory, K.J. (1976). Changing river basins. *Geographical Journal*, **142**, 237–247.

Gregory, K.J. (ed.) (1977a). *River Channel Changes*. John Wiley & Sons, Chichester.

Gregory, K.J. (1977b). Channel and network metamorphosis in northern New South Wales. In *River Channel Changes* (ed. Gregory, K.J.). John Wiley & Sons, Chichester, pp. 389–410.

Gregory, K.J. (1979). Hydrogeomorphology: how applied should we become? *Progress in Physical Geography*, **3**, 85–101.

Gregory, K.J., and Brookes, A. (1983). Hydrogeomorphology downstream from bridges. *Applied Geography*, **3**, 145–159.

Gregory, K.J., and Walling, D.E. (eds) (1986). *Man and Environmental Processes: A Physical Geography Perspective*. Dawson, Westview Press, Folkestone.

Griggs, G.B., and Paris, L. (1982). Flood control failure: San Lorenzo River, Cali-

fornia. *Environmental Management*, **6**, 407–419.
Grimes, D.J. (1975). Release of sediment-bound coliforms by dredging. *Applied Microbiology*, **29**, 109–111,
Griswold, B.L., Edwards, C.J., Woods, L.C., and Weber, E. (1978). *Some Effects of Stream Channelization on Fish Populations, Macroinvertebrates and Fishing in Ohio and Indiana*. Report No. FWS/OBS-77/46, Office of Biological Services, Fish and Wildlife Service, US Department of the Interior, Washington, DC.
Grizzell, R.A. (1976). Flood effects on stream ecosystems. *Journal of Soil and Water Conservation*, **31**, 283–285.
Groen, C.L., and Schmulbach, J.C. (1978). The sport fishery of the unchannelized Middle Missouri River. *Transactions of the American Fisheries Society*, **107**, 412–418.
Hails, J.R. (ed.) (1977). *Applied Geomorphology*. Elsevier Scientific Publishing Company, Amsterdam.
Ham, S.F., Wright, J.F., and Berrie, M.D. (1981). Growth and recession of aquatic macrophytes on an unshaded section of the River Lambourn, England, from 1971–1976. *Freshwater Biology*, **11**, 381–390.
Hamilton, J.D. (1961). The effect of sand-pit washings on a stream fauna. *Verhandlungen der internationalen Vereinigung für theoretische und angewandte Limnologie*, **14**, 435–439.
Hammer, T. R. (1972). Stream channel enlargement due to urbanization. *Water Resources Research*, **8**, 1530–1540.
Handmer, J.W., and Penning-Rowsell, E.C. (1987). Overview: Lessons for Britain. In *Flood Hazard Management: British and International Perspectives* (ed. Handmer, J.). Geobooks, Norwich, UK, pp. 287–294.
Hansen, D.R. (1971a). *Effects of Stream Channelization on Fishes and Bottom Fauna in the Little Sioux River, Iowa*. Report No. ISDWRRI-38, Iowa State Water Resources Research Institute, Ames, IA.
Hansen, D.R. (1971b). Stream channelization effects on fishes and bottom fauna in the Little Sioux River, Iowa. In *Stream Channelization: A Symposium* (eds Schneberger, S., and Funk, J.L.). Special Publication No. 2, North Central Division, American Fisheries Society, Bethesda, MD, pp. 29–51.
Hansen, D.R., and Muncy, R.J. (1971). *Effects of Stream Channelization on Fish and Bottom Fauna in the Little Sioux River, Iowa*. Iowa State Water Resources Research Institute, Ames, IA.
Hansen, E.A. (1973). In-channel sedimentation basins—a possible tool for trout habitat management. *The Progressive Fish Culturist*, **35**, 138–141.
Hansen, E.A., Alexander, O.R., and Dunn, W.H. (1983). Sand sediment in a Michigan trout stream. Part I. A technique for removing sand bedload from streams. *North American Journal of Fisheries Management*, **3**, 355–364.
Harvey, M.D., and Watson, C.C. (1988). Channel response to grade-control structures on Muddy Creek, Mississippi. *Regulated Rivers*, **2**, 79–92.
Harvey, M.D., Watson, C.C., and Schumm, S.A. (1982). A geomorphic approach to channel rehabilitation. *Geological Society of America Abstracts (with programs)*, **15**, 509.
Haslam, S.M. (1978). *River Plants*. Cambridge University Press, Cambridge.
Haslam, S.M., and Wolseley, P.A. (1981). *River Vegetation: its identification, assessment and management*. Cambridge University Press, Cambridge.
Hautum, F. (1957). Über die Ufergestaltung von Stömen und Flüssen. In *Naturnaher Ausbau von Wasserläufen* (eds Olschowy, G., and Kohler, H.). Hiltrup bei Münster, pp. 39–46.
Hawkes, H.A. (1975). River zonation and classification. In *River Ecology* (ed. Whitton, B.A.). Blackwell, Oxford, pp. 312–375.

Hay, R.C., and Stall, J.B. (1974). *History of Drainage Channel Improvement in the Vermilion Watershed, Wabash Basin*. Illinois State Water Survey Report No. 90, Illinois.

Haynes, J.M., and Makarewi, J.C. (1982). Comparison of benthic communities in dredged and undredged areas of the St. Lawrence River, Cape Vincent, New York, *Ohio Journal of Science*, **82**, 165–170.

Headrick, M.R. (1976). Effects of stream channelization on fish populations in the Buena Vista Marsh, Portage County, Wisconsin. Unpublished MS thesis, University of Wisconsin, WI.

Hedman, E.R. (1965). Effects of spur dikes on flow through contraction. *Journal of the Hydraulics Division, American Society of Civil Engineers*, **91**, 155–165.

Heede, B.H. (1980). *Stream Dynamics: an overview for land managers*. Technical Report No, 72, Forest Services, US Department of Agriculture, Rocky Mountain Forest and Range Experimental Station, Fort Collins, CO.

Henderson, F.M. (1961). Stability of alluvial channels. *Journal of the Hydraulics Division, American Society of Civil Engineers*, **87**, 109–138.

Henderson, F.M., and Shields, F.D. (1984). *Environmental Features for Streambank Protection Projects*. Technical Report E-84–11, US Army Engineers Waterways Experiment Station, Vicksburg, MS.

Heneage, A. (1951). *Land Drainage in England and Wales*. Report of the Land Drainage Subcommittee of the Advisory Water Subcommittee. Ministry of Agriculture, Fisheries and Food, Her Majesty's Stationery Office, London.

Herbkersman, C.N. (1981). *A Guide to the George Palmiter River Restoration Techniques*. Institute of Environmental Science, Miami University, Ohio.

Hermansen, H., and Krog, C. (1984). A review of brown trout, *Salmo trutta* L. spawning beds, indicating methods for their re-establishment in Danish lowland rivers. Paper presented at *EIFAC Symposium of habitat modification and freshwater fisheries*, Århus, Denmark, May 1984.

Hermens, L.C.M. (1978). Grune Bache in Limburg. In *Proceedings of the European Weed Research Society 5th Symposium on Aquatic Weeds*, pp. 212–218.

Herr, L.A. (1973). Stream channelization in the Federal Highway Program. In *Panel on stream channelization and trout fisheries* (ed. White, R.J.). *Trout Magazine*, **14**, 25,26.

Heuvelmans, M. (1974). *The River Killers*. Stackpole Books, Harrisburg, PA.

Hey, R.D. (1974). Prediction and effects of flooding in alluvial systems. In *Prediction of Geological Hazards* (ed. Funnell, B.M.). Miscellaneous paper No 3. Geological Society, pp. 42–56.

Hey, R.D. (1978). Determinate hydraulic geometry of river channels. *Journal of Hydraulic Engineering, American Society of Civil Engineers*, **104**, 869–885.

Hey, R.D. (1986). River mechanics. *Journal of the Institute of Water Engineers and Scientists*, **40**, 139–158.

Hey, R.D., and Thorne, C.R. (1986). Stable channels with mobile gravel beds. *Journal of the Hydraulics Division, American Society of Civil Engineers*, **112**, 671–689.

Hill, A.R. (1976). The environmental impacts of agricultural land drainage. *Journal of Environmental Management*, **4**, 251–274.

Hillman, E.C. (1936). The effect of flood relief works on flood levels below such works. *Journal of the Institute of Civil Engineers*, **2**, 393.

Hinge, D.C., and Hollis, G.E. (1980). *Land Drainage, Rivers, Riparian Areas and Conservation*. Discussion papers in Conservation No. 37, University College London, London.

Hjorth, E.S., and Tryk, I. (1984). *Vandløbsrestaurering af Vrold, Vestermose bæk*. Horsens Ingeniorskøle, Horsens, Denmark.

Hogger, J.B., and Lowery, R.S. (1982). The encouragement of freshwater crayfish populations by attention to the construction and maintenance of waterways. *Journal of the Institute of Water Engineers and Scientists*, **36**, 214–220.

Holder, T.W. (1970a). *Disappearing Wetlands in Eastern Arkansas*. Arkansas Planning Commission, Little Rock, AR.

Holder, T.W. (1970b). The destruction of our most valuable wildlife habitat. *Proceedings of the Annual Conference South-east Association Game and Fish Commission*, **23**, 13–18.

Hollis, G.E. (1976). River management and urban flooding. In *Conservation in Practice* (eds Warren, A., and Goldsmith, F.B.), John Wiley & Sons, Chichester, pp 201–216.

Hollis, G.E. (ed.) (1979). *Man's Influence on the Hydrological Cycle in the UK*. Geobooks, Norwich, UK.

Holm, T.F. (1984). *Strømkoncentratorers virkning pa vandløbets morfologi og sedimentdynamik*. Teknisk rapport No. 2. Miljøstyrelsens Ferskvandslaboratorium, Silkeborg, Denmark.

Holmes, N.T.H. (1986). *Wildlife Surveys of Rivers in Relation to their Management*. Report ER 1292-M, Water Research Centre, Medmenham, Marlow, Bucks.

Holtz, D.D. (1969). The ecology of the channelized and unchannelized Missouri River, Nebraska, with emphasis on the life history of the flathead catfish. Unpublished MS thesis, University of Missouri, Columbia, MO.

Hook, D.D. (ed.) (1987). *The Ecology and Management of Wetlands*, 2 volumes. Croom Helm, London.

Hook, J.M. (1977). The distribution and nature of changes in river channel patterns: The example of Devon. In *River Channel Changes* (ed. Gregory, K.J.), John Wiley & Sons, Chichester, pp. 265–280.

Hortle, K.G., and Lake, P.S. (1982). Macroinvertebrate assemblages in channelized and unchannelized sections of the Bunyip River, Victoria. *Australian Journal of Marine and Freshwater Research*, **35**, 1071–1082.

Hortle, K.G., and Lake, P.S. (1983). Fish of channelized and unchannelized sections of the Bunyip River, Victoria, *Australian Journal of Marine and Freshwater Research*, **34**, 441–450.

Huang Chung (1972). A new chapter in taming the Yellow River. In *China Tames her Rivers*. Foreign Language Press, Peking, China, pp. 1–14.

Hubbs, C.L., Greenly, J.R., and Tarzwell, C.M. (1932). Methods for the improvement of Michigan trout streams. *Michigan Institute of Fisheries Research Bulletin*, **1**, 54.

Huet, M. (1962). Influence du courant sur la distribution des poissons dans les eaux courantes. *Schweizerische Zeitschrift für Hydrologie*, **24**, 413–432.

Huggins, D.G., and Moss, R.E. (1974). Fish population structure in altered and unaltered areas of a small Kansas stream. *Transactions of the Kansas Academy of Science*, **77**, 18–30.

Hughes, D.A. (1966). Mountain streams of the Barberton area, eastern Transvaal. Part II. The effect of vegetational shading and direct illumination on the distribution of stream fauna. *Hydrobiologia*, **27**, 401–438.

Humenik, F.J., Bliven, L.F., Overcash, M.R., and Koehler. F. (1980). Rural nonpoint source water quality in a south eastern watershed. *Journal of the Water Pollution Control Federation*, **52**, 29–43.

Hung, Y.H., and Gaynor, R.K. (1977). Effects of stream channel improvement on downstream floods. *Research Report Pr-102 Kentucky Water Resources Research Institute*, Lexington, KY.

Hunt, R.L. (1969). Effects of habitat alteration on production, standing crops and yield of brook trout in Lawrence Creek, Wisconsin. In *Symposium on Salmon and Trout in Streams* (ed. Northcote, T.G.), Institute of Fisheries, University of British Columbia, Vancouver pp. 281–312.

Hunt, R.L. (1971). *Responses of a Brook Trout Population to Habitat Development in Lawrence Creek*. Technical Bulletin No. 48, Wisconsin Department of Natural Resources, Madison, WI.

Hunt, R.L. (1976). A long-term evaluation of trout habitat development and its relation to improving mangement-related research. *Transactions of the American Fisheries Society*, **105**, 361–365.

Hunt, W.A., and Graham, R.J. (1972). *Preliminary Evaluation of Channel Changes Designs to Restore Fish Habitat*. Department of Civil Engineering, Montana State University, Bozeman, MT.

Hussey, M.M., and Zimmerman, H.L. (1953). Rate of meander development as exhibited by two streams in Story County, Iowa. *Proceedings of the Iowa Academy of Science*, **60**, 390–392.

Hydraulics Research (1987). *Morphological Effects of River Works: a review of current practice*. Report for the Ministry of Agriculture, Fisheries and Food, Report No. SR116, Hydraulics Research Ltd, Wallingford, Oxfordshire, UK.

Hynes, H.B.N. (1960). *The Biology of Polluted Waters*. Liverpool University Press, Liverpool, UK.

Hynes, H.B.N. (1961). The invertebrate fauna of a Welsh mountain stream. *Archives of Hydrobiology*, **57**, 344–388.

Hynes, H.B.N. (1968). Further studies on the macroinvertebrate fauna of a Welsh mountain stream. *Archives of Hydrobiology*, **65**, 360–379.

Hynes, H.B.N. (1970). *The Ecology of Running Waters*. Liverpool University Press, Liverpool, UK.

Hynes, H.B.N. (1973). The effects of sedimentation on the biota in running water. In *Proceedings of the 9th Canadian Hydrology Symposium: Fluvial Processes and Sedimentation*. Inland Waters Directorate, Canadian Department of the Environment, Canada, pp. 653–662.

Hynson, J.R. (1985). *Environmental Features for Streamside Levee Projects*. Technical Report E-85–4, US Army Engineer Waterways Experiment Station, Vicksburg, MS.

Ibad-Zade, J.A. (1973). Flood control measures leveeing and bottom deepening. In *Flood Investigation*, vol. II. *Proceedings of the International Association of Hydraulic Research. Symposium on River Mechanics*. Asian Institute of Technology, Bangkok, pp. 499–510.

Ingle, R.M. (1952). Studies of the effect of dredging operations upon fish and shellfish. *Technical Service, Report No. 5, Florida Conservation Board*, FL.

Institute of Water Engineers (1961). *Manual of British Water Engineers Practice*, 3rd edition. London.

Irizarry, R. (1969). The effects of stream alteration in Idaho. Job completion Report, Project No. F-55-R-2. Idaho Fish and Game Department, ID.

Iwamonto, R.N., Salo, E.O., Madej, M.A., McCormas, R.L., and Rulfison, R.L. (1978). *Sediment and Water Quality: a review of the literature including a suggested approach for water quality criteria with summary of workshop and conclusions and recommendations*. Technical report No. EPA 910/9–78–048, US Environmental Protection Agency.

Jackson, H.B., and Bailey, R.A. (1979). Some practical aspects of river regulation in England and Wales. *Journal of the Institute of Water Engineers and Scientists*, **33**, 183–199.

Jackson, W.L., and Beschta, R.L. (1984). Influences of increased sand delivery on the morphology of sand and gravel channels. *Water Resources Bulletin*, **20**, 527–533.

Jackson, W.L., and Van Haveren, B.P. (1984). Design for a stable channel in coarse alluvium for riparian zone restoration. *Water Resources Bulletin*, **20**, 695–703.

Jaeggi, M. (1984). Problems related to important changes in river regime following intensive river training works in channels and channel control structures. In *Proceedings of the First International Conference on Hydraulic Design in Water Resources Engineering*. Southampton, UK, April 1984, pp. 603–612.

Jaffrey, L.J. (1967). The use of longitudinal embankments or *levees* as flood protection measures with particular reference to the River Trent in Great Britain. *International Commission of Irrigation and Drainage, 4th Congress*, **14**, 49–55.

James, L.D. (1965). Nonstructural measures for flood control. *Water Resources Research*, **1**, 9–24.

Jansen, P., Bendegom, L., Berg, J. de Vries, M., and Zanen, A. (1979). *Principles of River Engineering—the non-tidal alluvial river*. Pitman, London.

Januszewski, H., and Range, W. (1983). Ecologically based river regulation in small steps: the regulation of the Kahl in Lower Franconia. *Garten und Landschaft*, **2**, 99–102.

Jenkins, R.A., Wade, K.R., and Pugh, E. (1984). Macroinvertebrate-habitat relationships in the R. Teifi catchment and the significance to conservation. *Freshwater Biology*, **14**, 23–42.

Joglekar, D.V. (1971). *Manual on River Behaviour Control and Training*. Publication No. 60, Central Board of Irrigation and Power, New Delhi, India.

Johnson, E.A.G. (1966). Land drainage in England and Wales. In *River Engineering and Water Conservation Works* (ed. Thorn, R.B.). Butterworth, London, pp. 29–46.

Johnson, R.H., and Paynter, J. (1967). The development of a cutoff on the River Irk at Chadderton, Lancashire. *Geography*, **52**, 41–49.

Johnson, R.R. (1970). Tree removal along southwestern rivers and effects on associated organisms. *American Philosophical Society Yearbook*, pp. 321–322.

Johnson, R.R., and McCormick, J.F. (1978). *Strategies for Protection and Management of Floodplain Wetlands and other Riparian Ecosystems*. Technical Report WO-12, Forest Service, US Department of Agriculture, Washington, DC.

Jones, J.W., and King, G.M. (1950). Further experimental observations on the spawning behaviour of the Atlantic Salmon (*Salmo salar* Linn.). *Proceedings of the Zoological Society, London*, **120**, 317–323.

Joyal, R. (1970). Description de la torbiere a sphaignes Mer Bleue pres d'Ottawa. I. Vegetation. *Canadian Journal of Botany*, **48**, 1405–1418.

Kanaly, J. (1975). *Stream Improvement Evaluation in the Rock Creek Fishway, Carbon County (Addendum)*. Administrative Report for Project 5075–08–6602, Fish Division, Wyoming Game and Fish Department, Cheyenne, WY.

Karaki, S., Mahmood, K., Richardson, E.V., and Simons, D.B. (1974). *Highways in the River Environment, Hydraulic and Environmental Design Consideration*. Colorado State University, Fort Collins, CO.

Karr, J.R., and Schlosser, I.J. (1978). Water resources and the land water interface. *Science*, **201**, 229–234.

Kashef, A.I. (1981). Technical and ecological impacts of the High Aswan Dam *Journal of Hydrology*, **53**, 73–84.

Keller, E.A. (1975). Channelization : a search for a better way. *Geology*, **3**, 246–248.

Keller, E.A. (1976). Channelization : environmental, geomorphic and engineering aspects. In *Geomorphology and Engineering* (ed. Coates, D.R.). George Allen & Unwin, London, pp. 115–140.

Keller, E.A. (1978). Pools, riffles and channelization. *Environmental Geology*, **2**, 119–127.

Keller, E.A., and Brookes, A . (1984). Consideration of meandering in channelization projects: selected observations and judgements. In *River Meandering, Proceedings of Conference Rivers '83*. American Society of Civil Engineers, Vicksburg, MS, pp. 384–398.

Keller, E.A., and Hoffman, E.K. (1976). Sensible alternative to stream channelization. *Public Works*, 70–72.

Keller, E.A., and Hoffman, E.K. (1977). Urban streams: sensual blight or amenity. *Journal of Soil and Water Conservation*, **32**, 237–242.

Keller, E.A., and Swanson, F.J. (1979). Effects of large organic material on channel form and fluvial process. *Earth Surface Processes*, **4**, 115–140.

Keller, E.A., and Tally, T. (1979). Effects of large organic debris on channel form and fluvial processes in the coastal redwood environment. In *Adjustments to the Fluvial System* (eds Rhodes, D.D., and Williams, G.P.). Dubuque, Kendall-Hunt, IA, pp. 169–179.

Kellerhalls, R., Church, M., and Bray, D.I. (1976). Classification and analysis of river processes. *Journal of the Hydraulics Division, American Society of Civil Engineers*, **102**, 813–829.

Kendle, E.R. (1970). *The effects of channelization in the Missouri River on fish and fish food organisms*. Nebraska Game and Parks Commission, Nebraska.

Kennedy, G.J.A., Cragg-Hine, D., Strange, C.D., and Steward, D.A. (1983). The effects of a land drainage scheme on the salmonid populations of the River Camowen, Co. Tyrone. *Journal of Fish Management*, **14**, 1–16.

Kennedy, R.G. (1895). The prevention of silting in irrigation canals. *Proceedings of the Institute of Civil Engineers*, **119**, 2826

Keown, M.P. (1981). *Field Investigation of the Fisher River Channel Realignment Project Near Libby, Montana*. Inspection Report No. 11, Section 32 Program, US Army Engineer Waterways Experiment Station, Vicksburg, MS.

Keown, M.P., Oswalt, N.R., Perry, E.B., and Dardeau, E.A. (1977). *Literature Survey and Preliminary Evaluation of Streambank Protection Methods*. Technical Report No. H-77–9, US Army Engineer Waterways Experiment Station, Vicksburg, MS.

Keown, M.P., Oswalt, N.R., Perry, E.B., and Dardeau, E.A. (1981). *Streambank Erosion Control Evaluation and Demonstration Act of 1974, Section 32, Public Law 93–251, Appendix A. Literature Survey*. US Army Engineers Waterways Experiment Station, Vicksburg, MS.

Kern-Hansen, U. (1978). The drift of *Gammarus pulex* I. in relation to macrophyte cutting in four small Danish lowland streams. *Verhandlungen der internationalen Vereinigung für theoretische und angewandte Limnologie*, **20**, 1440–1445.

Kern-Hansen, U. (1980). The effect of macrophyte-cutting on the invertebrate fauna in a Danish lowland stream. Poster lecture presented at XXXI SIL Congress, Kyoto, Japan.

Kern-Hansen, U., and Holm, T.F. (1982). Aquatic plant management in Danish streams. *Proceedings of the European Weed Research Society 6th Symposium on Aquatic Weeds*, pp. 123–131.

Kern-Hansen, U., Holm, T.F., Thyssen, N., Mortensen, E., Hunding, C., and Bostrup, V. (1983). *Vedligeholdelse og restaurering af vandløb*. Tekniske anvisninger, Miljøstyrelsens Ferskvandslaboratorium, Silkeborg, Denmark.

Keutner, C. (1935). *Die Verwendung von Drahtnetzkorpen in Wasserbau*. Research Institute for River Control, Munich, Germany.

Kiel, W.H., Hawkins, A.S., and Perret, N.G. (1972). *Waterfowl Habitat Trends in the Aspen Parkland of Manitoba*. Report No. 18, Canadian Wildlife Service, Canada.

King, D.L., and Ball, R.C. (1964). The influence of highway construction on a stream. *Research Report of the Michigan State Agricultural Experimental Station*.

King, L.R., and Carlander, K.D. (1976). *A Study of the Effects of Stream Channelization and Bank Stabilization on Warmwater Sport Fish in Iowa. Subproject No. 3. Some effects of short-reach channelization on fishes and fish food organisms in central Iowa warmwater streams*. Report No. FWS/OBS 76/13, Office of Biological

Services, Fish and Wildlife Service, US Department of the Interior, Washington, DC.

Kite, D.J. (1979). Ecological study of land drainage works—River Stort navigation. Unpublished summary report for period April 1979–March 1980, University College London, London.

Klawitter, R.A. (1965). Woodland drainage in the south east. *Journal of Soil and Water Conservation*, **20**, 181–182.

Klawitter, R.A. (1970). Water regulation of forest land. *Journal of Forestry*, **68**, 338–342.

Klingeman, P.C., and Bradley, J.B. (1976). *Williamette River Basin Streambank Stabilization by Natural Means: Corvallis, Oregon*. Water Resources Research Institute, Oregon State University, OR.

Knight, D.W. (1981). Boundary shear in smouth and rough channels. *Journal of the Hydraulics Division, American Society of Civil Engineers*, **107**, 839.

Knight, D.W., and Demetriou, J.D. (1983). Flood plain and main channel flow interaction, *Journal of Hydraulic Engineering, American Society of Civil Engineers*, **109**, 1073–1092.

Knight, D.W., and Hamed, M.E. (1984). Boundary shear in symmetrical compound channels. *Journal of Hydraulic Engineering, American Society of Civil Engineers*, **110**, 1414–1430.

Knight, D.W., and Sellin, R.H.J. (1987). The SERC Flood Channel Facility. *Journal of the Institution of Water and Environmental Management*, **1**, 198–204.

Komura, S. (1970). Effect of channel straightening on the movement of flood waves on the Boyer River, Unpublished MS thesis, Iowa State University, Ames, IA.

Komura, S., and Simons, D.B. (1967). River bed degradation below dams. *Journal of the Hydraulics Division, American Society of Civil Engineers*, **93**, 1–14.

Krause, A. (1977). On the effect of marginal tree rows with respect to the management of small lowland streams. *Aquatic Botany*, **3**, 185–192.

Kuenzler, E.J., Mulholland, P.J., Ruley, L.A., and Sniffen, R.P. (1977). *Water Quality in North Carolina Coastal Plain Streams and Effects of Channelization*. Report No. 127, Water Resources Research Institute, University of North Carolina, NC.

Kuprianov, V.V. (1977). *Hydrologicheskie aspecty urbanizatsii*. Leningrad, Gidrometeoizdat.

Lacey, G. (1930). Stable channels in alluvium, *Proceedings of the Institute of Civil Engineers*, **229**, 259–292.

Lagasse, P.F. (1976). A geomorphic analysis of riverine dredging problems. *Proceedings of the World Dredging Conference (WODCON) VII*. San Francisco, CA, 9–13 July 1976.

Lagasse, P.F., and Simons, D.B. (1976). Impact of dredging on river system morphology. In *Proceedings of Conference Rivers '76*. American Society of Civil Engineers, pp. 434–458.

Lagasse, P.F., Winkley, B.R., and Simons, D.B. (1980). Impact of gravel mining on river system stability. *Journal of the Waterways, Port, Coastal and Ocean Division, American Society of Civil Engineers*, **106**, 389–404.

Lambrechts, C. (1983). Impact studies. *Naturopa*, 29–30.

Lamplugh, G.W. (1914). Taming of streams. *Geographical Journal*, **43**, 651–656.

Lane, E.W. (1947). The effects of cutting off bends in rivers. *Proceedings of the Third Hydraulics Conference*, Bulletin No. 31. University of Iowa, IA.

Lane, E.W. (1955a). Design of stable channels. *Transactions of the American Society of Civil Engineers*, **120**, 1234–1279.

Lane, E.W. (1955b). The importance of fluvial morphology in hydraulic engineering. *Proceedings of the American Society of Civil Engineers*, **81**, 1–17.

Larimore, R.W., and Smith, P.W. (1963). The fishes of Champaign County, Illinois, as affected by 60 years of stream changes. *Bulletin Illinois State Natural History Survey*, **28**, 299–382.

Lauder, T.D. (1830). *An Account of the Great Floods of August 1829 in the Province of Moray and Adjoining Districts*, 2nd edn. Adam Black, Edinburgh, Scotland.

Lee, G.F. (1976). *Environmental Impact of Dredging, Dredged Material Disposal and Dredged Material Research in the United States*. Occasional Paper No. 10, Centre of Environmental Study, University of Texas, TX.

Leentvaar, P. (1978). Aquatic weed control related to nature management. In *Proceedings of the European Weed Research Society 5th Symposium on Aquatic Weeds*, pp. 83–89.

Leliavsky, S. (1955). *Irrigation and Hydraulic Design*. Chapman and Hall, London.

Leliavsky, S. (1966). *An Introduction to Fluvial Hydraulics*. Dover Publications, New York.

Leopold, L.B. (1969). *Quantitative Comparison of some Aesthetic Factors among Rivers*. Circular 620, Geological Survey, US Department of the Interior, Washington, DC.

Leopold, L.B. (1977). A reverence for rivers. *Geology*, **5**, 429–430.

Leopold, L.B., and Maddock, T. (1954). *The Flood Control Controversy*. Ronald Press, New York.

Leopold, L.B., and Marchand, M.O. (1968). On the quantitative inventory of the riverscape. *Water Resources Research*, **4**, 707–717.

Leopold, L.B., Clarke, F.E., Hanshaw, B.B., and Balsley, J.R. (1971). *A Procedure for Evaluating Environmental Impact*. Circular 645, Geological Survey, US Department of the Interior, Washington, DC.

Lester, H.H. (1946). Stream Bank erosion control. *Agricultural Engineering*, **27**, 407–410.

Lewin, J. (1976a). Environmental Impact Statements. *Area*, 127–128.

Lewin, J. (1976b). Initiation of bedforms and meanders in coarse-grained sediment. *Bulletin of the Geological Society of America*, **87**, 281–285.

Lewis, G., and Williams, G. (eds) (1984). *Rivers and Wildlife Handbook: a guide to practices which further the conservation of wildlife on rivers*. Royal Society for the Protection of Birds, Bedfordshire, and the Royal Society for Nature Conservation, Lincoln, UK.

Lewis, K. (1973a). The effect of suspended coal particles on the life forms of the aquatic moss *Eurhynchium ripariodes* Hedw. I. The gametophyte plant. *Freshwater Biology*, **3**, 251–258.

Lewis, K. (1973b). The effect of suspended coal particles on the life forms of the aquatic moss *Eurhynchium ripariodes* Hedw. II. The effect on spore germination and regeneration of apical tips. *Freshwater Biology*, **3**, 391–396.

Lindley, E.S. (1919). Regime channels, *Proceedings of the Punjab Engineering Congress*, 7, 63–74

Lindner, C.P. (1952). Diversions from alluvial streams. *Transactions of the American Society of Civil Engineers*, **118**, 245–269.

Linsley, R.K., Kohler, M.A., and Paulhus, H.L.H. (1949). *Applied Hydrology*. McGraw-Hill, New York.

Lisle, T.E. (1981). Roughness elements: a key resource to improve anadromous fish habitat. In *Proceedings of the Propagation, Enhancement and Rehabilitation of Anadromous Salmonid Populations and Habitat in the Pacific North west symposium* (ed. Hassler, T.J.). Humboldt State University, Arcata, CA, pp. 93–98.

Little, A.D. (1973). *Channel modification: an environmental, economic and financial assessment*. Report to the Council on Environmental Quality, Executive Office of the President, Washington, DC.

Lohmeyer, W., and Krause, A. (1974). Uber den Geholzbewuchs an kleinen Fliessgewassern Norwest Deutschlands und seine Bedeutung für den Uferschutz. *Natur und Landschaft*, **49**, 323–330.

Luey, J.E., and Adelman, I.R. (1980). Downstream natural areas as refuges for fish in drainage development watersheds. *Transactions of the American Fisheries Society*, **109**, 332–325.

Lund, J.A. (1976). *Evaluation of Stream Channelization and Mitigation on the Fishery Resources of the St. Regis River, Montana*. Report No. FWS/OBS 76–07, Office of Biological Services, Fish and Wildlife Services, US Department of the Interior, Washington, DC.

Lynn, M.A. (1970). *Arterial Drainage in Ireland*. Office of Public Works, Dublin.

McCall, J.D., and Hyer, R. (1975). 'New Look' in channel work. *Soil Conservation*, February, pp. 20–21.

McCall, J.D., and Knox, R.F. (1978). Riparian habitat in channelization projects. In *Proceedings of the Symposium on Strategies for Protection and Management of Floodplain Wetlands and Other Riparian Ecosystems*. Technical Report No. GTR-WO-12, Forest Service, US Department of Agriculture, Washington, DC.

McCart, P. (1969). Digging behaviour of *Oncorhynchus nerka* spawning in streams at Babine Lake, British Columbia. In *Symposium on Salmon and Trout in Streams* (ed. Northcote, T.G.). Institute of Fisheries, University of British Columbia, Vancouver, pp. 39–51.

McCarthy, D.T. (1975). *The Effects of Drainage on the Flora and Fauna of a Tributary of the River Boyne*. Fishery Leaflet No. 68. Department of Agriculture and Fisheries, Dublin, Ireland.

McCarthy, D.T. (1977). The effects of drainage on the Trimblestown river. I. Benthic invertebrates and flora. *Irish Fisheries Investigations*, Series A, No. 16.

McCarthy, D.T. (1980). *Impact of Arterial Drainage on Fisheries*, MAB & NBST Symposium, Dublin, Ireland.

McCarthy, D.T.(1981). The effects of arterial drainage on the invertebrate fauna and fish stocks of Irish rivers. *Proceedings of the Second British Freshwater Fisheries Conference*, pp. 208–213.

McCarthy, D.T., and Glass, B. (1982). *Pilot Environmental Impact Statement, Drainage Ireland West. Appendix II. Hydrobiology*. Department of Fisheries and Forestry, Fisheries Research Centre, Abbotstown, Co. Dublin, Ireland.

McClellan, T.J. (1974). *Ecological Recovery of Realigned Stream Channels: Portland, Oregon*. Technical Report. Federal Highways Administration, US Department of Transportation, Portland, OR.

McConnell, C.A., Parsons, D.R., Montgomery, G.L., and Gainer, W.L. (1980). Stream renovation alternatives : the Wolf River Story. *Journal of Soil and Water Conservation*, **35**, 17–20.

McHarg, I.L. (1969). *Design with Nature*. Doubleday, Garden City, New York.

McLeod, G. (1970). Flooding (South Wales valleys). In *Problems of the South Wales Valleys*. Institute of Civil Engineers, London, UK, pp. 75–79.

MacBroom, J.G. (1980). Applied fluvial morphology in open channel design. *Water Management Conference, American Society of Civil Engineers*, Washington, DC.

Maccaferri Gabions of America (1968). *Stream Improvement Handbook*. Publication No. MN-604, Flushing, New York.

Macdonald, S.M. (1983). The status of the otter (*Lutra lutra*) in the British Isles. *Mammal Reviews*, **13**, 11–23.

Macdonald, S.M., and Mason, C.F. (1982). Some factors influencing the distribution of otters (*Lutra lutra*). *Mammal Reviews*, **13**, 1–10.

Macdonald, S.M., Mason, C.F., and Coghill, I.S. (1978). The otter and its conservation in the River Teme catchment. *Journal of Applied Ecology*, **15**, 373–384.

Mackin, J.H. (1948). Concept of the graded river. *Geological Society of America Bulletin*, **59**, 463–512.

Maddock, T., Jr. (1976). A primer on floodplain dynamics. *Journal of Soil and Water Conservation*, **31**, 44–47.

Madsen, B.L. (1983). *Den nye vandløbslov*. Miljøstyrelsens, Copenhagen, Denmark.

Makin, T.E., Weber, A.J., Hazel, D.W., Hunter, D.C., Hyberg, B.T., Flinchum, D.M., Lollis, J.P., Rognstand, J.B., and Gregory, J.D. (1980). *Effects of Stream Channelization on Bottomland and Swamp Forest Ecosystems*. Report No. 147, Water Resources Research Institute, North Carolina, NC.

Malhotra, S.L. (1951). *Effects of Barrages and Weirs on the Regime of Rivers*. International Association of Hydraulics Research, Bombay, India.

Manners, I.R., and Mikesell, M.W. (eds) (1974). *Perspectives on Environment*. Association of American Geographers Publication No. 13.

Margalef, R. (1963). On certain unifying principles in ecology. *American Naturalists*, **97**, 357–374.

Marsh, G.P. (1864). *Man and Nature or Physical Geography as Modified by Human Action*. Charles Scribner, New York.

Marsh, P.C., and Waters, T.F. (1980). The effects of agricultural drainage development on benthic invertebrates in undisturbed downstream reaches, *Transactions of the American Fisheries Society*, **109**, 213–223.

Marshall, E.J.P. (1981). The ecology of a land drainage channel. I: Oxygen balance. *Water Research*, **15**, 1075–1085.

Marshall, E.J.P. (1984). The ecology of a land drainage channel. II: Biology, chemistry and submerged weed control. *Water Research*, **18**, 817–825.

Marshall, E.J.P., Wade, P.M., and Clare, P. (1978). Land drainage channels in England and Wales. *Geographical Journal*, **144**, 254–263.

Marsland, N. (1966). The design and construction of earthern floodbanks. In *River Engineering and Water Conservation Works* (ed. Thorn, R.B.). Butterworth, London, pp. 361–392.

Martin, E.C. (1969). Stream alteration and its effects on fish and wildlife. *Proceedings of the 23rd American Conference. South-east Association of Game and Fish Commission*. Mobile, Alabama.

Martvall, S., and Nilsson, G. (1972). *Experimental Studies of Meandering: the transport and deposition of material in curved channels*, Report No. 20, Institute of Physiography, Uppsala University, Sweden.

Marzolf, R.G. (1978). *The Potential Effects of Clearing and Snagging on Stream Ecosystems*. Report No. FWS/OBS-78/14, Office of Biological Services, Fish and Wildlife Service, US Department of the Interior, Washington, DC.

Mason, C.F., and Bryant, R.J. (1975). Changes in the ecology of the Norfolk Broads, *Freshwater Biology*, **5**, 257–270.

Massachusetts Water Resources Commission (1971). *Neponset River Basin Flood Plain and Wetlands Encroachment Study*. Massachusetts Water Resources Commission, Boston, Massachusetts.

Matthes, G.H. (1949). Mississippi river cutoffs. *Transactions of the American Society of Civil Engineers*, **113**, 1–39.

Maughan, O.E., Nelson, K.L., and Ney, J.J. (1978). *Evaluation of Stream Improvement Practices in Southeastern Trout Streams*, Virginia Polytechnic Institute and State University, Blacksburg, Project No. VP1-VWRRC-BULL 115, prepared for Office of Water Research and Technology, US Department of the Interior, Washington, DC.

May, B. (1975). *Evaluation and Mitigation Measures in Fisher River, Wolf Creek and Fortine Creek, 1969–1972*. Job Progress Report and Completion report for Contract No. DACW 67-70-C-0001, Fisheries Division, Montana Department of Fish and Game, in Cooperation with US Army Corps of Engineers, Helena, MT.

Mayer, J.R. (1978). Aquatic weed management by benthic semi-barriers. *Journal of Aquatic Plant Management*, **16**, 31–33.

Maynard, S.T. (1978). *Practical Riprap Design*. Miscellaneous Paper No. H-78-7. US Army Engineer Waterways Experiment Station, Vicksburg, MS.
Medrington, N. (1965). Some land drainage works in the area of the Lee Conservancy Catchment Board. *Association of River Authorities Year Book*, pp. 77–90.
Medrington, N. (1968). Thrust-bored culverts for watercourses. *Association of River Authorities Year Book*, pp. 51–54.
Meehan, W.R. (1971). Effects of gravel cleaning on bottom organisms of three southeast Alaska streams. *Progressive Fish Culturist*, **33**, 107–111.
Melhorn, W.N., Keller, E.A., and McBane, R.A. (1975). *Landscape Aesthetics Numerically Defined (LAND System): Application to fluvial environments*, Technical Report No 37, Water Resources Research Center, Purdue University, West Lafayette, IN.
Miers, R.H. (1977). *The Role of Vegetation in Land Drainage. Report of study tour undertaken in West Germany*. Ministry of Agriculture, Fisheries and Food, London.
Miers, R.H. (1979). Land drainage: its problems and solutions. *Journal of the Institute of Water Engineers and Scientists*, **33**, 547–579.
Mifkovic, C.S., and Petersen, M.S. (1975). Environmental aspects—Sacremento bank protection. *Journal of the Hydraulics Division, American Society of Civil Engineers*, **101**, 543–555.
Mih, W.C. (1978). A review of restoration of stream gravel for spawning and rearing of salmon species. *Fisheries*, **3**, 16–19.
Mih, W.C. (1979). *Hydraulic Restoration of Stream Gravel for Spawning and Rearing of Salmon Species*. Albrook Hydraulic Laboratory, Report to Washington State Water Research Center, Washington State.
Mih, W.C., and Bailey, G.C. (1981). The development of a machine for the restoration of stream gravel for spawning and rearing of salmon. *Fisheries*, **6**, 16–20.
Miles, W.D. (1976). Land drainage and weed control. In *Aquatic Herbicides*. British Crop Protection Council Monograph No. 16, pp. 7–13.
Miller, J.N., and Simmons, R. (1970). Crisis on our rivers. *Readers Digest*, December, 97, pp. 584–590.
Mills, G.A. (1981). The spawning of roach *Rutilis rutilis*, L. in a chalk stream. *Fisheries Management*, **12**, 49–54.
Milner, N. (1984). Fish. In *Rivers and Wildlife Handbook: a guide to practices which further the conservation of wildlife in rivers* (eds Lewis, G., and Williams, G.). Royal Society for the Protection of Birds, Bedfordshire, UK, pp. 51–53.
Milner, N.J., Scullion, J., Carling, P.A., and Crisp, D.T. (1981) The effects of discharge on sediment dynamics and consequent effects on invertebrates and salmonids in upland rivers. *Advances in Applied Biology*, **VI**, 153–220.
Minikin, R.C.R. (1920). *Practical River and Canal Engineering*. Griffin & Co., London.
Ministry of Agriculture, Fisheries and Food (1979). *Code of Practice for the Use of Herbicides and Weeds in Watercourses and Lakes*. MAFF, London.
Ministry of Agriculture, Fisheries and Food (1985). *Investment Appraisal of Arterial Drainage Schemes*. Land and Water Services, River and Coastal Engineering Group, MAFF, London.
Mitchell, D.S. (1974). *Aquatic Vegetation and its Use and Control*. UNESCO, Paris.
Möller, H.M., and Wefers, K. (1983). The restoration of cutoff river channels: as illustrated by the lower reaches of the Krückau in the District of Pinneberg. *Garten und Landschaft*, **2**, 107–108.
Moore, W.L., and Morgan, C.W. (eds) (1969). *Effects of Watershed Changes in Streamflow*. Water Research Symposium No. 2. University of Texas, Austin, TX.
Morisawa, M. (1976). Readjustment of an 'improved' stream channel. *Geological Society of America Abstract (with programs)*, **6**, 877.

Morisawa, M. (1985). *Rivers: Form and Process*. Geomorphology Texts. Longman, London.

Morisawa, M., and Vemuri, R. (1975). *Multiobjective Planning and Environmental Evaluation of Water Resource Systems*, Technical Project C-6065, Final Report. Office of Water Research, US Department of the Interior, Washington, DC.

Morris, L.A., Langemeirs, R.M., Russell, T.R., and Witt, A. (1968). Effects of main stem impoundments and channelisation upon the limnology of the Missouri River, Nebraska. *Transactions of the American Fisheries Society*, **97**, 380–388.

Mortensen, E. (1977). Density-dependent mortality of trout fry (*Salmo trutta*) and its relationship to the management of small streams. *Journal of Fish Biology*, **11**, 613–617.

Mosley, M.P. (1975). Meander cutoffs on the River Bollin, Cheshire, in July 1975. *Revue de Geomorphologie Dynamique*, **24**, 21–31.

Mosley, M.P. (1981). The influence of organic debris on channel morphology and bedload transport in a New Zealand forest stream. *Earth Surface Processes*, **6**, 571–580.

Moss, B. (1980). Further studies on the palaeolimnology and changes in the phosphorus budget of Barton broad, Norfolk. *Freshwater Biology*, **10**, 261–279.

Moyle, P.B. (1976). Some effects of channelization on the fishes and invertebrates of Rush Creek, Modoc County, California. *California Fish and Game*, **63**, 179–186.

Mrowka, J.P. (1974). Man's impact on stream regimen and stream quality. In *Perspectives in Environment* (eds Manners, I.R., and Mikesell M.W.). Publication No. 13, Association of American Geographers, pp. 79–104.

Muckleston, K.W. (1976). The evolution of approaches to flood damage reduction. *Journal of Soil and Water Conservation*, 53–59.

Müller, K. (1953). Produktionsbiologische Untersuchungen in Nordschwedischen Fliessgewässern. Teil 1. Der Einfluss der Flössereiregulierungen auf den quantitativen und qualitativen Bestand der Bodenfauna. *Reports of the Institute of Freshwater Research, Drottningholm*, **34**, 90–121.

Müller, K. (1955). Produktionsbiologische Untersuchungen in Nordschwedischen Fliessgewässern. Teil 3. Die Bedeutung der Seen and Stillwasserzonen für die Produktion in Fliessgewässern. *Reports of the Institute of Freshwater Research, Drottningholm*, **36**, 148–162.

Müller, K. (1962). Limnologisch-Fischereibiologische Untersuchungen in regulierten Gewässern Schwedisch Lapplands. *Oikos*, **13**, 125–154.

Mumford, L. (1931). *The Brown Decades: a study of the arts in America, 1865–1895*. Dover Publications, NY.

Munn, R.E. (1979). *Environmental Impact Assessment: Principles and Procedures*. Scientific Committee on Problems of the Environment (SCOPE), Report No. 5 (second edn). John Wiley & Sons, Chichester.

Munroe, D.A. (1967). The prairies and the ducks, *Canadian Geographical Journal*, **75**, 2–13.

Myers, C.T., and Ulmer, R.L. (1975). *Streambank Stabilization Measures in Mississippi*. Paper No. 75-2517. American Society of Agricultural Engineers, Michigan. Prepared for annual meeting of American Society of Agricultural Engineers, Chicago, Illinois.

NEDECO (1965). *Development and Maintenance Dredging on Rivers: a study of hydraulic engineering considerations*. NEDECO, The Hague, Netherlands.

Neller, R.J., and Broughton, W.C. (1981). Modifications to stream channels in the Brisbane Metropolitan area, Australia. *Environmental Conservation*, **8**, 299–307.

Nelson, K.L., and Weaver, E. (1981). Criteria for the design and evaluation of stream excavation projects in North Carolina. *Fisheries*, **6**, 7–10.

Nelson, R.W., Horak, G.C., and Olsen, J.E. (1978). *Western Reservoir and Stream Improvements Handbook*. Report No. FWS/OBS-78/56. Office of Biological Services, Fish and Wildlife Service, US Department of the Interior, Washington, DC.

Neuhold, J.H. (1981). Strategy of stream ecosystem recovery. In *Stress Effects on Natural Ecosystems* (ed. Barrett, G.W., and Rosenburg, R.). (Environmental Monographs and Symposia, convenor and general editor N. Polunin). John Wiley & Sons, New York.

Newbold, C., Purseglove, J., and Holmes, N. (1983). *Nature Conservation and River Engineering*. Nature Conservancy Council, Shrewsbury, UK.

Newson, M.D. (1986). River basin engineering—fluvial geomorphology. *Journal of the Institution of Water Engineers and Scientists*, **40**, 307–324.

New York State Court of Claims (1966). *Robert H. Barnes* vs *State of New York Attorney General, Open File Decision*, 22 June, 1966.

New York State Supreme Court (1957). *Paul Demoski* vs. *State of New York* Claim No. 33940. *New York Supreme Court 12*, Misc. 2nd 416, Oct. 1.

Nielsen, M.B. (1985). Grødeskaeringsforsøg i Surbæk. *Vand and Miljø*, **2**, 120–124.

Nixon, M. (1966). Flood regulation and river training. In *River Engineering and Water Conservation Works* (ed. R.B. Thorn). Butterworth, London, pp. 293–297.

Noble, E.L., and Palmquist, R.C. (1968). Meander growth in artificially straightened streams. *Iowa Academy of Science Proceedings*, **75**, 234–242.

North West Water Authority (1978). *Environmental Appraisal: Haweswater, Borrow Beck, Morecambe Bay, Hellifield*. Report prepared by Land Use Consultants Ltd, London.

Norton, S.E., Timbol, A.S., and Parrish, J.D. (1978). *Stream Channel Modification in Hawaii. Part B. Effect of channelization on the distribution and abundance of fauna in selected streams*. Report No. FWS/OBS 78/17 Office of Biological Services, Fish and Wildlife Service, US Department of the Interior, Washington, DC.

Nunnally, N.R. (1978a). Stream renovation: an alternative to channelisation. *Environmental Management*, **2**, 403–411.

Nunnally, N.R. (1978b). Improving channel efficiency without sacrificing fish and wildlife habitat: the case for stream restoration. *Proceedings of National Symposium on Strategies for Protection and Management of Riparian Ecosystems*. Calloway Gardens, GA, pp. 394–399.

Nunnally, N.R. (1985). Application of fluvial relationships to planning and design of channel modifications. *Environmental Management*, **9**, 417–426.

Nunnally, N.R., and Beverly, L. (1983). Morphologic effects of Lower Mississippi River Dike Fields. *Proceedings of Conference Rivers '83, American Society of Civil Eengineers*, Vicksburg, MS. pp. 418–429.

Nunnally, N.R., and Keller, E.A. (1979). *Use of Fluvial Processes to Minimise Adverse Effects of Stream Channelisation*. Report No. 144, Water Resources Research Institute, University of North Carolina, NC.

Nunnally, N.R., and Shields, F.D. Jr (1985). *Incorporation of Environmental Features in Flood Control Channel Projects*. Technical Report E-85-3, Environmental and Water Quality Operational Studies, US Army Engineer Waterways Experiment Station, Vicksburg, Mississippi.

Nunnally, N.R., Shields, F.D., Jr and Hynson, J. (1987). Environmental considerations for levees and floodwalls *Environmental Management*, **11**, 183–191.

Nuttall, P.M. (1972). The effects of sand deposition upon the macroinvertebrate fauna of the River Camel, Cornwall. *Freshwater Biology*, **2**, 181–186.

O'Riordan, T., and More, R.J. (1969). Choice in water use. In *Water, Earth and Man* (ed. Chorley, R.J.). Methuen, London, pp. 547–573.

Ockerson, J.A. (1898). Dredgers and dredging on the Mississippi River. *Transactions of the American Society of Civil Engineers*, **XL**, 215–310.

Office, Chief of Engineers, US Department of the Army (1982a). *Planning, Project*

Purpose Planning Guidance. Engineer Regulation No. 1105-2-20, Washington, DC.

Office, Chief of Engineers, US Department of the Army (1982b). *Planning, Economic Considerations*. Engineer Regulation No. 1105-2-40, Washington, DC.

Olschowy, G. (1957). Bepflanzung von Bachläufen und Gräben. In *Naturnaher Ausbau von Wasserläufen*, (eds Olschowy, G., and Köhler, H.). Hiltrup bei Münster, pp. 86–94.

Olschowy, G., and Köhler, H. (eds) (1957). *Naturnaher Ausbau von Wasserläufen*. Vorträge, Aussprachen und Ergebrusse der gleichartigen Arbitstagung auf Bundesebene vom 10-12.10.1956 in Würzburg, Landwirtschaftsverlag, Hiltrup bei Münster.

Osakov,Y.A., Kirikov, S.V., and Formozov, A.N. (1971). Land Game, In *Natural Resources of the Soviet Union: their use and renewal* (eds Gerasimov, I.P., Armand, D.L. and Yefron, K.M.). W.H. Freeman & Co., San Francisco, CA, pp. 251–292.

Osborne, P.L., and Moss, B. (1977). Palaeolimnology and trends in the phosphorous and iron budgets of an old man-made lake, Barton Broad, Norfolk. *Freshwater Biology*, **7**, 213–233.

Otis, M.B. (1958). *Guide to Stream Improvement*, Information leaflet. State Conservation Department, New York.

Palmer, L. (1976). River management criteria for Oregon and Washington. In *Geomorphology and Engineering* (ed. Coates, D.R.). George Allen & Unwin, London, pp. 329–346.

Park, C.C. (1981). Man, river systems and environmental impacts. *Progress in Physical Geography*, **5**, 1–31.

Parker, D.J., and Penning-Rowsell, E.C. (1980). *Water Planning in Britain*. George Allen & Unwin, London.

Parker, D.J., Green, C.H., and Thompson, P.M. (1987). *Urban Flood Protection Benefits: a project appraisal guide*. Gower Technical Press, Farnborough, UK.

Parker, G., and Andres, D. (1976). Detrimental effects of river channelization. *Proceedings of Conference Rivers '76, American Society of Civil Engineers*, pp. 1248–1266.

Parrish, J.D., Maclolek, J.A., Timbol, A.S., Hathaway, C.B., and Norton, S.E. (1978). *Stream Channel Modification in Hawaii. Part D: Summary Report*. Report No FWS/OBS-78/19, Office of Biological Services, Fish and Wildlife Service, US Department of the Interior, Washington DC.

Parsons, D.A. (1965). Vegetative control of streambank erosion. In *Proceedings of the Third Federal Interagency Sedimentation Conference*. Miscellaneous Publication 970, US Department of Agriculture, pp. 130–136.

Patrick, R. (1971). *The Effects of Channelization on the Aquatic Life of Streams*. Academy of Natural Science of Philadelphia, PA.

Paynting, T. (1982). Flood scheme reconciles conservation and alleviation. *Surveyor*, 14–16.

Pearson, R.G., and Jones, N.V. (1975). The effects of dredging operations on the benthic community of a chalk stream. *Biological Conservation*, **8**, 273–278.

Pearson, R.G., and Jones, N.V. (1978). The effects of weed-cutting on the macroinvertebrate fauna of a canalised section of the River Hull, a northern English chalk stream. *Journal of Environmental Management*, **7**, 91–97.

Peltier, W.H., and Welch, E.B. (1969). Factors affecting growth of rooted aquatics in a river. *Weed Science*, **17**, 412–416.

Penning-Rowsell, E.C., and Chatterton, J.B. (1977). *The Benefits of Flood Alleviation: a manual of assessment techniques*. Saxon House, Gower Technical Press, Farnborough.

Percival, E., and Whitehead, H. (1929). A quantitative study of the fauna of some types of stream bed. *Journal of Ecology*, **17**, 282–314.

Perry, E.W. (1974). The effect of stream improvement structures on the sport fishery in a channelized section of the Olentangy River, Unpublished MS thesis, Ohio State University, Columbus, OH.

Peters, J.C., and Alvord, W. (1964). Man-made alterations in thirteen Montana streams and rivers. *Transactions of North American Wildlife Conference*, **29**, 93–102.

Petersen, M.S. (1964). Hydraulic aspects of Arkansas river stabilization. *Symposium on Channel Stabilisation Problems*. Technical Report No. 1. US Army Corps of Engineers,

Petryk, S., and Bosmajian, G. (1975). Analysis of flows through vegetation. *Journal of the Hydraulics Division, American Society of Civil Engineers*, **101**, 871–884.

Petts, G.E. (1979). Complex response of river channel morphology subsequent to reservoir construction. *Progress in Physical Geography*, **3**, 329–362.

Petts, G.E. (1980). Morphological changes of river channels consequent upon head-water impoundment. *Journal of the Institute of Water Engineers and Scientists*, **34**, 374.

Petts, G.E. (1984). *Impounded Rivers: perspectives for ecological management*. John Wiley & Sons, Chichester.

Petts, G.E., and Greenwood, M. (1981). Habitat changes below Dartmoor reservoirs, *Reports and Transactions, Devonshire Association for the Advancement of Science*, **113**, 13–27.

Pfeffer, H. (1978). Lebendbau und Landschaftspflege am Beispiel Untere Vils. *Garten und Landschaft*, **1**, 31–36.

Phillipson, J. (1956). A study of factors determining the distribution of the blackfly *Simulium ornatum*. Mg. *Bulletin of Entomology Research*, **47**, 227–238.

Pickles, G.W. (1931). Run-off investigations in Central Illinois. *University of Illinois Bulletin*, **29**, 3.

Pickles, G.W. (1941). *Drainage and Flood Control Engineering*, 2nd edn. McGraw-Hill, New York.

Piest, R.F., Elliot, L.S., and Spomer, R.G. (1977). Erosion of the Tarkio Drainage System, 1845–1976, *Transactions of the American Society of Agricultural Engineers*, **20**, 458–488.

Pitlo, R.H. (1978). Regulation of aquatic vegetation by interception of daylight. In *Proceedings of the European Weed Research Society 5th International Symposium on Aquatic Weeds*, 5–8 September 1978, pp. 91–99.

Platts, W.S. (1981). Streamside management to protect bank-channel stability and aquatic life. In *Proceedings of the Interior West Watershed Management Symposium* (ed. Baumgartner, D.M.). Washington State University, Pullman, Washington, DC, pp. 245–255.

Platts, W.S., Shirazi, M.A., and Lewis, D.H. (1979). *Sediment Particle Sizes used by Salmon for Spawning and Methods for Evaluation*. Report No. EPA 600/3-79-043, US Environmental Protection Agency, Washington, DC.

Ponce, V.M., and Mahmood, K. (1976). Meandering thalwegs in straight alluvial channels. *Proceedings of Conference, Rivers '76, American Society of Civil Engineers*, pp. 1418–1441.

Porter, M.E. (1977). Effects of stream channelization on the macrobenthic and fish communities of the Little Auglaize River, Ohio, Unpublished PhD thesis, Ohio State University, Columbus, OH.

Posewits, J. (1967). Ravage the river. *Montana Wildlife*, 18–21.

Possardt, E.E. (1975). The effects of stream channelization on aquatic and riparian wildlife in the White River watershed, Vermont, Unpublished MS thesis, University of Massachusetts, Amherst, MA.

Possardt, E.E., and Dodge, W.E. (1978). Stream channelization impacts on songbirds and small mammals in Vermont. *Wildlife Society Bulletin*, **6**, 18–24.
Priestly, J. (1831). *Historical Account of the Navigable Rivers, Canals and Railways of Great Britain*. London.
Prudhomme, P. (1975). Development of the lower course of the River Aar and protection of its aquifer. An example of controlled gravel extraction (in French). *Houille Blanche*, **2–3**, 145–153.
Public Law 566 (1954). Watershed Protection and Flood Prevention Act. *US Statutes at Large*, **68**, 1.
Purseglove, J.J. (1983). River engineering and nature conservation in England. *Garten und Landschaft*, **2**, 119–122.
Raleigh, R.F. (1982). *Habitat Suitability Index Models: Brook trout*. Report No. FWS/OBS-82/10-24, Office of Biological Services, Fish and Wildlife Service, US Department of the Interior, Washington DC.
Ramser, C.E. (1929). Flow of water in drainage channels, *US Department Agriculture Technical Bull*., 129.
Rankin, D. (1980). Trees and rivers. *Journal of the Soil Conservation Service of New South Wales*, **36**, 129–133.
Raudkin, A.J. (1976). *Loose Boundary Hydraulics*, 2nd edn. Pergamon Press, Oxford.
Raven, P.J. (1985). Ecological effects of two-stage flood relief channels on the River Roding, Essex, England. Unpublished PhD thesis, University of London, London.
Raven, P.J. (1986). Changes of in-channel vegetation following two-stage channel construction on a small rural clay river. *Journal of Applied Ecology*, **23**, 333–345.
Rayment, A.F., and Cooper, D.J. (1968). Drainage of Newfoundland peat soils for agricultural purposes. In *Proceedings of the 3rd International Peat Congress*, Quebec, Canada, pp. 345–355.
Rees, W.H. (1959). Effects of stream dredging on young salmon and bottom fauna. *Washington Department of Fish, Fisheries Research Papers*, **2**, 53–65.
Report of the Royal Commission on Land Drainage in England and Wales (1927). Her Majesty's Stationery Office, London.
Ribaut, J.-P. (1982). Nature and environment. *Naturopa*, **42**, 4–6.
Rich, K. (1969). Channelization severely altering fish populations. *Missouri Game and Fish*, March-April.
Richards, B.D. (1950). *Flood Estimation and Control*. Chapman & Hall, London.
Richards, K.S. (1979). Channel adjustment to sediment pollution by the china clay industry in Cornwall, England, In *Adjustments of the Fluvial System* (eds. Rhodes, D.D., and Williams, G.P.). Kendall-Hunt, Dubuque, IO, pp. 309–333.
Richards, K.S. (1982). *Rivers: form and process in alluvial channels*, Methuen, London.
Richardson, E.V., and Christian, M. (1976). Channel improvements of the Missouri River. In *Proceedings of the Third Federal Interagency Sediment Conference*, pp. 113–124.
Richardson, E.V., Stevens, M.A., and Simons, D.B. (1975). The design of spurs for river training. In *Proceedings of the 16th Congress of the International Association of Hydraulics Research, Sao Paulo*, vol. 2, pp. 382–388.
Richardson, E.V., Simons, D.B., Karaki, S., Mahmood, K., and Stevens, M.A. (1975). *Highways in the River Environment: Hydraulic and environmental design consideration. Training and design manual*. Federal Highways Administration, US Department of Transportation, Washington, DC.
Ritter, D.F. (1979). The effects of channelization on a high-energy river. *Environmental Geology*, **3**, 29–38.
Robertson, H.R. (1875). *Life on the Upper Thames*. Virtue, Spalding & Co., London.

Robinson, G.W. (1969). The use of herbicides in the maintenance of land drainage channels on Romney Marsh. *Journal of the Institute of Water Engineers*, **23**, 159–169.

Robinson, G.W. (1971). Practical aspects of chemical control of weeds in land drainage channels in England and Wales. In *Proceedings of the European Weed Research Society, 3rd Symposium on Aquatic Weeds*, pp. 297–303.

Robinson, G.W., and Leeming, J.B. (1969). The experimental treatment of some waters in Kent with Diuron to control aquatic weed growth. *Association of River Authorities Year Book*, pp. 58–64.

Robinson, J.W. (1969). Twilight for two rivers. *Montana Conservationist*, **30**, 4–5.

Robinson, M., and Blyth, K. (1982). The effect of forestry drainage operations on upland sediment yields: a case study. *Earth Surface Processes and Landforms*, **7**, 85–90.

Robson, T.O. (1973). *The Control of Aquatic Weeds*. Bulletin, No. 194, Ministry of Agriculture, Fisheries and Food, London.

Robson, T.O. (1977). Perspectives of biological control of aquatic weeds in Temperate climatic zones. *Aquatic Botany*, **3**, 125–131.

Robson, T.O., and Fearon, J.H. (1976). *Aquatic Herbicides*. British Crop Protection Council, London.

Rolt, L.T.C. (1950). *The Inland Waterways of England*. Allen & Unwin, London.

Rolt, L.T.C. (1969). *Navigable Waterways*. Longmans, London.

Rouse, H. (1950). *Engineering Hydraulics*. John Wiley & Sons, New York.

Royal Society for the Protection of Birds (1983). *Land Drainage and Birds in England and Wales: an interim report*. RSPB Conservation Planning Department, The Lodge, Sandy, Bedfordshire, UK.

Ruhe, R.V. (1970). Stream regimen and man's manipulation. In *Environmental Geomorphology* (ed. Coates, D.R.). First Annual Geomorphology Symposium, State University of New York, pp. 9–23.

Ryckborst, H. (1980). Geomorphological changes after river-meander surgery. *Geologie en Mijnbouw*, **59**, 121–128.

Sammell, E.A., Baker, J.A., and Brackley, R.A. (1966). *Water Resources of the Ipswich River Basin, Massachusetts*. Water Supply Paper 1826, Geological Survey, US Department of the Interior, Washington DC.

Sanders, D.F. (1976). Effects of stream channelization on aquatic macroinvertebrates, Buena Vista Marsh, Portage County, Wisconsin. Unpublished MS thesis, University of Wisconsin, Stevens Point, WI.

Santema, P. (1966). Influence of flood protection works on physical and biological environments. In *Scientific Problems of the Humid Tropic Zone Deltas and their Implications*. UNESCO, Paris, pp. 333–339.

Sato, N. (1971). Changes in river bed in three main rivers in Nishi-Ou District, north-east Honshu. *Geographical Review of Japan*, **44**, 356–365.

Sato, N. (1975). On the changes of river bed in the Oyodo River and its influence on the drainage basin. *Japanese Journal of Limnology*, **36**, 33–47.

Saunders, J.W., and Smith, M.W. (1962). Physical alterations of stream habitat to improve brook trout production. *Transactions of the American Fisheries Society*, **91**, 185–188.

Scarnecchia, D.L. (1988). The importance of streamlining in influencing fish community structure in channelized and unchannelized reaches of a prairie stream, *Regulated Rivers*, **2**, 155–166.

Schamberger, M., and Krohn, W.B. (1982). Status of the habitat evaluation procedures, *Transactions of the 47th North American Wildlife and Resources Conference*. Wildlife Management Institute, Washington, DC.

Schlosser, I.J., and Karr, J.R. (1981). Water quality in agricultural watersheds: impact of riparian vegetation during baseflow. *Water Resources Bulletin*, **17**, 233–240.

Schmal, R.N., and Sanders, D.F. (1978). *Effects of Stream Channelization on Aquatic Macro-invertebrates, Buena Vista Marsh, Portage County, Wisconsin.* Report No. FWS/OBS-78/92, Office of Biological Services, Fish and Wildlife Service, US Department of the Interior, Washington DC.

Schmidtke, R.F. (1987). Project appraisal, resource allocation and public involvement. In *Flood Hazard Management: British and international perspectives* (ed. Handmer, J.). Geobooks, Norwich, UK, pp. 262–278.

Schneberger, E., and Funk, J.L. (eds) (1971). *Stream Channelization—a Symposium.* North Central Division, American Fisheries Society Special Publication No. 2, Washington, DC.

Schnick, R.A., Mortom, J.M., Mochalski, J.C., and Beall, J.T. (1982). *Mitigation and Enhancement Techniques for the Upper Mississippi River System and Other Large River Systems.* Resource Publication No. 149. Office of Biological Services, Fish and Wildlife Service, US Department of the Interior, Washington, DC.

Schoof, R. (1980). Environmental impact of channel modification. *Water Resources Bulletin*, **16**, 697–701.

Schroeder, R. (1982). *Habitat suitability index models: Yellow headed blackbird.* Report No. FWS/OBS-82/10.26, Office of Biological Services, Fish and Wldlife Service, US Department of the Interior, Washington, DC.

Schumm, S.A. (1969). River metamorphosis. *Journal of the Hydraulics Division, American Society of Civil Engineers*, **95**, 255–273.

Schumm, S.A. (1977). *The Fluvial System.* Wiley-Interscience, New York.

Schumm, S.A., Harvey, M.D., and Watson, C.C. (1984). *Incised channels: morphology, dynamics and control.* Water Resources Publications, Littleton, CO.

Scott, D. (1958). Cover on river bottoms. *Nature*, **188**, 76–77.

Seaburg, K. (1971). The stream that used to be. *Montana Outdoors*, 26–29.

Seibert, P. (1960). Naturnahe Querprofilgestaltung bei Anbau von Wasserlaufen. *Natur und Landschaft*, **35**, 12–13.

Seibert, P. (1968). Importance of natural vegetation for the protection of banks of streams, rivers and canals. *Council of Europe Nature and Environment Series 2, Freshwater*, pp. 35–67.

Seppala, K. (1969). Post-drainage growth rate of Norway spruce and Scots pine on peat. *Acta Forestalia fennica*, **93**, 77–88.

Seymour, R. (1978). Trees and a fishery. *Angling*, February, pp. 32–34.

Sherlock, R.L. (1922). *Man as a Geological Agent.* Witherby, London.

Sherlock, R.L. (1923). The influence of man as an agent in geographical change. *Geographical Journal*, **65**, 258–273.

Shetter, D.S., Clark, O.H., and Hazzard, A.S. (1946). The effects of deflectors in a section of Michigan trout stream. *Transactions of the American Fisheries Society*, **76**, 248–278.

Shields, F.D. (1982a). Environmental features for flood control channels. *Water Resources Bulletin*, **18**, 779–784.

Shields, F.D. (1982b). *Environmental Features for Flood Control Channels.* Environmental and Water Quality Operational Studies Technical Report E-82-7, US Army Engineer Waterways Experiment Station, Vicksburg, MS.

Shields, F.D. (1983a). Design of habitat structures for open channels. *Journal of Water Resources Planning and Management, American Society of Civil Engineers*, **109**, 331–344.

Shields, F.D. (1983b). Environmental guidelines for dike fields, *Proceedings of Conference Rivers '83, American Society of Civil Engineers*, pp. 430–441.

Shields, F.D., and Nunnally, N.R. (1984). Environmental aspects of clearing and snagging. *Journal of Environmental Engineering, American Society of Civil Engineers*, **110**, 152–165.

Shields, F.D., and Sanders, T.G. (1986). Water quality effects of excavation and diversion. *Journal of Environmental Engineering, American Society of Civil Engineers*, **112**, 211–228.

Shirazi, M.A., and Seim, W.K. (1979). *A Stream Systems Evaluation: an emphasis on spawning habitat salmonids*. Report No. EPA-600/3-79-109, US Environmental Protection Agency, Washington DC.

Simmons, C.E., and Watkins, S.A. (1982). *The Effects of Channel Excavation on Water Quality Characteristics of the Blackwater River and on Groundwater Levels, Near Dunn, North Carolina*. Water Resources Investigations 82-4083, Geological Survey, US Department of the Interior, Raleigh, NC.

Simons, D.B. and Lagasse, P.F. (1976). Impact of dredging on river system morphology. *Proceedings of Conference Rivers '76, American Society of Civil Engineers*, pp. 435–457.

Simons, D.B., and Senturk, F. (1977). *Sediment Transport Technology*. Water Resources Publications, Fort Collins, CO.

Simons, D.B., and Albertson, M.L. (1960). Uniform water-conveyance channels in alluvial material. *Journal of the Hydraulics Division, American Society of Civil Engineers*, **86**, 33–71.

Simons, D.B., Chen, Y.H., and Swenson, L.J. (1984). *Hydraulic Test to Develop Design Criteria for the use of Reno Mattresses*. Report prepared by the Civil Engineering Department, Colorado State University and Simons, Li. Associates for Maccaferri Steel Wire Products Ltd, Ontario, Canada.

Simons, D.B., Li, R.M., Alawady, M.A., and Andrew, J.W. (1979). *Report on Connecticut River Streambank Erosion Study, Massachusetts, New Hampshire and Vermont*. Report No. CSU-213. Colorado State University Research Institute, Fort Collins, CO.

Simpson, P.W. (1981). *Manual of Channelization Impacts on Fish and Wildlife*. Contract No. 14-16-0009-80-066, Office of Biological Services, Fish and Wildlife Service, US Department of the Interior, Washington, DC.

Sincock, J.L., Smith, M.M., and Lynch, J.F. (1964). Ducks in Dixie. In *Waterfowl Tommorrow* (ed. Lindusda, J.P.). US Fish and Wildlife Service, Washington, DC, pp. 94–124.

Smith, B. (1910). Some recent changes in the course of the Trent. *Geographical Journal*, **XXXV**, 568–579.

Smith, M.F. (1975). *Environmental and Ecological Effects of Dredging: a bibliography with abstracts*. National Technical Information Service Report for 1964–1975, Springfield, VA.

Smith, P.W. (1968). An assessment of changes in the fish fauna of two Illinois rivers and its bearings on their future. *Transactions of the Illinois State Academy of Science*, **6**, 3–45.

Snell, E.L. (1968). Fibreglass protection to river banks. *Journal of the Institute of Water Engineers*, **22**, 72.

Soil Conservation Service (1947). *Handbook of Channel Design for Soil and Water Conservation*. Technical Report No. 61, US Department of Agriculture, Washington, DC.

Soil Conservation Service (1970) *North Fork Broad River (Project) Watershed, Georgia, Watershed Evaluation Study*, US Department of Agriculture, Soil Conservation Service, Athens, GA.

Soil Conservation Service (1977a). *Planning and Design of Open Channels*. Technical Release No. 25, US Department of Agriculture, Washington, DC.

Soil Conservation Service (1977b). Compliance with the National Environmental Policy Act, 1969; Use of channel modification as a means of water management; and guide for environmental assessment. *Federal Register*, **42**, 40119–40122.

Soil Conservation Service (1977c). *Guide for Environmental Assessment*. Soil Conservation Service, US Department of Agriculture, Washington DC.

Soulsby, P.G. (1974). The effect of a heavy cut on the subsequent growth of aquatic plants in a Hampshire chalk stream. *Journal of the Institute of Fish Management*, **5**, 49–52.

Spence, L. (1968). A river fights back. *Montana Wildlife*, 12–13.

Spieker, A.M. (1970). *Water in Urban Planning. Salt Creek Basin, Illinois*. Water Supply Paper 2002, Geological Survey, US Department of the Interior, Washington, DC.

Spillett, P.B. (1981). The use of groynes and deflectors in river management. In *Proceedings of Conference on Aquatic Weeds and their Control*. Association of Applied Biologists, Oxford, UK, pp. 189–198.

Spillett, P.B., and Armstrong, G.S. (1984). Ameliorative methods to reinstate fisheries following land drainage operations. Paper No. 37 presented at *Symposium of habitat modification and freshwater fisheries*. European Inland Fisheries Advisory Commission, Thirteenth Session. Århus, Denmark.

Sprules, W.M. (1947). *An ecological investigation of stream insects in Algonquin Park, Ontario*. University of Toronto, Canada, Studies in Biology ser. 56, 181.

Stall, J.B., and Herricks, E.E. (1982). Evaluating aquatic habitat using stream network structure and streamflow predictions. In *Applied Geomorphology*. Proceedings of the Annual Binghamton Geomorphology Symposium, Ohio (eds Craig, R.G. and Craft, J.L.). George Allen & Unwin, London, pp. 240–253.

Stalnaker, C.B. (1980). The use of habitat structure preferenda for establishing flow regimes necessary for maintenance of fish habitat. In *The Ecology of Regulated Streams* (eds Ward, J.V. and Stanford, J.A.). Plenum, New York, pp. 212–337.

Statzner, von, B., and Stechman, D.H. (1977). Der Einfluß einer mechanischen Entkrautungsmaßnahme auf die Driftraten der Makro-Invertebraten im Unteren Schierenseebach. *Faunistische ökologische Mitteilungen* **5**, 93–109.

Stevens, M.A., Simons, D.B., and Schumm, S.A. (1975). Man-induced changes of the Middle Mississippi River. *Journal of the Waterways, Harbors and Coastal Engineering Division, American Society of Civil Engineers*, **101**, 119–133.

Stewart, A.J.A., and Lance, A.N. (1983). Moor-draining: a review of impacts on land use. *Journal of Environmental Management*, **17**, 81–99.

Stott, B. (1977) On the question of the introduction of grass-carp (*Ctenopharyngodon idella* Val) into the United Kingdom. *Fisheries Management*, **8**, 63–71.

Stott, B., and Robson, T.O. (1970). Efficiency of grass carp (*Ctenopharyngodon idella* Val) in controlling submerged water weeds. *Nature*, **226**, 870.

Strauser, C.N., and Long, N. (1976). Discussion of 'Man-induced changes of the Middle Mississippi River'. *Journal of the Waterways, Harbors and Coastal Engineering Division, American Society of Civil Engineers*, **102**, 281.

Strom, H.G. (1962). *River Improvement and Drainage in Australia and New Zealand*. State Rivers and Water Supply Commission, Melbourne, Australia.

Stroud, R.H. (1971). Stream destruction by channelization. *Sport Fishing Institute Bulletin*, **226**, 1–3.

Stuart, T.A. (1959). The influence of land drainage works, levees, dykes, dredging etc. on the aquatic environment and stocks. In *Proceedings of the International Union for the Conservation of Nature, Technical Meeting, Athens*, vol. 4, pp. 337–345.

Stuart, T.A. (1960). Land drainage in Scotland and its effect on fisheries. *Salmon and Trout Magazine*, **158**, 44–62.

Sukopp, H. (1973). Conservation of wetlands in Central Europe. *Polish Archives Hydrobiology*, **20**, 223–228.

Sumner, F.H., and Smith, O.R. (1939). *A Biological Study of the Effect of Mining Debris Dam and Hydraulic Mining on Fish Life in the Yuba and American Rivers in California*, Stanford University, CA.

Swales, S. (1979). Effects of river improvements on fish populations. *Proceedings of*

the First British Freshwater Conference. University of Liverpool, UK, pp. 86–89.
Swales, S. (1980). Investigations of the effects of river channel works on the ecology of fish populations. Unpublished PhD thesis, University of Liverpool, UK.
Swales, S. (1981). Fisheries conservation and land drainage improvement in *Proceedings of the F.A.O. Technical Consultation on the Allocation of Fishery Resources*, Vichy, France, April 1980.
Swales, S. (1982a). A 'before and after' study of the effects of land drainage works on fish stocks in the upper reaches of a lowland river. *Fish Management*, **13**, 105–113.
Swales, S. (1982b). Environmental effects of river channel works used in land drainage improvements. *Journal of Environmental Management*, **14**, 103–126.
Swales, S. (1982c). Impacts of weed cutting on fisheries: an experimental study in a small lowland river. *Fisheries Management*, **13**, 125–137.
Swales, S., and O'Hara, K. (1980). Instream habitat devices and their use in freshwater fisheries management. *Journal of Environmental Management*, **10**, 167–179.
Swicegood, W.R., and Kriz, G.J. (1973). Physical effects of maintaining drainage channels in North Carolina's coastal area. *Journal of Soil and Water Conservation*, **28**, 266–269.
Szilagy, J. (1932). Flood control on the Tisza river. *Military Engineer*, **24**, 632.
Tarplee, W.H., Louder, D.E., and Weber, A.J. (1971). *Evaluation of the Effects of Channelization on the Fish Populations in North Carolina's Coastal Streams*. North Carolina Wildlife Resources Commission, NC.
Tarzwell, C.M. (1932). Trout stream improvement in Michigan, *Transactions of the American Fisheries Society*, **61**, 48–57.
Tarzwell, C.M. (1937). Experimental evidence on the value of trout stream improvement in Michigan. *Transactions of the American Fisheries Society*, **66**, 177–187.
Tarzwell, C.M. (1938). An evaluation of the methods and results of stream improvement in the south-west. *Transactions of the North American Wildlife Conference*, pp. 339–364.
Task Committee (1978). Environmental effects of hydraulic structures. *Journal of the Hydraulics Division, American Society of Civil Engineers*, **104**, 203–221.
Taylor, J. (1864). Description of the River Tees and of the works upon it connected with navigation. *Minutes and Proceedings of the Institute of Civil Engineers*, **XXIV**, 62–103.
Tesch, F.W., and Albrecht, M.L. (1961). Uber den Einfluss verschiedener Umweltfaktoren auf Wachstum und Bestand der Bachforelle (*Salmo trutta fario* L.) in Mittelgebirgsgewasser. *Verhandlungen der internationalen Vereinigung für theoretische und angewandte Limnologie*, **14**, 763–768.
Thames Water Authority (1988). *Lower Colne Study: Environmental Impact Assessment*. Report by Land Use Consultants Ltd for Thames Water, Rivers Division, Reading, UK.
Thames Water Authority (1988). *Manual of Appraisal*. Thames Water, Rivers Division, Reading, UK.
Thienemann, A. (1950). Vereitungsgeschichte der Susswassertierwelt Europas. *Binnengewasser*, 18.
Thienen, C.W. (1971). *Stream Channelization*. Mimeographed Report of the Ad Hoc Committee on Stream Channelization, North Central Division, American Fisheries Society, Washington DC.
Thomas, W.L. (1956). *Man's Role in Changing the Face of the Earth*. University of Chicago Press, Chicago, IL.
Thorn, R.B. (1966). The hydraulic design of river control structures. In *River Engineering and Water Conservation Works* (ed. Thorn, R.B.). Butterworths, London, pp. 272–308.
Timbol, A.S., and Maciolek, J.A. (1978). *Stream Channel Modification in Hawaii. Part A: Statewide inventory of streams; habitat factors and associated biota*. Report

No. FWS/OBS-78/16, Office of Biological Services, Fish and Wildlife Service, US Department of the Interior, Washington DC.

Toms, R. (1975). The environmental impact of land drainage work. In *Conservation and Land Drainage Conference*. Water Space Amenity Commission, London, pp. 6–15.

Toner, E.D., O'Riordan, A., and Twomey, E. (1965). The effects of arterial drainage works on the salmon stock of a tributary of the River Moy. In *Irish Fisheries Investigations*, Series A, No. 1, pp. 36–55.

Townsend, C.R. (1980). *The Ecology of Streams and Rivers*. Studies in Biology, No. 122. Edward Arnold, London.

Trautman, M.B. (1939). The effects of manmade modifications on the fish fauna in Lost and Gordon Creeks, Ohio, between 1887–1938. *Ohio Journal of Science*, **39**, 275–288.

Trautman, M.B., and Gartman, D.K. (1974). Re-evaluation of the effects of manmade modifications on Gordon Creek between 1887 and 1973 and especially as regards its fish fauna. *Ohio Journal of Science*, **74**, 162–173.

Tuckfield, C.G. (1980). Stream channel stability and forest drainage in the New Forest, Hampshire. *Earth Surface Processes*, **5**, 317–379.

United Nations (1951). *Methods and Problems of Flood Control in Asia and the Far East*. Flood Control Series 2, United Nations.

United States Army Corps of Engineers (1955). *Susquehanna River Flood Control Project*. Conklin-Kirkwood, New York, US Army District Office.

United States Army Corps of Engineers (1963–65). *Committee on Channel Stabilisation. Symposium on channel stabilization problems*. US Army Engineers, Vicksburg, MS.

United States Army Corps of Engineers (1966). *Symposium on Channel Stabilization Problems*. Report of the Committee on Channel Stabilization, Technical report No. 1, US Army Engineers, Vicksburg, MS.

United States Army Corps of Engineers (1969). *State of Knowledge of Channel Stabilization in Major Alluvial Rivers*. Report of Committee on Channel Stabilization, Technical report No. 7, US Army Engineers, Vicksburg, MS.

United States Bureau of Reclamation (1952). *Canals and Related Structures: design and construction manual*. Design Supplement No. 3. US Government Printing Office, Washington DC.

United States Department of Agriculture (1970). *North Fork Broad River (Project) Watershed, Georgia, Watershed Evaluation Procedure*. US Department of Agriculture, Athens, GA.

United States Fish and Wildlife Service (1974). *Stream Channel Alteration Guidelines*. Fish and Wildlife Service, US Department of the Interior, Washington, DC.

United States Fish and Wildlife Service (1976). *Habitat Evaluation Procedures: for use by the Division of Ecological Services in evaluating water and related land resource development projects*. Mimeo report, Fish and Wildlife Service, US Department of the Interior, Washington, DC.

United States Fish and Wildlife Service (1980a). *Habitat as a Basis for Environmental Assessments*. ESM 101, Fish and Wildlife Service, US Department of the Interior, Washington, DC.

United States Fish and Wildlife Service (1980b). *Habitat Evaluation Procedures (HEP)*. ESM 102, Fish and Wildlife Service, US Department of the Interior, Washington, DC.

United States Fish and Wildlife Service (1980c). *Human Use and Economic Evaluation (HUEE)*. ESM 104. Fish and Wildlife Service, US Department of the Interior, Washington, DC.

United States Fish and Wildlife Service(1981). *Standards for the Development of*

Habitat Suitability Index Models. ESM 103. Fish and Wildlife Service, US Department of the Interior, Washington, DC.

United States Geological Survey (1969). *Map of Water Resources Development*. Geological Survey, US Department of the Interior, Washington, DC.

United States Water Resources Council (1976). *A Unified National Program for Floodplain Management*. US Government Printing Office, Washington, DC.

Van Bendegom, L. (1973). *Natuurlijke waterlopen*. Delft Technical University.

Van Der Voo, E.E. (1962). Danger to scientifically important wetlands in The Netherlands by modification to the surrounding environment. In *Project MAR. The Conservation and Management of Temperate Marshes, Bogs and Other Wetlands*, Vol. 1. New Series No. 3, International Union for the Conservation of Nature, pp. 224–277.

Vandre, W.G. (1975). Effects of channel dredging on wildlife and wildlife habitat of Buena Vista Marsh. Unpublished MS thesis, University of Wisconsin, WI.

Vannote, R.L., Minshall, G.W., Cummins, K.W., Sedell, J.R., and Cushing, C.E. (1980). The river continuum concept *Canadian Journal of Fisheries and Aquatic Science*, **37**, 103–137.

Vanoni, V.A. (ed.) (1975). *Sedimentation Engineering*. Manuals and reports on engineering practice No. 54, American Society of Civil Engineers, New York.

Vaughan, R. (1610). *Water Workes*, George Eld, London.

Wade, P.M. (1977). The dredging of drainage channels : its ecological effects. Unpublished PhD thesis, University of Wales Institute of Science and Technology, Cardiff, UK.

Wade, P.M. (1978). The effects of mechanical excavators on the drainage channel habitat. In *Proceedings of the European Weed Research Society 5th Symposium on Aquatic Weeds*, pp. 333–342.

Wade, P.M., and Edwards, R.W. (1980). The effect of channel maintenance on the aquatic macrophytes of the drainage channels in the Monmouthshire Levels, South Wales, 1870–1976. *Aquatic Botany*, **8**, 307–322.

Walker, J.J. (1979). Environmental considerations in engineering design. In *The Mitigation Symposium: a national workshop on mitigating losses of fish and wildlife habitats*. Technical report RM-65, Forest Service, US Department of Agriculture, Fort Collins, CO.

Wallen, I.E. (1951). The direct effect of turbidity on fishes. *Bulletin of the Oklahoma Agricultural Experimental Station*, **28**, 1–27.

Ward, R.C. (1978). *Floods—A Geographical Perspective*. Macmillan, London.

Warner, K., and Porter, I.R. (1960). Experimental improvement of a bulldozed trout stream in Maine. *Transactions of the American Fisheries Society*, **89**, 59–63.

Warner, R.E. (1979) *Proceedings of a workshop on fish and wildlife resource needs in riparian ecosystems*. Fish and Wildlife Service, US Department of the Interior, Kenneysville, W.VI.

Warner, W.L., and Preston, E.H. (1973). *Review of Environmental Impact Assessment Methodologies*. Battelle Columbus Laboratory, Columbus, OH.

Warner, W.L., Moore, J.L., Chatterjee, S., Cooper, D.C., Ifreadi, C., Lawhon, W.T., and Reyners, R.S. (1974). *An Assessment Methodology for the Environmental Impact of Water Resource Projects*. Report No. EPA 600/5–74–016, US Environmental Protection Agency, Washington, DC.

Water Space Amenity Commission (1980a). *Conservation and Land Drainage Guidelines*. WSAC, London.

Water Space Amenity Commission (1980b). *Working Party Report on Conservation and Land Drainage*. WSAC, London.

Watson, R.M. (1981). The Stokesley Flood Relief Scheme. *Journal of the Institute of Water Engineers and Scientists*, **35**, 143–150.

Weber, E.C. (1977). Angler use and success in two channelized warmwater streams.

Unpublished MS thesis, Ohio State University, Columbus, OH.
Weeks, K.G. (1982). Conservation aspects of two river improvement schemes in the River Thames catchment. *Journal of the Institute of Water Engineers and Scientists*, **36**, 447–458.
Welker, B.D. (1967). Comparison of channel catfish populations in channeled and unchanneled sections of the Little Sioux River. *Proceedings of the Iowa Academy of Science*, **74**, 99–104.
Wene, G., and Wickliff, E.L. (1940). Modification of a stream bottom and its effect on the insect fauna. *Canadian Entomologist*, **72**, 131–135.
Wentz, F. (1983). The management of tree and shrub belts along rivers: as illustrated by the examples of the Glonn and the Amper in Upper Bavaria. *Garten und Landschaft*, **2**, 117–118.
Wessex Water Authority (1976). *Environmental and Conservation Aspects Relating to River Works*. Wessex Water, Bristol.
Westlake, D.F. (1968a). The biology of aquatic weeds in relation to their management. In *Proceedings of the Ninth British Weed Control Conference*, pp. 371–379.
Westlake, D.F. (1968b). The weight of water weed in the River Frome. *Association of River Authorities Year Book*, pp. 59–68.
Westlake, D.F. (1968c). Methods used to determine the annual production of reed swamp plants with extensive rhizomes. In *Methods of Productivity Studies in Root Systems and Rhizosphere Organisms* (eds Shilarov, M.S., Kovda, V.A., Novichkova, L.N., Rodin, L.E., and Sveshnikova, V.M.). Acad. Science, Leningrad, USSR, pp. 226–234.
Westlake, D.F, and Dawson, F.H. (1982). Thirty years of weed cutting on a chalk stream. In *Proceedings of the European Weed Research Society 6th Symposium on Aquatic Weeds*, pp. 132–140.
Westlake, D.F., and Dawson, F.H. (1986). The management of *Ranunculus calcareus* by pre-emptive cutting in southern England. In *Proceedings of the European Weed Research Society 7th Symposium on Aquatic Weeds*, pp. 395–400.
Wharton, C.H. (1970). *The Southern River Swamp: a multiple-use environment*. Bureau of Business and Economic Research School of Business Adminstration, Georgia State University, GA.
Whipple, W., and Dilouie, J. (1981). Coping with increased stream erosion in urbanizing areas. *Water Resources Research*, **17**, 1561–1564.
Whistleblower (1974). How to kill a river by 'improving' it. *Observer Magazine*, 9 June.
Whitaker, G.A., McCuen, R.H., and Brush, J. (1979). Channel modification and macro-invertebrate community diversity in small streams. *Water Resources Bulletin*, **15**, 874–879.
White, R.J. (1968). *So baut man Forellenunterstände: Ein Schwerpunkt der Bachpflege* (Creating shelters for strout: a stream management). Verlag Paul Parey, Hamburg and Berlin.
White, R.J. (1972). Trout population responses to habitat change in Big Roche-a-Cri Creek, Wisconsin. Unpublished PhD thesis, University of Wisconsin, Madison, WI.
White, R.J. (1975). Trout population responses to streamflow fluctuation and habitat management in Big Roch-A-Cri Creek, Wisconsin. *Verhandlungen der internationalen Vereingung für theoretische und angewandte Limnologie*, **19**, 2469–2477.
White, R.J. (1979). Stream habitat management. In *A Manual of Wildlife Conservation* (ed. Teague, R.D.). The Wildlife Society, Washington, DC.
White, R.J., and Brynildson, O.M. (1967). *Guidelines for the Management of Trout Steam Habitat in Wisconsin*. Technical Bulletin No. 39, Division of Conservation, Wisconsin Department of Natural Resources, Madison, WI.
Whitehead, E. (1976). *A Guide to the Use of Grass in Hydraulic Engineering Practice*.

Technical Note 71, Construction Industries Research and Information Association, London.
Whitely, J.K. (1964). Control of weeds in farm ponds *Proceedings of the North Central Weed Control Conference*, **20**, 31–33.
Whitney, G.R. and Bailey, J.E. (1959). Detrimental effects of highway construction on a Montana stream. *Transactions of the American Fisheries Society*, **88**, 72–73.
Whitney, G.R., Burdett, R.C., and Hunt, D.E. (1972). The anticipated effects of A.R.D.A. funded drainage proposals on wetland system in Wellington, Grey, Dufferin and Simcoe Counties. Unpublished report, Conservation Authorities Branch, Ministry of Natural Resources.
Whitton, B.A. (ed.) (1975). *River Ecology*. Blackwell, Oxford.
Wilcock, D.N. (1977). The effects of channel clearance and peat drainage on the water balance of the Glenullin Basin, Co. Londondery. *Proceedings of the Royal Irish Academy*, section B, pp. 253–267.
Willan, T.S. (1936). *River Navigation in England, 1600–1750*. Oxford University Press, Oxford.
Willeke, G.E. (1981). *The George Palmiter River Restoration Techniques*. Institute of Environmental Science, Miami University, Oxford, OH.
Williams, E.L. (1860). Account of the works recently constructed upon the River Severn, at the Upper Lode, near Tewkesbury. *Proceedings and Minutes of the Institute of Civil Engineers*, **XIX**, 527–545.
Williams, G. (1980). Swifter flows the river. *Birds*, **8**, 119–122.
Williams, G., Henderson, A., Goldsmith, L., and Spreadborough, A. (1985). The effects on birds of land drainage improvements in the North Kent Marshes. *Wildfowl*, 34.
Williams, M. (1970). *The Draining of the Somerset Levels*. Cambridge University Press, Cambridge.
Williamson, K. (1971). A bird census study of a Dorset Dairy Farm. *Bird Study*, **18**, 80–96.
Wilson, K.V. (1968). Flood-flow characteristics of a rectified channel, Jackson, Mississippi. In *Geological Survey Research 1968*, Professional Paper 600-D, Geological Survey, US Department of the Interior, Washington, DC, pp. D57–D59.
Wilson, K.V. (1973). Changes in flood-flow characteristics of a rectified channel caused by vegetation, Jackson, Mississippi. *Journal of Research, United States Geological Survey*, **1**, 621–625.
Wingate, P.J., and Weaver, E. (1977). *Environmental Assessment, Chicod Creek Watershed Work Plan*, Raleigh, NC.
Winger, P.V., Bishop, C.M., Glesne, R.S., and Todd, R.M. (1976). *Evaluation Study of Channelization and Mitigation Structures in Crow Creek, Franklin County, Tennessee, and Jackson County, Alabama*. Final Report for contract No. AG47 SCS-00141. Soil Conservation Service, US Department of the Interior, Nashville, TN.
Winget, R.N. (1984). Methods for determining successful reclamation of stream ecosystems. In *Successful Reclamation of Stream Ecosystems* (ed. Gore, J.A.). Butterworth, London.
Winkley, B.R. (1976). Response of the Mississippi River to the cutoffs. In *Proceedings of Conference Rivers '76*, American Society of Civil Engineers, pp. 1267–1284.
Winkley, B.R. (1977). *Man-made Cutoffs on the Lower Mississippi River: Conception, Construction and River Reponse*. Report 300–2, Potamology Investigations, US Army Engineer District, Vicksburg, MS.
Winkley, B.R. (1982). Response of the Lower Mississippi to river training and realignment. In *Gravel-bed Rivers* (eds Hey, R.D., Bathurst, J.C., and Thorne, C.R.). John Wiley & Sons, Chichester, pp. 652–681.
Winkley, B.R., and Harris, P.C. (1973). *Preliminary Investigations of the Effects of*

Gravel Mining in the Mississippi River. US Army Corps of Engineers, Vicksburg, MS.

Wisdom, A.S. (1979). *Law of Watercourses*. Shaw & Sons, London.

Witten, A.L., and Bulkley, R.V. (1975). *A Study of the Effects of Stream Channelization and Bank Stabilization on Warmwater Sport Fish in Iowa: Subproject No. 2. A study of the impact of selected bank stabilization structures on game fish and associated organisms*. Contract No. 14–16–0008–745. Iowa Cooperative Fishery Research Unit, Iowa State University, Ames. Prepared for the Fish and Wildlife Service, US Department of the Interior, Washington DC.

Wojcik, D.K. (1981). Flood alleviation, conservation and fisheries : an experimental scheme on the River Roding. Unpublished MSc thesis, Department of Civil Engineering, City University, London.

Wolman, M. G. (1967). A cycle of sedimentation and erosion in urban river channels. *Geografiska Annaler*, **49A**, 385–395.

Woods, L.C. (1977). The effect of stream channelization and mitigation on warmwater macroinvertebrate communities. Unpublished MS thesis, Ohio State University, OH.

Wright, J.F., Hiley, P.D., Cameron, A.C., Wigham, M.E., and Berrie, A.D. (1983). A quantitative study of the macroinvertebrate fauna of five biotopes in the River Lambourn, Berkshire, England. *Archives of Hydrobiology*, **96**, 271–292.

Wyche, B. (1972). Stream channelization: the new flood control controversy. Unpublished thesis, Princeton University.

Yearke, L.W. (1971). River erosion due to channel relocation. *Civil Engineering*, **41**, 39–40.

Zahar, A.R. (1951). The ecology and distribution of black-flies (*Simuliidae*) in south-east Scotland. *Journal of Animal Ecology*, **20**, 33–62.

Zeller, J. (1967). Meandering channels in Switzerland. *International Association of Scientific Hydrology*, **75**, 174–186.

Zimmer, D.W. (1976). The effects of long-reach channelization on habitat and invertebrate drift in some Iowa streams. Unpublished PhD thesis, Iowa State University, Ames, IA.

Zimmer, D.W., and Bachmann, R.W. (1978). Channelization and invertebrate drift in some Iowa streams. *Water Resources Bulletin*, **14**, 868–883.

Zimmerman, R.C., Goodlett, J.C., and Comer, G.H. (1967). The influence of vegetation on channel form in small streams. *International Association of Scientific Hydrology*, **75**, 255–275.

Zon, J.C.J. van (1977). Grass carp (*Ctenopharyngidon idella* Val.) in Europe. *Aquatic Botany*, **3**, 143–155.

Zumberge, J.H. (1957). Land drainage and the water-table in southern Michigan and Northern Indiana. *Michigan Academy of Science Letters and Papers*, **42**, 105–113.

Author Index

Geographical Index

Subject Index